# Environmentalism and the mass media

By studying mass media in different countries it is possible to assess the ways in which some of the dominant concerns of contemporary societies are reflected and projected.

*Environmentalism and the Mass Media* throws new light on the way in which environmental ideas circulate in the United Kingdom and India. Using original research the book shows that the ideas of 'environmentalism' are too broad and imprecise to hold either meaning or currency, that the greater mass of the Indian people are not part of any global consensus on these issues, that their goal remains the attainment of development. If anything, 'environmentalism' is seen as neo-colonialism by other means. At a grass-roots level the public in developed Britain often react in similar ways. They may be more ready to recognise environmental problems, but these occur elsewhere – in Eastern Europe or developing countries.

*Environmentalism and the Mass Media* outlines the differing cultural, religious and political contexts of Britain and India, against which different world-views form. It considers the contrasting nature of their environments, and then, by in-depth interviews with journalists in both countries, and by the analysis of media content, a picture emerges of the environmental concerns the media convey. Public reaction is examined both by the use of focus groups in India and the UK, and through public opinion survey.

Whether or not environmentalism becomes a universal cause depends on how, and to what extent, the many sharply contrasting world-views can ever converge – in other words whether the world can 'know itself'. A simulation shows how a common world-view on environmental issues will not emerge in a world of unequal communication access.

**Graham Chapman** is Professor of Geography, Lancaster University; **Keval Kumar** is Reader in the Department of Communication and Journalism, University of Poona, India; **Caroline Fraser** is an Independent Media Consultant and Broadcast News Editor, Lyons, France; **Ivor Gaber** is Professor of Broadcast Journalism, Goldsmiths College, University of London.

# GLOBAL ENVIRONMENTAL CHANGE SERIES

Edited by Michael Redclift, *Wye College, University of London*; Martin Parry, *University College, London*; Timothy O'Riordan, *University of East Anglia*; Robin Grove-White, *University of Lancaster*; and Brian Robson, *University of Manchester*.

The *Global Environmental Change Series*, published in association with the ESRC Global Environmental Change Programme, emphasises the way that human aspirations, choices and everyday behaviour influence changes in the global environment. In the aftermath of UNCED and Agenda 21, this series helps crystallise the contribution of social science thinking to global change and explores the impact of global changes on the development of social sciences.

Also available in the series

**Argument in the Greenhouse**
The international economics of controlling global warming
Edited by Nick Mabey, Stephen Hall, Clare Smith and Sujata Gupta

**Environmental Change in South-East Asia**
People, politics and sustainable development
Edited by Michael Parnwell and Raymond Bryant

**The Environment and International Relations**
Edited by John Vogler and Mark Imber

**Politics of Climate Change**
A European perspective
Edited by Timothy O'Riordan and Jill Jäger

**Global Warming and Energy Demand**
Edited by Terry Barker, Paul Ekins and Nick Johnstone

**Social Theory and the Global Environment**
Edited by Michael Redclift and Ted Benton

# Environmentalism and the mass media

## The North–South divide

Graham Chapman, Keval Kumar,
Caroline Fraser and Ivor Gaber

**Global Environmental Change Programme**

**Indian Institute of Advanced Study, Shimla**

Routledge
Taylor & Francis Group

LONDON AND NEW YORK

First published 1997
by Routledge
2 Park Square, Milton Park, Abingdon, Oxon OX14 4RN

Simultaneously published in the USA and Canada
by Routledge
711 Third Avenue, New York, NY 10017

*Routledge is an imprint of the Taylor & Francis Group, an informa business*

Typeset in Times by
Florencetype Ltd, Stoodleigh, Devon

*British Library Cataloguing in Publication Data*
A catalogue record for this book is available from the British Library

*Library of Congress Cataloguing in Publication Data*
Environmentalism and the mass media: the North–South divide / Graham Chapman
 p. cm. – (Global environmental change series)
Includes bibliographical references and index.
 1. Mass media and the environment – Great Britain.
 2. Environmentalism – Great Britain. 3. Mass media and the
environment – India. 4. Environmentalism – India. 5. Economic
development. I. Chapman, Graham. II. Series.
 P96.E572G744 1997
333.7 – dc20             96–32316

ISBN 0–415–15504–5
ISBN 0–415–15505–3 (pbk)

# Contents

# Maps

# Figures

# Tables

# The authors and the project

**Graham Chapman** Professor of Geography at Lancaster University, UK, and former Fellow, Indian Institute of Advanced Study, Shimla, India

**Caroline Fraser** Independent Media Consultant and Broadcast News Editor, Lyons, France

**Ivor Gaber** Professor of Broadcast Journalism, Goldsmiths College, University of London, and Political Consultant, ITN, London, UK

**Keval Kumar** Reader, Department of Communication and Journalism, University of Poona, India

The project 'The Mass Media and Global Environmental Learning' funded by the ESRC Global Environmental Change Programme, Phase 2, was based at the South–North Centre for Environmental Policy, School of Oriental and African Studies, University of London, UK.

# Acknowledgements

This book is the result of a co-operative research project involving many people and institutions, in the end so numerous that although we can try to thank publicly as many as possible, we hope that it is understood why in some instances we thank groups of individuals.

The project would not have got off the ground at all had it not been for the funding received from the Economic and Social Research Council of the United Kingdom, who provided grant L320253059 under Phase II of their Global Environmental Change Programme. We thank them for their backing. We would also like to thank, through the Director Professor Mrinal Miri, the Indian Institute of Advanced Study at Shimla, which provided an ideal milieu and the support of a Fellowship for Graham Chapman, during which several parts of the work were completed, and the Royal Society for providing an air-fare to India.

We thank Charlotte Reisch for her incomparable appetite for work, inputting twice as much data as originally anticipated, for transcribing so many tapes, for organising the office and keeping track of money, and for her finely honed skills in sniffing out bargain air-fares; and for keeping her sense of humour throughout it all. We are not sure what would have been completed had she not accepted the post of project secretary. We thank Richard and Emma Insell-Jones for allowing their home in Shropshire to be turned into a focus group theatre with humour and cakes, and our coders in the UK, Sue Pearlman, Raine Lawson, Michelle Hodge, Paula Briggs and Ruth Reynolds for persisting with monitoring the airwaves.

We thank Professor A. Relton of Bishop Heber College, Tiruchirapalli, for co-ordinating the questionnaire-based surveys and the focus group discussions in Tamil Nadu, as well as for translating the focus group discussions (which were conducted in Tamil) into English. He was ably assisted by Professor A. Suresh Kumar of Madurai. We also thank Vishram Dhole, Seema Ginde and Shekar Nagarkar for assisting Dr Keval J. Kumar in organising the surveys and the focus groups in Maharashtra, and in translating the Marathi focus group discussions into English. They also assisted in coding the contents of television, radio and the press.

In Britain we thank the broadcasters and journalists who were interviewed, and Alison Anderson of the University of Plymouth for conducting many of the interviews so professionally. We thank the many participants of the focus groups, and we hope they enjoyed the sessions as much as we appreciated the results. In India we thank the journalists and broadcasters who gave of their time so freely, and the many participants there in the focus group sessions.

Finally, we thank Sarah Lloyd of Routledge for her incomparably good-humoured encouragement while getting the book to and through the press, and Matthew Smith of Routledge for his cheerful, prompt and efficient handling of the text.

Although we have done our level best to minimise them, the responsibility for the flaws and faults which remain lies squarely with the four of us.

Graham Chapman, Keval Kumar, Caroline Fraser, Ivor Gaber

# Foreword

As the title of this book implies, it is written on a topic which lies at the inter-section of three areas of academic enquiry – media studies, environmentalism, and development studies, the latter particularly at the global scale which looks at North–South inequalities. These three areas are, however, not just areas of acad-emic enquiry. The mass media have an impact on virtually all contemporary societies – so much so in Britain that the term 'politico-media complex' has recently been coined – nobody lives on earth without an environment, all people display some set of values towards their environment, and the global struggle for development is the substance of Development Decades, Lomé Conventions, the United Nations Development Programme, institutions such as the World Bank, and of course the 1992 UNCED conference at Rio de Janeiro. The momentum behind UNCED was undoubtedly driven at least in part by the success of scientists in persuading politi-cians in the North that there are serious, but potentially manageable, global climatic changes about to take place as a result of human activity. Assuming that they exist, that they are 'for real', unravelling problems of such magnitude would clearly involve many disciplines, for example linking physical and biological science to the social sciences and humanities. It would be concerned with the economic, political, legal, religious and philosophical processes needed to induce major changes in human behaviour. The Economic and Social Research Council of the UK decided to play its part in this process, and funded a programme on Global Environmental Change. Seeing the significance of the intersection we have identified, we were fortunate in receiving a grant under the GEC programme to see how the mass media in Britain and India treated environmental issues, to see how the public reacted to sample material, and to assess public attitudes and knowledge in these two contrasting countries, one developed the other developing, on environmental and related matters.

Here we introduce briefly our approach to the three topics. We start with the environment, since that is the central pivot of the GEC programme. The 'envi-ronment' is a categorising word, and as with any such word, there are two distinct ways (amongst many others) in which it can be used. One way is bottom-up, to think of all those things that can be grouped together into a set which carries the categorising label. For example, in one sense we can say that the National Trust, a long-established non-governmental organisation (NGO) in Britain which conserves historical houses and landscapes, is the sum total of the properties it owns. Most of the buildings have official Listed Status, and cannot be arbitrarily demolished. The other way is top-down. A label has a meaning, against which candidates for association can be assessed, so that the label becomes a lens and authority imposing meaning on the world. People visit the National Trust proper-ties, and they become aware of the type of inheritance the NT conserves, properties that satisfy the condition 'National-Trust-ness'. Now there are modern commercial buildings in Britain which are Listed, but which the National Trust could never take on. This is because these properties are screened out by testing them with the condition 'National-Trust-ness', which they fail. The two ways are of course linked, since it is usual for a new word to acquire the beginnings of meaning first by the bottom-up approach, but when it has developed some sense and some value, it may then be used increasingly top-down. If it becomes a well-known big word (or phrase) with publicity value of its own, then there will probably be fights over the right meanings and ownership of the word or phrase – such as 'sustainable'.

The environment is everything. Every physical act of every living thing has environmental significance, using energy, absorbing and converting chemicals, and changing the environment for other living things. The fact that the environment is everything means that it is very difficult to establish, as we try to do in this project, the amount of environmental programming on TV. In any drama, people may get in cars, pick flowers, turn a light switch on. None of these might be the semantic centre of the drama, but each one is also an illustration of the values associated with the use of the environment. Yet, if you ask the average man in the street in Britain if he has an opinion on environmental issues, he will give an answer, of whatever sort. Hardly anyone will say, 'I do not understand, please explain that word.' The word does have a meaning which many people have absorbed, modified, interpreted, and are prepared to apply in different contexts. What we have discovered in our research is that it is more likely that this acceptance has occurred in Britain than in India, and that its uses in Britain and India are different. In India if it is used it is difficult to separate it from 'development'. In Britain it has become an abstract idea which can stand on its own and with which any number of interest groups can play.

Why should this be the case? We put forward just one simple reason, out of many possible ones, why this may be so. On a regular daily basis, in Britain most people appropriate nothing directly from the environment, other than oxygen, which they inhale in their lungs and their vehicle engines, and visual and aural signals which may give pleasure (such as aesthetic landscapes, an increasingly important environmental demand) or pain. Everything else – water, food, clothing, and many visual and aural signals – comes indirectly. Neither do they directly return anything to the environment, apart from $CO_2$, some SoxNox (sulphur and nitrogen oxides), and maybe some litter and noise. The people who are actively and daily involved in appropriation, the farmers, fishermen, quarry workers, miners, are collectively known as primary producers, and constitute less than 2 per cent of the national workforce. To this may be added a small number of construction workers who dig foundations for roads and buildings. For the vast majority of people, therefore, even though they all live and act totally within their own environment, whatever that is, it is possible to conceptualise something called an environment which is far off, and separate, and abstract, from which their physical needs are satisfied indirectly. The water comes by pipes, energy by pipes and cables, the effluent is conveyed away by pipes, and food comes in packages from supermarket shelves.

In India more than 70 per cent of employment is provided by agriculture. For the majority of the population fuel is something collected by the family, be it wood or dung. The rural population is dependent on local wells, pumpsets and small reservoirs for water, collected in pitchers and head-loaded home. Half of the 25 per cent that constitute the urban population, and all of the 75 per cent rural population, because they have no toilets, use open ground, something readily discussed in the transcripts which follow. In other words the majority of the population is involved on a daily basis with physical exchange with a local lived-in environment.

For this and many other reasons which we will touch on in the book, we surmise that the idea of the environment will be different in these two cultures, and that dominant forms of knowledge will also be different. In Britain many people have some vague grasp of global warming, and the science that lies behind the predictions of the impact of carbon dioxide on the atmosphere. In India farmers know about their own land and plants, using their own calendars and explanations, few of which they would be able to explain in theories which satisfied the scientific community, but which represent sometimes acutely perspicacious valuable knowledge. Sometimes it is tacit, as Polanyi (1967) postulated; sometimes it is articulated, as Chapman (1983, 1984) has shown in the use of folklore by Bihari farmers. The few primary producers in Britain almost certainly have equivalent knowledge (Wynne, 1991).

Knowledge about the environment, particularly general public knowledge about distant places, is articulated and circulated by many routes, but one set of routes, known collectively as the mass media, has drawn increasing attention in modern society: it is perceived to be very influential and yet it is difficult to see how it can maintain objectivity. This is because, at the very least, the channels of communication have finite and therefore limited capacity, and have to be selective in terms of the information conveyed. In addition, of course, meaning is attached and changed by the essential acts of editorial compression and audience interpretation. Given both their power and their suspected power to 'distort', the media become an obvious candidate for social and political analysis, and for inclusion in this project. What we have in mind is a simple model, which we will use here to chart out the territory and our approach to it, but which can be and should be endlessly complexified. The first point to note is that the media are part of a circle which feeds back on itself. Indeed, the first defence of any producer accused of promoting violence in society is to deny it and maintain that (s)he merely is reflecting it. It is possible to enter any circle at any point. That people often enter this circle at the point of the channel gatekeepers is probably not just fortuitous but does reflect a latent and nagging suspicion that here lies an independent power source that can be manipulated. We enter the circle at this point, by interviewing the journalists, editors and television and radio executives who are important in Britain and India, and in some cases in international networks. We have tried to capture the opinions of our sample of the gatekeepers by taped and transcribed interview, using only the loosest of structuring. The questions we have asked look both ways round the circle – 'backwards' to their attitude towards sources and society and 'forwards' towards the material they select and how they view the audience. We have tried to capture the next stage, the messages sent through the media, by analysing the content of two UK and two Indian newspapers, the main terrestrial TV channels, and some radio channels in both countries. We used our own system of coding and analysis and our own teams of coders, who worked in both countries during the same six sample weeks, spread over one year. For general programmes and for the newspapers, we have collected data on environmental matters only: we have therefore had to acknowledge and use our own 'environment' lenses. For the broadcast News we have coded all items. We readily recognise that the idea of 'mass communications' and particularly of 'broadcasting' is historically bound – that, with burgeoning methods of point-to-point communication and expanding channel capacities of all sorts, the current role played by systems which send messages from a few sources to many receivers may be significantly diminished. The supply-side, which now seems dominant, may be replaced by a demand-driven information system – the World Wide Web being just one example.

The messages are then decoded by receivers – the audiences and readership in the two countries. There is no way of obtaining a comprehensive approach to the understanding and reaction of the public to what they hear and read. We have therefore used focus groups to see how they react to a sample of TV documentaries on the environment, some British and some Indian, the same material being shown in both countries. To the extent that Indian TV documentaries are rare in Britain and the kind of British TV documentary we used also rare in India, we are adding something to the simple model of the circle described above.

The messages received and decoded radiate out into society at large, producing multiple effects. This transition is the breeding ground for the endless and endlessly contentious studies and assertions about the impact of the media in general, but TV in particular, on social values. Some claim that it is obvious that the mass shootings in Hobart, Tasmania in 1996, were sparked by world coverage of the mass shootings in Dunblane, Scotland in the same year. Others (we are among them) would claim it is impossible to prove such a link – but we do not say one should not have pause for thought. We have, however, been looking at the less violent problem of the environment. We do not attempt to study the transition process between the received messages and the formation of ideas and values in society. We content ourselves with public opinion surveys in both countries, asking the same questions – general ones about the state of the world and more specific ones about public knowledge of the environment. We attempt to correlate this information with media consumption but, as with any correlation, which way the arrows of causality point can be guessed at but not proved. This stage of the model, like other stages, could and should be much more complex. There is no such thing as a singular coherent 'society': in any country it is a mosaic of different groups and interests, with most people simultaneously members of many different kinds of groups. Different elements of this mosaic absorb information from many sources other than the domestic media, so a complex model would have to include many other kinds of sources both internal and external. We have not had the resources to pursue such a complex approach, but we believe there is enough of value in our surveys to suggest a continuing thread between the different elements of our model, as well as the inevitable 'avenues for further research'. There is, though, a caveat, and that is that we have to confront the fact that half of the Indian population is not literate and that many people have no direct access to newspapers. Although the majority of the population has access to radio, only a minority has access to TV. How can we incorporate these people in our model? The honest answer is, with difficulty. We assume that there are multiplier effects, whereby those who are connected with the media may pass an interpretation of their received material to other people via the bush telegraph. But it still remains true that there may be many who are effectively disconnected – but from what, and with what consequences? They still have their own social networks and still form parts of their own society. At the end of the book we propose a simple model which will touch on the consequences of such disconnection for both sides – since it is no more possible for only one part of a communication pair to be disconnected than it is possible for one hand to clap.

How some activities of society become subjects for media attention is the step which closes the circle. With respect to environmental issues we have already partly touched on this in the interviews with media gatekeepers. But we could also have touched on it from the point of view of active campaigners who know how to publicise their message and who often try to 'manipulate' the media. Somewhere there had to be a limit to what we could attempt within the time and money at our disposal, and this is one aspect which remains untouched here, but it is not an area untouched by others (see Hansen, 1993) .

There remains the question of what source of 'energy' or 'stimulus' keeps the circle going. The simple answer is, the careers and profits that keep the media industry going like any other industry, but that does not say enough about how it can sell what it sells, through either the public or private sector. What we surmise is that the public need to debate, even in the most trivial of ways, the fundamental project(s) of society is the essence of the continual motion. The 'debates' may result in many different forms of communication, from commercialised media which pander to consumerism to public media which are part of the process of rejecting a past colonial subservience.

So we come to the last of the three intersecting spheres. This is the axis of 'development', which is concerned with global inequalities in wealth, health, demographics, resources, communications and power and the concomitant historical analyses, contemporary trends and contemporary political action. All of this spans any number of geographical scales. There are differences in all of the variables mentioned within both countries as well as between them. Thus, although in aggregate terms it seems clear that Britain belongs to the 'developed' group of nations and India to the 'developing', a sizeable part of Britain's current population lives in poverty, and there are whole groups of people and areas of India which are wealthy and could easily be classed as at least 'middle income', if not outright 'developed'. The distinctions within Britain do not feature much in this book, and Britain by contrast with India appears as much more homogeneous – partly because of its size, no more than the equivalent in population and area of an average state of the Indian Union, and dominated by a single language. The differences in India are important and significant, because they lead to a greater degree of differentiation of society, in income, education, language and 'publics', all of which impinge on our analysis of the relations between media and society. But the national aggregate scale remains important too, since it is at this scale that international political identity is expressed, and for all its heterogeneity, India has a self-image which influences its reaction to international negotiation and its own internal policies for 'development'. Much of this image is forged in response to a perceived Northern hegemony. Indeed, this book will end with a quote by a senior government official which makes just such a point.

'Development' is a multi-faceted term, appropriated at many scales by many disciplines spanning both physical and psychological dimensions along medical, social, political and economic axes, from developmental embryology to (human) identity formation, from language acquisition and the attainment of literacy to the pursuit of democracy and 'good governance', from improving individual savings rates to international aid and finance. But whatever the term is taken to connote, 'development' is embedded within human cultures, understood and interpreted by these cultures. We therefore start our book by talking not of the media, nor of the journalists and reporters, but of the two cultures within which our study is embedded.

Finally, we suggest that there are some themes which the reader can look out for as (s)he reads the book. Although we know that it is important to provide many signposts along the way, sometimes they are more implicit than explicit – so the readers can also provide some of their own. One important aspect to watch is the way in which 'publics' are constituted – what kinds of people form interest groups, how they define the knowledge they need and the sources that will satisfy it, and how they are seen by the media. Then, what the world-views of the various actors are, be they English-speaking broadcasting executives or Marathi- or Tamil-speaking peasant farmers. There may be few of the former and they may be powerful, but there are many of the latter, and their collective demands may compensate for their apparent individual 'powerlessness'.

## REFERENCES

Chapman, G.P. (1983) 'The folklore of the perceived environment in Bihar', *Environment and Planning* A, 15: 945–68.
—— (1984) 'The structure of two farms in Bangladesh', in Bayliss-Smith, Tim and Wanmali, Sudhir (eds), *Understanding Green Revolutions*, Cambridge: Cambridge University Press, pp. 212–52.
Hansen, A. (ed.) (1993) *The Mass Media and Environmental Issues*, Leicester: Leicester University Press.
Polanyi, M. (1967) *The Tacit Dimension*, London: Routledge.
Wynne, B. (1991) 'Knowledges in context', *Science, Technology and Human Values* 16(1): 111–21.

# Abbreviations

| | |
|---|---|
| AIR | All India Radio |
| CSE | Centre for Science and Environment |
| DD | Doordarshan (Indian State Television) |
| ESRC | Economic and Social Research Council |
| GATT | General Agreement on Tariffs and Trade |
| GEC | Global Environmental Change |
| GoI | Government of India |
| IMF | International Monetary Fund |
| INS | Indian Newspaper Society |
| NAM | Non-Aligned Movement |
| NATO | North Atlantic Treaty Organisation |
| NGO | Non-Governmental Organisation |
| PTI | Press Trust of India |
| SITE | Satellite Instructional Television Experiment |
| UNCED | United Nations Conference on Environment and Development |
| UNESCO | United Nations Educational, Scientific and Cultural Organisation |
| UNI | United News of India |
| WHO | World Health Organisation |
| WWF | World-Wide Fund for Nature |

# 1   The cultural context

## 1.1 INTRODUCTION

This book reports on the results of a large research programme into the extent and style of reporting on the environment by the mass media in the UK and India, and into what the public know and think about environmental issues. To do this we have used a number of different research methodologies, both in the collection of varied and large data sets and in their analysis. We could begin by explaining these – but we think it even more important if we begin by explaining something of the cultural and economic contexts of these two countries – which of course constitute the framework within which any understanding of our results has to be placed. Although these two countries, one from the South, one from the North, have been intertwined with each other for the last three centuries and have had a major impact on each other (and as a result both India and Britain have a civil service which operates in English (in India at the higher levels) and both drive on the left), the cultural baggage which each one of them carried into the era of their mutual encounter could hardly have been more different, and their experiences of the world around them also radically different. Britain began industrialising two centuries before India, and the former came to dominate the latter through its imperial adventures. In the latter half of the twentieth century one became known as 'developed', the other 'developing': Gandhi surmised that if Britain had to conquer half the world to become developed, there were not enough worlds left for India and the others to conquer – implying that that sort of development would be neither possible nor desirable for India. Most of the historical cultural baggage survives transformations that take place in each society even if it itself is also metamorphosed along the way. The right to monopolise the meaning of such words as 'development' and 'environment' leads to consequent censorship of some of the most important policy-making facing mankind. This can lead straight to the heart of political manipulation at both international and national scales. This is a theme to which we will repeatedly return. In India, Gandhi's cautions did not prevail. From pre-Independence days the idea of 'development' had already taken root (Zachariah, 1996).

## 1.2 THE PROBLEM OF ORIENTALISM

Much has been written in recent years criticising 'Orientalist' approaches in both academic and popular western publishing on 'The Orient', i.e. South Asia (India of old), Indo-China, China and Japan. The basis of the criticism is that 'Orientalism' has its origins in colonialism and the colonial subjugation of people who were known collectively as 'Orientals', a term which has a certain disparaging edge to it. As part of the process of subjugation, knowledge about and images of the subjugated cultures were framed in particular ways – often stressing the differences and 'otherness' of the people in a rather simplified and collective manner, and usually stressing differences which implicitly recognised the superiority of the European opposite pole of the difference: 'the orientalist paradigm was a persistent feature of social science which constructs the Orient (as stagnant, irrational and backward) as a contrast case to explain the Occident (as changeful, rational and progressive)' (Turner, 1994: 96). The use of the word 'Orientals' is of course not necessarily restricted to the written word – it can occur in any form of communication, in any medium in which discourse is possible. In the nineteenth century, for example, art prints by European artists might show neatly dressed and disciplined European troops standing near neat, classically styled colonial buildings, a short distance away from poor and shabbily dressed (but 'colourful') natives, outside semi-dilapidated hovels. In the twentieth century TV comedies (for example BBC TV's It Ain't Half Hot Mum! in which British soldiers are posted in India during the Second World War, affording plenty of opportunity for 'native' menials such as punkah-wallahs to wander in and out speaking in funny accents) as much as TV documentaries or advertisements put out in Britain by India's own tourist promotion board can perpetuate Orientalism at a popular level.

Modern Orientalism is seen as alive and well and manifest in new forms by some commentators. 'Today it returns in the guise of waffle about the "tiger economies" of the Pacific Rim' (Ascherson, 1996: 27). Most contemporary academic writing in the West now tries to come to terms with the Orientalist critique, but it is not clear that there is any way of avoiding other kinds of relativism which merely replace the original social construction of knowledge with another. We hope this book avoids obvious Orientalist traps, but that does not mean it will be free from some kind of relativism, nor indeed should it be free from Orientalism at another level. We are studying popular culture and popular imagery and communication – in the press, TV and radio, and the responses of ordinary people to that material. To the extent that that is 'Orientalist', we should say so.

Part of Said's (1978) original exposition of Orientalism was the idea that there was no equivalent Occidentalism. 'Orientalism was based on the fact that we [westerners/imperialists] know or talk about the Orientals, while they neither know themselves adequately nor talk about us' (Turner, 1994: 45). This is clearly true of the way in which Europeans have conceptualised the growth of knowledge about The World – it was apparently Columbus and not the Caribs who 'discovered' the Caribbean – but there might be no necessary theoretical reason why this has to remain true. There is evidence in our work of a growing kind of Occidentalism at a popular level – contrasting the pollution, social breakdown, alienation, greed and diseases of affluence of the West with the ecological sensitivity of Hinduism, the loyalty of the extended family, the modest levels of personal consumption, and the diets and proscriptions (on alcohol for example) which promote health and are part of a preventive approach to disease. Indeed, studies of the image of the West in Indian thought are beginning to emerge (for example, see Stasik, 1994).

## 1.3 DISPARITIES IN WEALTH

Although the British may disparage themselves and believe that they have slipped economically behind the USA, Japan and Germany, the United Kingdom today is fundamentally a wealthy nation, even if that wealth is not equally distributed and the last decades have seen a real increase in poverty. Basically, water, sanitation, electric power, telecommunications, are reliably available to all, and the level of illiteracy is nearly as low as the mental abilities of the populace will allow. The growth rate of the population hovers a little above zero – certainly not enough to be a major policy concern of the government. The population is in fact ageing, and more elderly people, living longer, are becoming dependent on the remaining work-force. Whether this shift ultimately helps in reducing unemployment levels from an unacceptable 10 per cent has yet to be seen. These 10 per cent have access to state benefits, which, though seen locally as meagre, mostly mean that people do not starve. Two per cent of the populace work in agriculture yet produce 50 per cent of the nation's food requirements. The British have an average income of £10,000 per year, and most families own a TV set, have a telephone, a washing machine, and a car. Meat forms a significant part of the diet, and if we count animal feed as part of what people actually eat (indirectly) then per capita consumption is over 9,000 calories per day. Much as the Indians do not eat beef, the British do not eat horsemeat. The inspection of vehicle maintenance standards has become quite rigorous, and by and large the process is not corrupt – something which contrasts with the current state of play in India. Although Britain has rather archaic political institutions, it is nevertheless a democracy.

India's population is nudging towards the 1 billion mark; it is certainly well over 900 million now. It is still growing at more than 2 per cent per year as a result of that process known as the demographic transition, during which death-rates fall quite fast, lifetime expectancy increases, yet birth-rates do not drop either as fast as death rates, or as soon. No one knows what the ultimate stable population of India will be. It is routinely stated that a stable point will be reached sometime late in the next century, and that by then the population may be somewhere near 1.7 billion – surpassing China's. The bulk of India's population lives in the North – indeed, the Ganges–Brahmaputra basins in South Asia (including therefore the population of Nepal, Bhutan and Bangladesh) together have some 500 million people – 10 per cent of the human race live in this river system alone. This 500 million includes 30 per cent of the world's poorest 1 billion people. In India as a whole, according to the 1991 census, 52 per cent of the population are literate, and between 70 and 75 per cent are rural, living in about half a million villages. The urban population is massive – about 250 million people, or five times the popula-tion of the UK – but it is concentrated in a few megacities and 250 cities with a population of over 100,000. Altogether, for the half million villages, there are only 4,000 urban places, most of them with very minor infrastructure, to demand produce from the countryside and to provide goods and services in return. Looked at geographically, there are two Indias: a 'high-speed' India in the northwest, west and parts of the south, where there have been constant gains in living standards, and 'low-speed' India in the northeast and east and other parts of the south, where standards have stagnated for decades (see Chapman, 1992, for an elabora-tion of this).

The diet is largely vegetarian, partly because of religious beliefs and cultural practices, but even those whose religion allows them to eat meat – Muslims and Christians, for example – actually derive very little calorific input from meat, because there is little of it and it is relatively expensive. Overall, therefore, the average indirect consumption of food per capita is almost the same as the direct consump-tion – somewhere around 3,000 calories per person per day. This is about the WHO minimum for a healthy life: since access to food is unequal, it means that there are many surviving on a calorific level below the minimum. Now follows an Orientalist observation. Strangely, for a vegetarian country, it seems to be overrun by cows – India has the largest bovine stock in the world. These are of course venerated – literally holy cows, but venerated or not, many are ill-fed and of little use. Strays are beaten from the countryside where they might thieve fodder from productive beasts, and end up in towns, browsing rubbish tips and eating cardboard packing cases. This is a fact of 'otherness' which is visually unavoidable to visitors from the West, who have not grown up accustomed to seeing cows wandering unattended down shopping streets or chewing their cud as they lie in the middle of crossroads at night. The 'differentness' of this experience is bound to attract TV film producers and tourist cameras alike, and so the image will be perpetuated back home.

It is often claimed that India's 250 million urban people constitute a new consumerist middle class but this is an exaggeration on two counts. First, the salary level of the employed middle classes is low by international comparisons – £100 (Rs 5000/-) a month being a good salary – and second, because many of the 250 million are extremely poor: somewhere around 30–40 per cent live in squatter settlements, and work in the 'domestic' or 'unorganised' or 'informal sector', with little job security. No one knows what the overall level of un- and under-employment is, but it is often quoted as at least 30 per cent in terms equivalent to western-style calculations. Nevertheless, India is ranked somewhere near the world's twelfth industrial power. It has nuclear power stations, exploded a nuclear device in 1974, builds space rockets, builds its own railway locomotives (and exports a lot of railway equipment in general) and has a rapidly growing automobile industry. It has the world's largest number of trained computer software engineers, and exports software globally. It has the world's largest film industry and a growing tele-communications sector, though one which heavily favours the largest urban places. It has a vibrant free press, with over 28,000 regular registered publications, in many Indian languages and in English.

For nearly fifty years, since Independence in 1947, and with one wobbly exception of dictatorial rule by Mrs Indira Gandhi during the 'Emergency' of 1975–77, India has functioned as the world's largest democracy, despite the illiteracy and linguistic and religious diversity. The armed forces have remained subordinate to civil political power throughout.

## 1.4 NATIONALISM AND THE IDEA OF DEVELOPMENT

Whenever we look out at the world around us we do so on the basis of unexamined assumptions. By this we do not mean that these assumptions are never examined, but they are not examined in the shorter term acts of experiencing our surroundings – including the messages that come via the media. The unexamined assumptions come at different hierarchical scales. We hold assumptions about trivial things, such as that we will have breakfast tomorrow morning, that the post will arrive. We assume less trivial things, that the students in class this morning will not have a mass shoot-out in which they kill each other. The whole of communication depends on shared assumptions: for example, news presenters in the UK assume that most of the audience will have some idea of what NATO means, and that they can distinguish between local and parliamentary elections. There is also an assumption that the audience will understand the conventions of how visual footage is edited, and how one scene is connected to the next, something which we will touch on later in the book, since the conventions in different cultures are not the same. Then at the higher hierarchical levels come the assumptions that are harder to see about shared cultural histories. It is assumed by TV presenters in Britain that the audience knows that Britain has not been invaded for 900 years, that it has a long history of democratic government, that it is a post-imperial middle-ranking power, which is quite wealthy and technically advanced in global terms but which in relative terms has slipped down international rankings quite a lot since the Second World War, and that it has an ambivalent attitude towards the rest of Europe – 'The Continent'. There is an assumption that, for all its problems, Britain is not a 'developing' country.

The word 'development' is, of course, part of the language, and it has many meanings in Britain. It is commonly used to talk about buildings – 'Then they put up that new development' – and about change in a news story – 'There has been a new development in the Peace Process in Northern Ireland today.' Academically it has many meanings too. If you look up the word 'development' in the periodicals index of a University Library, there are hundreds of entries. Some journals are about the development of embryos, some are about the development of the faculties of newborn children, some about the socialisation of young children, and some about education, psychology, physics, etc. Only a very few are about development economics and developing nations, and these are specifically directed at South case studies and how to 'improve things' – by bringing sanitation, education, health, etc. to the mass of the world's population who do not have the same standards of living as the North. (All of these meanings will surface in one of our chosen UK videos shown to both British and Indian focus groups.) The interesting thing is that quite clearly Britain is developing the whole time in the sense that there is change going on and that capital continues to accumulate and per capita incomes increase. (We are not arguing that this is necessarily the same as having a better quality of life, nor that it is sustainable.) But this is not instinctively thought of as 'development' since Britain is supposed to be a 'developed' country. At one point in this research project when we were attaching key words to broadcast news items it rapidly became apparent that for many items on Indian TV we needed a key word 'development' for government drinking-water schemes, primary education schemes, etc., but that this word did not fit stories about water quality in Britain or debates about education.

It is impossible to stress enough how different are the lenses through which India views itself and the world. For the same 900 years during which Britain has not been invaded, much of India has been under imperial rule by dynasties originating from outside India. They include the 300 years of the Afghan Sultanate of Delhi from AD 1100 onwards, the 300 years of the Moghuls from 1500 to 1800, and the 150–200 years of the British Raj, ending in Independence in 1947. India was not seen even by these rulers as a country – it was seen as an empire of nations in itself. The British were the first to bring the whole of South Asia under one hegemony, but they did not unite it as a single country since it was a hotchpotch of provinces treading slowly towards home rule, and a collection of princely states of the maharajahs and nawabs under British protection but internally autonomous. At Independence in 1947 the ramshackle edifice fell apart (the finger of blame is pointed in all directions) and gave birth to West and East Pakistan (now Bangladesh) and what became the Republic of India. Well over a million people died in communal massacres, nearly 20 million fled their homelands in a sort of ethnic cleansing, spontaneous but also instigated, such as has been seen more recently in the former Yugoslavia. Yet it is common to hear North Indians say that they do not understand the need for Pakistan. The people are the same, the culture is the same, everything, they say, is the same. With the same breath they will speak with incredulity about the habits of the people of the southern states, whose inhabitants speak non-Indo-European languages which use very different scripts. Since 1947, therefore, the central government of India has had as an absolute priority the instillation into the minds of each and every one of the citizens of the republic the idea that they are first and foremost Indian. It has had remarkable success – something from which Europeans have much to learn. But it is not total success, and the dark forces of Hindu–Muslim communalism, linguistic communalism and regional secession remain. Every day the schoolchildren all over India sing the national anthem – a hymn to all the states and regions of India. This may be called indoctrination or it may be called socialisation.

The first step was therefore to cultivate the idea of secular Indian-ness, but that alone is not enough to legitimise the Union (i.e. federal) Government. This central government must do something for its people: in modern political parlance it must pursue a 'great idea', so that they can all see how they as Indians are benefiting from *swaraj* – self-rule – in independent democratic India. It is also a truism that there would be greater cohesion amongst the citizens if they felt themselves to be some kind of 'us' against some kind of 'them'. 'They' were clearly the indus-trialised, imperialist countries, one of whom had ruled India for the last century and a half. This is not the place to enter into a debate about the demerits or merits of British rule, the laying of the railways (at profit to the British and at the cost of subsidies from the Indian tax-payer) and the development of irrigation. Suffice it to say that British rule was concerned first and foremost with the maintenance of 'law and order' (which will again surface in our contemporary public opinion surveys) so that trade could continue. It is true that new universities were founded, that democracy did painfully and haltingly advance in the British provinces, but for much of India very little changed. The level of industrialisation, even simple industrialisa-tion, was abysmally low (at Independence in 1947 India produced no bicycles indige-nously). The Prime Minister, Jawarhalal Nehru, viewed the India of 1947 as sunk in backwardness, ignorance, poverty and superstition.

Nehru had been party to the pre-Independence analysis of the route to modernity and 'true independence' (see Zachariah, 1996, cited in the Foreword); he believed that for the country to progress, planning by the government was essential, and that 'planning was science in action' (Vasudeva and Chakravarty, 1989: 417). In short, the government would seek to 'develop' India, and it readily acknowledged its place among the 'developing nations'. It took inspiration from the rapid devel-opment (and in the early years it was rapid) of the Soviet Union and evolved its own idea of five-year planning and public sector dominance in the commanding heights of the economy. Although diplomatically close to the Soviet Union, it helped promote the grouping of the non-aligned nations (NAM, Non-Aligned Movement), and for these two reasons and other geopolitical considerations as well it found itself not courted but spurned by the one democratic industrial power – the United States – which had been anti-imperialist and had championed India's freedom. This rift is relevant to the material contained in this book.

India turned its back on 'economic dependency' and became a closed economy. Its share of world trade collapsed dramatically and import substitution became a central goal. Public-sector investment was poured into heavy industry but the rural sector was not forgotten. Here the effort was as much administrative as financial. The structure of government that the British had inherited from the Moghuls, modified and then passed on to independent India, reached down to district level – a district being a subdivision of a province and commonly having 1 or 2 million inhabitants. For taxation and police purposes there were subdivisions of the district – but the district and its District Officer were paramount. In sweeping reforms in the late 1950s, villages were grouped into new blocks of about 100 villages to implement a programme known as Community Development. Each block was headed by a Block Development Officer who oversaw village-level workers. Substantial efforts were made in land reform to eliminate the excesses of feudal landlordism and to give more equal holdings to the peasantry. Nor did technology pass rural areas by. India accelerated the development of large-scale irrigation and began the construction of huge new dams – the temples of modern India as Nehru called them. The point is that the word 'development' intrudes wherever society and government intermesh. And the idea of science intrudes wherever development and society intermesh – new seed, new fertilisers, new health clinics, massive electrification programmes.

In the 1980s the small state-socialist economies of Africa and some of the large economies of Latin America have all succumbed to demands from the World Bank and the IMF for liberalisation and restructuring. Put simply, this means ending controlled currencies and allowing national currencies to find a true international value, massive reduction in anti-trade tariffs, and an increasing export-orientated industrial policy rather than an import-substitution industrial policy. These countries have often been massively in debt and unable to withstand the demands for change. In India's case, her size and the fact that, comparatively speaking, she has not had such a large overseas debt has meant that external pressures have not worked either so fast or so simply. But the realisation by Indian leaders and by different sectors of the new entrepreneurial classes that growth has been slow, that to be modern India's industry has to compete with the rest of the world, together with pressure from GATT (now the World Trade Organisation) and its North bosses, and that the government has inadequate revenues has indeed led to a policy of liberalisation and increasing openness. Now international investment is more likely courted, not spurned. The policy is not, however, completely firmly set, since vested interests in India are hurt by it. But it does seem that a total reversal is unlikely, even though after the Congress's first loss of a majority in the Union Government in May 1996 (apart from a short period after Mrs Gandhi's Emergency in 1975) it is clear that many compromises have to be made. These compromises will also surface in our analyses. *The Economist* (1996) points out that the many losers will continue to force a democratic government to subsidise the losses inflicted. At the risk of oversimplification, we may say that the emerging post-national classes (explained in Chapter 4) form the Indian version of a 'Northern' élite within India separated from the remaining 'Southern' masses.

Of course, there have always been dissenting critics in India of the whole modernist project, some of them decidedly Gandhian in outlook. Mostly they have objected to big cities and industrialisation as avoidable evils, something that is seen to destroy family life, and that has Kali Yuga quality (a Hindu concept that we are now in an age of destruction – see next section) about it, like Blake's 'dark, satanic mills' in England. They stress rural development with appropriate technology and institutional change to remove inequality in society. In this view, technology follows rather than leads, and institutions are built bottom-up. The Gandhian approach is revealed, in modern terminology, to be very environmentally aware, although during Gandhi's lifetime the environmentalist movement had not independently reached a peak which he could capture for his own purposes. The Gandhian dream is always honoured in India, but honoured more in the breach than the observance. The idea is kept alive as a kind of national conscience: utopian communities supported by private donations operate in all parts of India but they show no sign of becoming the dominant paradigm.

Independent India has much to be proud of. It has fed its burgeoning population, it has achieved an extensive industrialisation, although it has generated proportionately little employment. The rate of growth has not, however, eradicated massive poverty, particularly in rural areas, and so the continued need for 'development' is acutely felt.

This chapter was written in India, while Chapman was a Fellow at the Indian Institute of Advanced Study in Shimla (Simla, the hill station summer capital of British India), in Himachal Pradesh. Adjacent to the Institute is the University of Himachal Pradesh. In April 1995 the Department of Sociology there held a three-day national seminar on sustainable development. Chapman was invited, but could only go for one of the three days. He was the only non-Indian present on any of these three days. The dialogue at the seminar was framed nearly exclusively in terms of 'they' and 'we', or 'them and 'us'. 'They' and 'them' refer to the industrialised nations, 'us' to developing India. The majority of the delegates, members of India's intelligentsia, have never been outside India, for obvious economic reasons. They have little exposure to the 'reality' of the North, and know it through popular discourse, through contacts with migrants abroad in the Indian diaspora, through what is on TV and through the press. Their comments fell into two broad categories. First that the North tries to dictate – over nuclear non-proliferation, over the patenting of genes ('our neem', a tree which produces a natural pesticide, is being exploited by the 'multi-nationals'), over trade liberalisation, over the terms and costs of the transfer of technology. 'If they want our industries to be clean, they should give us that technology.' All this refers to a perceived 'unfairness' and reveals a strong suspicion of neo-colonialism at work. The second theme was how the North was an environmental disaster, each individual consuming 30 times the resource needs of an average Indian, with the inevitable result that pollution by the North was horrific. (Allied to this was the perception that in the North social life had been debased and all family values abandoned. The fact that the UK has the second lowest homicide rate in the world, coming after Iceland, did not feature.) The articles reproduced here are a recitation of some of these themes.

**The Tribune**
Chandigarh, 7 May 1995
Sunday Reading

### THE WHITE ARROGANCE

America reserves the right to impose its will on any nation in the world which does not conform to its conception of correct behaviour

says
G.S. Cheema

The Western media catering as it does to a vast mass audience simplifies and categorises everything into standard well-known stereotypes. For instance, the Muslim is almost invariably 'fanatical', the Indian 'superstitious', or 'poverty stricken', and the Chinese and Japanese 'inscrutable' and 'cruel'.

Such sweeping generalisations are mindlessly applied to 'Orientals' because they are 'different'. They are very seldom applied to fellow Caucasians, especially if they be Americans or Europeans. The division of the world's nations into 'us' and 'them' could not be starker.

**The Tribune**
Chandigarh, 1 June 1995

Editorial

**A Correct Stand**

India appears to be under tremendous pressure from various industrialised countries to open up its lucrative financial sector to international market forces ...

The country has long experience of discriminatory treatment by the industrialised nations. The USA, Japan and other European states have been taking unfair advantage of the weak economic and political leadership of the developing world. On all world issues, the league of the First World has been ganging together to brow beat the struggling and poverty stricken Third World. Whether it is nuclear proliferation, global warming, trade barriers or the elimination of poverty, the powerful group of industrial countries has taken a united stand to bulldoze its way. ... Only a resolute stand at the forthcoming negotiations on these issues at Geneva this month would ensure and protect the vital interests of the South in general and India in particular.

**The Tribune**
Chandigarh, 21 May 1995
Sunday Reading

**Running to Stay in Place**
By Gagan Dhir

Why is human hunger widespread in India but not in China, where there is half as much arable land? The pressure from a growing population is an obvious symptom but not the main cause of poverty, hunger and environmental degradation. The biggest contributors to the deteriorating environment are nations with stable or decreasing population.

... One quarter of the world's people living in developed countries consumes 75 per cent of the world's energy, 79 per cent of all commercial fuel and 85 per cent of all the wood. Each person in Europe, North America, Australia and Japan consumes, directly and indirectly, enough grain to feed four Indians. The average Swiss consumes 40 times more resources than the average Tanzanian.

Because they consume more, developed countries pollute more. Paul Shaw of the UN Population Fund calculated that in 1993 rich countries were producing approximately 1.9 tonnes of waste per capita. Poor countries were generating just 0.2 tonne per person and as little as .08 tonne in rural areas.

Says Shaw: 'This means that the biggest contributions to environmental degradation – measured by waste generation – come from countries where population growth has been stable, if not declining.'

*Note:* The article never gets around to answering the question it poses at the beginning about hunger in India.

Our point is not the truth or otherwise of these views. It is that again we come to unexamined assumptions. One of them in India is that a major part of India's predicament is that the global system disempowers the country and works against it. What was most interesting was that this dialogue could take place for the majority of the time with no dissenting views expressed, or even expressible, because 'they' were hardly present, since Chapman did not see his role as that of a North representative. To be fair, there are many in the business community who do support openness and the global trading system.

To conclude, in India an unexamined assumption that permeates every nook and cranny of the land is that there is such a thing as 'development'. Even the age of trade liberalisation is reinforcing the idea, since now 'development' is seen as internationally competitive, not just the private affair it remained during India's four decades of autarkic development.

## 1.5 SOCIETY, SCIENCE AND DEVELOPMENT

At the conclusion of the English civil war in the 1660s, England became a commonwealth, or in modern parlance, a republic. But it was short-lived, and the monarchy was restored remarkably quickly after Cromwell's death. The Restoration symbolises the incompleteness of the revolution, a fact mourned by significant political critics in contemporary England. The industrial revolution changed society radically, the peasants became members of a new working class – Marx's hope for the future – but the old aristocrats co-opted the new industrial barons into their own class, creating new peerages for them. The French Revolution and later Napoleon were kept at bay, and in this century two world wars failed to defeat the British establishment. The idea and ideals of class therefore remain more deeply ingrained in Britain than in any other European country.

Indian society is more deeply fragmented by religion than most of contemporary British society. The term 'communalism' is used to refer to the manifestation of the struggle between Hindus and Muslims. Although India's constitution proclaims it a secular state, it still does not have a common Civil Code, so that customary marriage, divorce and property law differs between Hindu and Muslim communities. 'Communalism' will surface in this book in our surveys of public opinion.

The next two paragraphs are about caste, and since they are brief yet deal with a highly complex topic, they will again tread dangerously close to the Orientalist tradition. India was, and to a large extent still is, despite urbanisation, a caste society. A caste is an endogamous marriage group into which a person is born. Most marriages are still arranged, and arranged within the marriage group – newspapers carry marriage columns in the classified ads in which the caste affiliation of the proposed bride or groom is part of the classification. Caste status is therefore not achieved but ascribed, just as social rank still is with the hereditary British aristocracy. Caste groups traditionally in the past, and to some extent even now but in more subtle guises, followed different occupations. These occupations have conferred a degree of ranked social stratification, and the inherited social distinctions are still important. One of the biggest current political issues in India is 'reservation', a system most recently reinforced by the Mandal (named after its chairman) Commission, by which groups or castes in some areas who are thought to have been particularly backward or suppressed are guaranteed a certain quota of places in public institutions, higher education and public employment – that is, the less equal are made more than equal in order to be equal. It is akin to US government policies of 'affirmative action'.

In traditional theology the highest caste group is the Brahmins, the priests, or pandits after whom the British call pontificating political commentators 'pundits'. Brahmins are thinkers, who eschew physical work – that is for the lowest castes, the Shudras, who till the earth and reap the crops. Here is not the place to explore this theme in depth but there clearly are resonances from this division in modern

India. There are, as noted above, many software developers in India – it suits the abstract mind that the culture has developed. Nor is India devoid of empiricists – anyone who studies local knowledge in farming is well aware of the depth and subtlety of local ecological knowledge (see Chapman, 1983, 1984). But it was in European societies that Baconian experimental science fused theory and empiricism and produced the industrial revolution.

India has enough physicists to produce an atomic bomb, but it is not just the number of scientists that will determine whether India can be at the forefront of applied research. The Scientific Policy Resolution adopted by the Lok Sabha (parliament) in 1958 under Nehru's leadership (Vasudeva and Chakravarty, 1989) gave science and scientists a greatly enlarged responsibility for changing society, and the institutions and funds to help in this aspiration. This would create what came to be known as the new 'scientific temper' of the country. It was an inherently ahistorical view of science as an objective activity outside society. The causality was therefore supposedly unidirectional. Such an attitude ignored the fact that science could lock itself into its own privileged empires of obfuscation and mystique.

> In keeping with the Indian philosophical tradition, the imposition of a foreign way of thinking did not penetrate beyond the surface levels of the culture. What resulted was a division which contributed further to the contradictory nature of the society. The introduction of modern technology and thought processes brought the empiricist tradition of observation and analysis but suited the Indian admirably since it enabled him to pursue science as an observational activity without interfering with his traditional world picture. The materialist thrust forward evident today is not at odds with the Hindu way of life which uncouples the temporal from the intellectual.
>
> (Reddi, 1989: 402)

So agricultural scientists conduct excellent pioneering work in research stations across the country, but agricultural extension remains a very poor cousin, and the gulf between the scientists and the peasant farmers often remains unbridged. Often the latter have to experiment with the new chemicals that spasmodically hit the markets, unaware of the proper treatments that have been devised, and of course unaware of the results published in academic papers, usually abroad. The gap between thinker and doer remains wide in most spheres. Speculation then remains unchecked by experience and can result in sweeping generalisations. If one of the hills is deforested and degraded, then all the Himalayas are degraded and deforested. If reforestation has been successful here, then it can be successful everywhere. The practical boundaries of application get scant regard.

In Britain it is perhaps only in the last decade that there has been a growing apprehension that maybe the passage of time, the march of progress, the growth of material wealth and the pursuit of happiness are not coterminous. Individuals can have their doubts about the impact of motor cars and commuting on the ideals of community living, about the impact of cars on urban atmospheres, on growing stress and insecurity in highly paid but highly demanding work. There may be a sense that people are beginning to think that perhaps we should get off the good ship *Progress*, even though the voyage is not finished, or at least pause a while and look at the charts to see more clearly where we are going. We are even beginning to wonder how we started on the voyage in the first place. Clearly part of an answer to this question lies in the development of science in post-Renaissance Europe, but that does not say enough about the cultural context that allowed the application of that science.

## 1.6 THE UNDERLYING COSMOLOGIES

The most ancient stories of the most ancient societies are stories of creation. Most of these stories are in a modern scientific sense objectively wrong, though nearly all contain some surprising aspects of 'the truth', which hint at a deeper understanding of the world than is apparent at first sight. The extent to which our values and attitudes may be guided by an unacknowledged, older and wrong 'truth' rather than a modern scientific new 'truth' is only clearly apparent when we start looking at the larger patterns of human behaviour. In contemporary Britain this means seeking out the relationships between the Christian past of the country and its current cultural values. In the Judaeo-Christian account of creation given in Genesis in the Bible, and uncontested in the public mind until Darwin's ideas took root less than 100 years ago, man and woman were created last and put in the garden of Eden.

> And God said, let us make man in our image, after our likeness: and let them have dominion over the fish of the seas, and over the fowl of the air, and over the cattle, and over all the earth, and over every creeping thing that creepeth upon the face of the earth.

The earth that God had created was therefore seen as essentially unchanging, except to the extent that man could and should, in the name of the Lord, tame the wilderness and establish a settled Christian civilisation. It is a story of individual morality – each individual is born on earth once and once only, and looks forward to a continued individual existence in the afterlife, the quality of which depends on his or her lifetime's actions. The attainment of perfection, not of harmony, is the ideal. No limits are conceived of the earth's capacity to sustain society, and not until Thomas Malthus did people begin seriously to wonder about the carrying capacity of the earth, and even then only from the viewpoint of its impact on man, not of the reciprocal impact of man on the carrying capacity of the earth. Thus the earth had no mechanisms inherent in itself for ending itself. An end was assumed, but it would come through a divine act on Judgement Day.

Hindu cosmology could hardly be more different. The idea of cycles and repetition dominates. There are many universes, new ones forming out of the collapse of older ones. They go through phases in which creation exceeds destruction, as order emerges out of chaos, to be followed by the descent again into chaos. Of the many divisions and subdivisions of time, one, the Kalpa, is most important – it represents one day in the life of Brahma, the life force of the universe. It is equivalent to 4,320 million years – close to the age of the earth.

Brahma is the one and only God – not a god in the sense of an anthropoid projection like the Christian 'Our Father' but rather energy, *the force* in everything in the universe, the force revealed, like a hologram, in fragments and in its totality in everything that lives (all animals and plants) or exists (all rocks and rivers). Man is in no sense a privileged animal, even though he has intellect. The pantheistic aspect of Hinduism is simply a representation of the manifold revelations of the life force – so there are many gods representing the many faces of the universe. The god Vishnu represents the Creator, and the god Shiva the Destroyer. Both are necessary, since the cycle of creation and destruction is unending, but at different times different aspects are dominant. The Kalpa is subdivided into Maha Yugas, each of these being subdivided into a further four yugas, of which the last stage is a Kali, or Black, Yuga in which destruction is uppermost and the descent into chaos occurs. According to sages we are in one now. That this is so can be confirmed – earthquakes strike, bridges collapse, droughts are followed by floods, dams collapse, the plague attacks us – and that is the way of things.

For most Hindus the belief in reincarnation is as deep and instinctive as the assumption for both believers and atheists in the West that each individual lives on this earth just once. For the atheists in the West, this means that there is no point in not engaging with the world here and now. For the believers it means that they have one chance to prove their worth and that if they fail in their actions in this world, then at least there is a chance of forgiveness and redemption. In Hinduism it is incumbent on the individual to fulfil the duties that birth, i.e. caste, has prescribed for him or her. (S)he will be reborn to a status that reflects that fulfilment, since there is no equivalent of forgiveness and redemption. The result is less of a willingness to engage the world fully, certainly less willingness to change it – which would be beyond 'karma'. Hindus have therefore been described by some as fatalistic, by others as 'cosmically lonely', with eyes set on the next world – i.e. the next incarnation.

In India, therefore, not only are thinkers separated from doers, but the thinkers are less predisposed to assume 'progress' in the first place. Perhaps this is why the idea of 'development' has to be cultivated so assiduously, but it is applied more carelessly because of a tacit cynicism about its value in a Kali Yuga, which will envelop mankind as surely as the tide enveloped Canute. To a westerner, one of the shocks of India is not just the number of horrific traffic accidents, the leading cause of death in many parts of the country, it is also the fatalism with which it is accepted. It almost seems that there is no belief in the possibility of preventive measures through better standards of driving or of maintenance (apart from the obvious economic statements about the cost of maintaining good tyres – which is a relative value statement anyway, an opportunity cost and not just a straight economic cost). That vehicle safety checks should be completely subverted by corruption does not really surprise anyone. That factory safety checks and pollution emission tests should similarly be subverted seems equally not really to surprise anyone. Yet clearly there is a degree of conscience about such violations, since the press constantly reports them.

Even now the majority of Americans assume that there can be a technical fix to just about any problem. The Indian attitude of fatalism may not ultimately be healthy in itself, but at least it is a healthy antithesis to western optimism, something which the West is slowly beginning to realise. Does Hinduism offer no way forward at all? Theoretically, is it desirable that India should reject its cultural baggage, if it could, much as Nehru seemed to hope? There are those, such as Gadgil and Guha (1992), who see in Hinduism and caste the possibility (and also the actuality, in a 'golden past' which we think is highly questionable) of a civilisation which is much more in harmony with nature, much more sustainable. Tomalin (1994) provides some contemporary support for the view that symbolic engagement with the environment is still very powerful, although not necessarily from a caste position. She points out that Lakshmi, the Goddess of Wealth, whom many Hindu business people revere, is actually celebrated perhaps more by the majority of rural women, in the form of festivals widespread in rural India which honour dung, the means of returning fertility to mother earth. But we will not go into this topic further here, partly because it something which has emerged on the research agenda only very recently.

## 1.7 THE LINGUISTIC MAP OF INDIA

This section applies to India alone. In Britain the use of the English language is overwhelmingly dominant. There are still pockets of Celtic speakers in Wales, served by a Welsh-language broadcasting system, and even smaller pockets of a closely related Celtic language in the Scottish Isles and Highlands, but these are not of a scale to constitute major political problems for central government – even if Scottish regionalism does (Maps 1.7 a and b).

India's population of more than 900 million embraces more linguistic diversity than any other nation on earth. Depending on the definition used and the extent to which a dialect becomes another language, there are between 225 (1931 Census) and 1,652 (1961 Census) languages spoken in India, but obviously not all can have equal official status. Fifteen of the languages are scheduled, that is to say recognised for official use in education and the law courts, etc. These fifteen have different scripts, and although for some of the northern languages derived from Sanskrit the scripts are very similar to the Devnagri used by Hindi, there are at least ten completely different calligraphic systems. Officially Hindi is the sole lingua franca, but attempts in the past to impose it in preference to English have met with stiff resistance, particularly from the South of the country. Thus English, which is not a scheduled language, is nevertheless used and accepted as a lingua franca in higher education, in business and in government. Proportionately, the number of people who speak English as a mother tongue is negligible, but the number of people who learn to speak it with some degree of competence is of course higher. Ancient India was noted for its scientific achievements (Basham, 1973), in particular its contribution to mathematics, but anecdotal evidence suggests that contemporary Indian languages have poorer scientific vocabularies than English. In terms of scientific publication in India, English is overwhelmingly preferred.

The population inhabits 22 states and 9 union territories – smaller enclaves such as the capital district of Delhi or the ex-French port of Pondicherry. Out of the 22 states, 14 are major states, defined since a significant boundary reorganisation starting in 1956 which based most of its decisions on the criterion of language. This move reduced the number of minority languages at the states' level, and enhanced the size of majorities speaking the dominant language of each state. This has had the effect of pushing the use of local languages higher up the educational hierarchy. Thus, although English remains the language of higher degrees, its use at undergraduate level has declined in favour of regional languages. Nevertheless, English is still the passport to the élite in administration. Candidates for the Indian Administrative Service (the Union level service, successor to the Indian Civil Service of the British period, known as the 'iron frame' of India) may take their exams in any of the scheduled languages, but less than 5 per cent of the successful candidates use a language other than English.

Although average literacy rates for the whole country are now (1991 Census) 52 per cent, there is considerable disparity between males at 64.1 per cent and females at 39.3 per cent. The disparity is exaggerated in rural areas, where 57.9 per cent of males are literate compared with 30.1 per cent of females. For both sexes urban literacy rates are much higher, though there is still a residual disparity in favour of males (81.1 per cent) over females (64.1 per cent).

Quite clearly, although language has been a major political issue in post-independence India, although it is not now as important on that agenda, the problem of communication between language groups has to play a prominent role in a study such as ours into the impact of the mass media on the general public, since we are

Legend:
- ⊠ >50% Welsh Speaking, 1981
- ▨ >50% Gaelic Speaking, 1991

100km

Source: Census of Great Britain, 1981 and 1991.

*Map 1.7a* The geography of language in Britain and location of focus groups

not talking about an undifferentiated public. It is even less undifferentiated than in Europe, where at least all the languages are Indo-European and share a similar script. Language will therefore surface as an issue in this book in several different ways. One is clearly the simple need for publishing and broadcasting in many

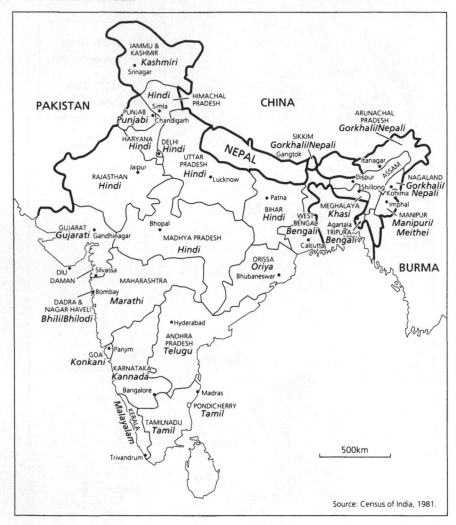

*Map 1.7b* The geography of language in India

vernaculars, but another is to do with the capacity of different languages to express concepts about science or about the environment. The British in the nineteenth century rejected the use of vernacular languages for higher education in India because 'the vernaculars of the country do not as yet afford the materials for conveying instruction of the comparatively high order' (Naik, 1963, quoting government documents from 1860, cited in Khubchandani, 1983: 20). Depressingly, Khubchandani seems to suggest that, according to one model of language development, the vernaculars will never catch up, since the target languages (possibly international variants of English) are acquiring new ideas and terms faster than they can be translated or assimilated by the vernaculars. Thus they are chasing a moving target – but that begs the question of why this particular target has been chosen. This underlies the contemporary British and European fear of 'Asian Tigers' that the target may not be so narrowly defined, but that it is whichever cultural group is accumulating superior 'knowledge'. It is also the case that for political reasons some languages may be manipulated to mould the messages communicated within them – as Ahmed (1996) has claimed of the Urdu press in Pakistan and Gupta and Sharma (1996) have claimed of the Hindi press in India.

## 1.8 THE MASS MEDIA

### Britain

By mass media we mean the press, particularly newspapers on sale for general circulation, radio and TV, particularly those channels which are transmitted terrestrially, and cinema. We do make passing comments on satellite TV, but we do not include it in any of our content analyses. Our analysis of cinema is limited to discussion of the Government of India's Films Division, one of whose producers we interview and two of whose films we use for our focus groups.

The oldest of these media is of course the press. The origin of newspapers in Britain is curiously tied up with the development of trade with the tropics. When coffee, and then tea, became available in London, coffee-houses became fashionable talking shops, a bit like pubs but definitely much more élite and for the wealthy intelligentsia. People wishing to bend the ear of this audience began to publish pamphlets, which rapidly evolved into single-sided broadsheets, and from then on into mainstream newspapers as we know them now. This is easy to say in a sentence, but just how the mixture of scandal, commercial news on ship movements and cargoes, and political opinionising developed into a structured system, a profession with its own values and standards is a long and complex tale. The success of the idea of journalism grew with the spread of literacy and education and the expansion of the market for all types of news. In the nineteenth century the first cable was laid between London and Brussels in 1851 by Mr Julius Reuters, and by 1867 a cable had reached India. It was used initially to report the prices at the Calcutta jute auctions to the jute barons in Dundee. For the first time messages could speed globally faster than the commercial ships carrying the cargoes, to which they might refer. It was a kind of coming of age in the separation of global news flow from freight transport – even if both can still be covered by the term 'communications'.

The daily press in Britain today is divided into different categories by its perceived journalistic weightiness, by its size and by its circulation, as well as, to some extent, by its political inclinations. The so-called quality press publishes in broadsheet format: there is nothing to suggest a serious daily could not be tabloid but the cultural association between tabloid and low-brow is too strong. The broadsheet newspapers have a lower circulation than the tabloids, and fairly clear marketing niches. *The Times* is no longer what it once was – the national arbiter – but tends towards conservative values, in both senses of the word. The *Daily Telegraph* has often been seen as the mouthpiece of the Conservative Party, but is less inclined to play that role at the moment. The *Financial Times* is normally expected to be pro-Conservative, but in the last general election backed Labour at the last minute, and may well do so again in the next one. The *Guardian* is left of centre and has claims to be the most intellectual. *The Independent* started in recent years as just that, but it is now part of Mirror Group and may come to take a more pro-Labour line. The founding philosophy was to represent some kind of uncommitted pragmatic centre, and a more educational and objective stance. This is the paper we have chosen to study as our representative of the 'quality' press in the UK. The tabloid press runs the whole gamut from the genteel *Daily Express* and the middle-class gossip articles of the *Daily Mail*, which has a modest news coverage given its page length, to the gutter press, which, though published daily, is more interested in soft pornography and sleaze than news. The worst is the *Daily Star*. Two tabloids vie with each other as the paper of the common man: one is left wing, the *Daily Mirror*, and takes its role as a purveyor of news and opinion seriously – it clearly has its own standards. The other, the *Sun*, has been staunchly pro-Conservative because of its ownership, and wrapped up its pro-government messages in a mixture of tawdry scandal and a modicum of soft pornography. We have taken the *Daily Mirror* as the paper for our content analysis of a mass distribution paper. There are other regional daily papers: for example, in Scotland the *Scotsman* is an influential and viable paper. But within Britain the major dailies we have mentioned dominate.

The press in Britain has rarely been interfered with by legislation or censorship. This does not mean that it does not come in for criticism, and over recent cases of the invasion of privacy (many to do with the endless goings-on of royal affairs) the threat of legislation has been used more than once to shake the industry's own self-regulation into more effective action. But the press is of course owned, and with different patterns of ownership come different editorial pressures. At the moment the most significant debate in Britain, as to some extent in India, is over the issue of cross-ownership between broadcasting networks and the press.

The broadsheet papers have their correspondents in significant capitals around the world, and they have their own freelance correspondents. They also all use the major news agencies – Reuters, Agence France Presse, Associated Press, etc. – and use the minor agencies less often. On occasion Tanjug, based in Belgrade, might have been used, but that is now effectively defunct. There is a structural bias, therefore, against receiving stories from the South which South nationals or agencies have originated.

With the invention of radio, the British Government was quick to act to keep broadcasting a state monopoly, anxious that the power of the new medium (which is a restricted resource in the electromagnetic spectrum) should be harnessed in the national interests. This did not mean that the government directly dictated the contents of radio, but with the establishment of the British Broadcasting Corporation (BBC) as the successor to the British Broadcasting Company, it defined a charter and a board of governors that ensured the style and content was acceptable. The legendary Lord Reith, the first Director General, defined the mission of the BBC to be to 'Inform, Educate, and Entertain' in that order, and for many years that was the order of priorities, the whole being paid for by a licence fee levied on all receivers. This was radically different from the USA, where radio from the beginning was commercial and became used as an extension of what would otherwise have been dying – the music hall. There, from the beginning, radio was essentially music, entertainment and advertising.

The independence that the BBC had, the style and gravity with which it developed its news culture and journalistic standards, all combined to give it a stature world-wide which is still largely intact even today. Domestically the radio organisation has expanded the number of national channels it runs to five, which cover popular entertainment (Radio 1) – mostly pop music, Radio 2, a middle-brow mostly musical channel, Radio 3 which is used for classical music and some drama, Radio 4 which is speech-orientated and includes many news analysis programmes and documentaries as well as drama, and Radio 5 Live, now a 24-hours news and sports network (although when we did the content analysis for this project it was still

Radio 5, a curious blend of sport, music, children's programmes and other things). Radio 4's morning programme Today running from 6.30 to 9 a.m., is thought to be the kernel of early morning political life – the issues and stories of the day are substantially defined there. No minister turns down the opportunity to be interviewed, no matter how terrible the grilling usually is. Commercial radio, known as independent radio, was first allowed to operate local stations (the BBC does too), but now includes national radio FM channels with specific remits. One is a classical music channel, another is a phone-in channel.

After the Second World War, when attempts at maintaining objective standards in broadcasting had made the BBC more powerful in Europe than any other broadcasting station, it was able to expand into the new medium of television with renewed vigour – but the same Reithian ethos remained. Legislation prohibited high levels of foreign (i.e. American) imported programmes, and information and education played a large part in programme output. Even entertainment was often fairly highbrow, with dramatisations of classical literature, a tradition in which the BBC has remained pre-eminent. This period when the BBC was not operating for profit or watching the bottom line enabled a culture of production to grow which stressed the quality of the output. It educated the public to expect high standards – something which is relevant to some of the discussions in later chapters. In 1955 commercial television was licensed for the first time, with companies operating on a largely regional basis, though co-operating in networking most programmes, particularly their news programmes. Independent Television News (ITN) was established as the first national alternative to the BBC's broadcast news monopoly, and has since achieved a status which some commentators think is superior to the BBC's. These commercial channels were livelier and provided something of a competitive shock to the BBC across a whole range of programmes. They did not import more US programmes – they were not allowed to do so – but it seemed as if they did, because frequently their quota would be scheduled at peak viewing hours only. But interestingly, the quality of BBC programmes undoubtedly had an impact on ITV, whose own quality was pulled upwards. There are now two BBC channels, BBC1 which is the more entertainment-orientated and popular channel, and BBC2, which is more highbrow information and education orientated, and two ITV channels – the regional companies of Channel 3, and the national Channel 4, orientated towards specialist audience groups, and often educationally highbrow. It started with subsidies from Channel 3 but is showing that commercial channels do not all have to regress to some awful lowest common denominator as they have done in the USA. Despite the advent of satellite – popular because it has bought up most big sport – and cable, the terrestrial channels are still very strong, and satellite penetration has remained below 20 per cent. A fifth terrestrial channel is due to start in 1997.

## India

Although English language newspapers were founded first in the late eighteenth century in India, the vernacular press soon followed. Most, but by no means all, of the English language press was in British hands before Independence in 1947, the remainder and much of the vernacular press is in Indian hands. Much of the press, including some that was British-owned, was to some extent anti-government, fostering the pursuit of Home Rule, and throughout the first half of this century the imperial government passed and implemented legislation aimed at curbing literature which would encourage disaffection against the crown, or enmity and hatred between different classes, or which would promote mutiny in the armed services. Censorship was not actually very fierce, except when the Gagging Act and the Official Secrets Act were imposed, and certainly did not include prior submission of material for approval. Since 1947 the press has frequently been accused of not knowing where it is going, and of being too close to government. However, in a country where newsprint was in short supply and rationed, and where some newspapers were owned by conglomerates which needed licences for larger industrial projects, it was not always easy to maintain a distance from government. Early after her assumption of the premiership in India, Indira Gandhi asked: 'How much freedom can the press have in a country like India fighting poverty, backwardness, ignorance, disease and superstition?' When she proclaimed a State of Emergency in 1975, assumed dictatorial powers, and immediately put the press under close censorship, most of the press simply caved in: '[They] crawled when they were only asked to bend' (Kumar, 1994: 69). After the emergency the new Janata government moved to reinstate the Press Council and to re-establish the independence of the major national news agencies. But the criticism remains, that the press is too reliant on government hand-outs, and that it is too dominated by politics at the expense of everything else. It certainly concentrates overwhelmingly on Indian affairs, and has a very limited coverage of other parts of the world. Facets of this will become evident in the content analysis in later chapters. In terms both of stories and readership it is also a press which covers the metropolitan cities first, with both English and vernacular papers, then the smaller urban places mostly with regional languages, and then finally and only in a limited way, the rural areas. Several of the more significant papers published in English from one city originally are now published in several cities in vernacular languages. One curious feature which is unlike anything in the current western press is the rather cavalier attitude towards time. *The Times of India*, published in Delhi on Wednesday, becomes *The Times of India* published in Chandigarh on Thursday, whence it is shipped to Shimla. But there is a local paper in Chandigarh – the *Tribune* – which subscribes to some of the national agencies and will carry the same main national stories as the Delhi *Times of India* on the same day. So for the stories that overlap one finds oneself reading today what one has already read yesterday.

In most mainstream papers the news values are similar to western ones, concentrating on 'timelines, immediacy, proximity, oddity, conflict, mystery, suspense, curiosity and novelty' (Kumar, 1994: 60), which, in the eyes of development journalists, 'promotes sensationalism, élitism and conservatism, and thus suppresses the voice of the silent and oppressed minority'. This is very different from the approach advocated by development journalists who would like the press to be pro-liberation, pro the poor and marginalised groups of society, pro the environment, and to adopt longer-term less sensational perspectives. There are, however, significant differences between the press in vernacular languages and the English language press. We look at this in detail in the following chapters.

For this project we chose to study the contents of two newspapers from the mainstream press. These are the *Express* (published in English) since the early 1930s and the *Navbharat Times*, established in 1950 and published in Hindi. The first is published from 17 cities across the country, but the second, for language reasons, has a more restricted geographical spread, being published in Bombay, Delhi, Patna and Jaipur. Both belong to large Indian publishing houses – the first to the Indian Express Group of Newspapers, and the second to The Times of India Group, which is proprietor of the much older English language *The Times of India*. The various editions of the papers from different cities carry substantially the same material –

the same editorials, the same national and international news, while the 'City' page, usually page 3, varies to reflect local interest. The other pages are transmitted from Delhi or Bombay to be printed locally. The English press (this shorthand is common usage and is to be taken to mean 'the English-language press') constitutes 17 per cent of the total Indian press, and the Hindi press nearly 33 per cent. More is said about circulation figures for the press in Chapter 4.

Beside the major newspapers, there is a large press involved in producing magazines for specialised audiences. According to the INFA Yearbook (1995) there are more than 28,400 newpapers and magazines in more than 85 languages. Kumar (1994) says that the explosion in the last five years has been particularly significant. The more specialist the topic and the more the readership is nationwide, the greater the necessity of publishing in English. In environmental topics excellent magazines such as *Down to Earth*, published by the Centre for Science and Environment in New Delhi (circulation 8,000), *Sanctuary* (published from Bombay) and *Science*, published by Publications and Information Directorate (New Delhi) exemplify this particular trend.

## Radio

Radio broadcasting began commercially in India in the 1920s, but became a government monopoly in the 1930s. The development of the BBC in Britain naturally formed some kind of reference point – but there was (and is) no way in which broadcasting in India could be paid for by a licence fee. All India Radio (AIR) was the new name used for the Indian State Broadcasting Service from June 1936. Despite growth during the war, mostly to combat foreign propaganda, at Independence in 1947 AIR had only six stations and 18 transmitters covering 11 per cent of the population. By 1985 AIR, directly under the authority of the Ministry of Information and Broadcasting, had extended this to 88 stations and 167 transmitters, covering 90 per cent of the population. By now transmission coverage should be nearly complete, but listening figures are lower: 64 per cent in urban and 46 per cent in rural areas in 1989. Much of the programme output is music, poetry reading and drama, the latter in regional languages. All regions carry the news in English, Hindi and the regional language. There are also documentaries, some associated with scientific developments, but anecdotal evidence (e.g. Malik, 1989) always seems to suggest that popular music is what most listeners want from radio. An important advantage of radio sets compared with TV sets is that they are easier to run off a battery.

## Television

Television in India started in 1959 with help from UNESCO, very much with an eye to educating and improving the audience. A single station in Delhi broadcast to 180 teleclubs, which had been given free receivers. This was an experiment to see what TV could achieve in community development and formal education (Kumar, 1994: 154). In 1961 educational TV programmes on science for school teachers started – and only in 1965 did the first entertainment transmissions begin, under pressure from the manufacturers of receiving sets. In 1967 farmers' teleclubs were started in rural areas near Delhi, to spread word of new agricultural techniques (the Green Revolution). From these small beginnings TV developed into a much more general public service broadcasting agency, broadcasting programmes in all the usual categories, including entertainment and film. The idea of community TVs was not abandoned and headmen in an expanding number of villages have been given such sets for community purposes, although there are competing privately-owned sets which are easier to see when the audience is less than ten rather than more than 40. It is also now common for the family of a bridegroom to ask for a TV set as part of the dowry from his wife, so clearly the potential for viewing is going up. However, power shortages and power failures are endemic in India, even more in rural than in urban areas. Viewing is likely to be disrupted, which means that the attraction of the box is commensurately reduced.

Starting in August 1975 the famous SITE (Satellite Instructional Television Experiment) used satellite direct broadcasting to receivers placed in 2,400 villages (out of India's 550,000) in parts of Orissa, Madhya Pradesh, Bihar, Rajasthan, Andhra Pradesh and Karnataka. Because the satellite carried only one video channel and two voice channels, it was possible to beam programmes to only two linguistic regions simultaneously. SITE had high hopes of spreading basic education in 'modernisation' and in science for schoolchildren, and in agricultural modernisation for farmers, and in health and family planning for families in general. Practical problems showed up fast – within a month only 70 per cent of sets were working, and by the end of the project in some areas less than 30 per cent. The follow-up evaluation suggested that results had been somewhat meagre, although not worthless. Partly this reflected technical problems, but it also reflected the top-down nature of the broadcasting, with little involvement by local development officials, who could have been briefed to lead discussion after programmes had finished. The SITE programme was followed in 1986 by six terrestrial transmitters continuing the same idea in some selected centres. Mostly the state governments have been unable or unwilling to take these over, so the majority are now with Doordarshan (see next paragraph). Beside the SITE, Doordarshan also makes programmes for educational television, broadcasting in school hours, and concentrating on science. Part of the reasoning is that few schools have laboratories for practical science work, and that TV provides an alternative means of demonstration. Take-up has been poor, in part because of language problems, when narrators may use vernacular languages but not local dialects. Partly it is organisational, since education is a local state government subject, whereas TV is still centrally run. Since 1984 in another development the University Grants Commission Higher Education Project launched a Countrywide Classroom scheme, beaming educational programmes, part domestic, part foreign, to localised audiences. These have a viewership of 19 m. (Kumar, 1994) with a majority of watchers being non-students.

In 1976 TV was split from AIR as the separate Doordarshan (DD). This became more heavily involved with entertainment and it also began to take advertising and programme sponsorship. The revenue taken exploded, from 7.7 million rupees in the first year to 3,500 million in 1992. This did not mean only entertainment for its own sake. Although Indian movies and live sport (cricket) topped audience figures, in 1984 a socially aware soap opera Hum Log ('we people', or 'us'), initially focusing on contemporary issues such as birth control and inter-community (religious) relationships, and inter-caste relationships, became a considerable success. The audience is still more urban than rural, and with 40.6 million sets in 1994 for a population of 900 million, the reach of TV is still limited, possibly to less than 25 per cent overall. Banerjee (1996) gives a figure of 49.5 m. TV sets for 1995, 30 m. of them urban and 19.5 rural (for a rural population of 600 m.). The latest development is the establishment of commercial channels (principally Zee TV) in the major cities, a second channel for Doordarshan in major cities, and of course satellite competition from Star TV (Agarwal, 1994). The latter has a small share of the audience – its programming in English with non-Indian accents must restrict its viewership, in addition to the obvious cost problems of dishes or local (illegal but frequent) cable distribution. However, it takes a very high share of advertising, a testimony to the élite nature of the small audience it finds. There is no doubt that this competition has shaken Doordarshan to improve on its drab and fusty image. The Pransar Harati Bill in 1989 gave broadcasting a greater degree of autonomy from government, but Banerjee cautions: '[In the age of liberalisation] the Indian broadcasting authority will have to maintain the right to censorship of news and information progammes. This is because there are very sensitive religious and ethnic issues in the country' (1996: 108). DD continues to retain a monopoly of terrestrial channels, and will continue to do so, although it does buy programmes from private producers. It remains sensitive to India's regional problems and diversity. The national network broadcasts 70 hours a week, while the regional component broadcasts 25 hours a week. The national network programmes are 70 per cent in Hindi, 20 per cent in English, and 10 per cent in other languages, while the regional network programmes are 90 per cent in the local language, 10 per cent in others. Cable TV is mostly a ramshackle affair of unregulated local distribution of satellite channels, but enforcement of an Act of Parliament of 1994 is beginning to impose some kind of order. Banerjee (1996) reckoned that in 1995 it reached 12.5 m. homes.

Broadcasting is transforming itself from a serious but limited attempt at 'improvement' as part of a government-led 'development crusade' with a 'scientific temper' to a more multifarious enterprise mostly sponsored by advertising (although the press retains the lion's share of advertising revenues: $842 m. in 1994 as opposed to $271 m. for TV (Banerjee, 1996: 104)). It more and more reflects the consumer behaviour and consumer power of the emergent urban middle classes of industrialising and liberalising India. It is also consistent with the world-wide philosophy of the late 1980s, of privatisation, government retrenchment and trade liberalisation. To many Indian critics all this represents the triumph of western-based materialism, at considerable environmental cost, and the final abandonment of any Gandhian dream of rural development based on appropriate technology. (But no one has yet suggested seriously a national Radio Gandhi.)

## Foreign broadcasting to India

In this project we have concentrated in our content analysis of broadcasting in India on the state radio and TV networks. It is common now to talk about the satellite invasion of India and of a foreign cultural hegemony seeping out of The Box, but for reasons we have spelt out in several places in this book, this can easily be exaggerated. Most of it reaches only a small élitist audience. We have also almost completely ignored one other much more significant, older and well-established route for foreign media invaders: broadcasts in vernacular languages as well as English by foreign radio stations. Pre-eminent amongst these is the BBC, which broadcasts not just in English on the World Service but also in Hindi, Bengali and several other Indian languages. The listenership for these broadcasts is very large, and for the Hindi news bulletins may make a significant challenge to the listenership for AIR Hindi bulletins. Simply for lack of resources, we have not been able to include such broadcasting – which of course also comes from many other nations too – in our analyses. This is simply a point which should be borne in mind.

**1.9 THE ENVIRONMENT AS EVERYTHING**

With respect to mass media, it is obviously important to understand the frameworks that are used for the representation of the environment, and how those frameworks come to be dominant. But before we reach that point, it is perhaps even more important to realise that the 'environment' will exist in all other frameworks on all other topics. For the moment, consider solely the nature of TV programmes. Every programme ever shown has to be about the environment, if viewed in that way. A thief's motor bike fleeing a rifled bank is a statement about transport technology and the use of fossil fuels, urban air pollution, noise pollution and much else. A cricket match at Lord's is tacit approval of fertiliser, motorised rollers and water sprinkler systems, besides much else. Shannahan (1993: 185) makes the same point: 'Most television is devoid of explicit environmental messages, although all television implicitly says something about the environment,' adding that 'One of the most typical messages of television entertainment is that there is, in fact, no environment as we know it, since so much television is set in sanitised studio settings.' The latter claim can be modified, in that the electric lights of the studio are a statement about power consumption, and the consumptive materialism of prizes offered is obvious. If we were to use some broad concept of implicit environment in a content analysis of TV or the press, clearly we would expect that nearly 100 per cent of programmes and stories would be included in the 'environment' category, even if they were also (in a cover-set rather than partitional classification) in other categories such as economics or politics. Routinely, however, analyses suggest that 'the environment' features little in media content. For this to be the case, there clearly has to be a restricted meaning to the concept of 'environment'.

A few decades ago most newspapers and broadcast channels had no environment correspondent, and the Department of the Environment in London was concerned with local government green belts and new bypasses (amongst other things). TV showed films about 'natural history' and 'animal behaviour'. Until very recently, no major newspaper in the UK was without an 'Environment' Correspondent. (They are again now a threatened species of journalist.) The 'environment', internationalised by concern over global warming, has become the umbrella *category* that has captured conservation, sustainability, biodiversity, India's forests, the Bengal tiger and the Assamese one-horned rhinoceros, ozone depletion, climatic change, greenhouse gases, desertification, flush toilets, plastic bags, green belts, green revolutions, village ponds and village bypasses – to name but a very few. This would suggest that 'environment' is a category used of objects and issues.

However, despite such a list, the rise of the environmental journalist clearly reflects a vaguely if intuitively restrictive alternative understanding of what the 'environment' comprises. Journalists and producers have their own categorising frameworks for different programme or reporting formats. The bank robber might fit into a 'crime' framework (see Dunwoody and Griffin, 1993: 25) or a 'human interest' framework. A story about bypasses could fit into a local politics 'NIMBY' (Not In My Back Yard) framework in a local UK paper, or an 'environmental' framework linking urban traffic with asthma. This therefore implies that in the media the word 'environment' is actually a classificatory umbrella for a frame of reference used to view and report particular issues. It is not necessarily a classificatory word for the issues or objects themselves. For example, to a farmer a tract of cultivated land he has sprayed with insecticide may be a field, while to a 'green' or perhaps an environmental journalist, it is 'the environment'. To this extent the only discoverable meaning of the word 'environment' is that it is a perceptual filter, which by definition has to have some minimum of polemical intent.

There is no doubt that some sort of 'environmental frame' has now been established in the media, but its establishment and perpetuation is clearly not unproblematic, nor will the frame automatically be the same in different countries. This is in part because of all the ambiguities in the topic and partly because, as a result of the ambiguities and complexities (Dunwoody and Griffin, 1993), it is difficult to put over adequately a news story within the time slot allowed for it, or a full programme which offers a solution within the normal programme time allowed. For this reason, news stories 'about the environment' may be placed within a different frame, such as 'natural disaster' for floods, or 'accident' for a wrecked oil tanker. However, once the framework has been established, the problem and issue 'definers' (Anderson, 1993 and Hall *et al.*, 1978) can set to work to mould the media 'environment' agenda.

## 1.10 SUMMARY

The historical and cultural circumstances of Britain and India are radically different. Both may be sovereign states – a construct and a value which both do seem to accept – on the same planet, but that is a far cry from saying that either they can or they must see the world in the same way. We can answer none of the basic questions that this project investigates without keeping these differences in mind. To what extent do different media, in different languages, inform and educate the public on environmental issues, and how are those issues defined as environmental? To what extent is the perceived environmental crisis something which is seen and understood in terms of synonymous categories across different languages and cultures? In India specifically, to what extent is the perceived environmental crisis acknowledged by the development debate and actually incorporated within it? In Britain is the environmental debate incorporated within some equivalent grand 'project' and, if not, how does it relate to society, through the Judaeo-Christian account of Genesis for example? The media are often suspected of setting the agenda – and we will investigate to what extent that may or may not be true of the environmental agenda – but in neither society do the media operate context-free, whether we care to define context in terms of cultural norms, in terms of national and international ownership, of state regulation and subsidy, or in terms of changing technologies.

It is unrealistic to imagine that we will ever get close to a full appreciation of all these issues, least of all with one project and one book. But we hope the readers will find many contributions to the answers to these questions in the following chapters.

## REFERENCES

Agarwal, Amit (1994) 'Doordarshan: opening up at last', *India Today* 19(7): 108–19.
Ahmed, K. (1996) 'One nation, two discourses', *Himal*, August.
All India Radio (1995) *Facts and Figures*, New Delhi: Audience Research Unit, AIR.
Anderson, A. (1993) 'Source–media relations: the production of the environmental agenda', in Hansen, Anders (ed.) *The Mass Media and Environmental Issues*, Leicester: Leicester University Press, pp. 51–68.
Ascherson, Neal (1996) 'Sentimentality count: zero', review of Buruma, I. (1996) *Love*

and *War in East and West*, London: Faber, in *The Independent on Sunday; The Sunday Review*, 29.4.96, p. 27.

Basham, A.L. (1973) *The Wonder That Was India*, 3rd revised edn, London: Sidgwick Jackson.

Banerjee, I. (1996) 'India', in International Institute of Communications, *Media Control and Ownership in the Age of Convergence*, London: International Institute of Communications.

Chapman, G.P. (1983) 'The folklore of the perceived environment in Bihar', *Environment and Planning A*, 15: 945–68.

—— (1984) 'The structure of two farms in Bangladesh', Ch. 11 in Bayliss-Smith, Tim and Wanmali, Sudhir (eds) *Understanding Green Revolutions*, Cambridge: Cambridge University Press, pp. 212–52.

—— (1992) 'The changing geography of India', Ch. 2 in Chapman, G.P. and Baker, K.M. (eds) *The Changing Geography of Asia*, London: Routledge.

Dunwoody, Sharon and Griffin, Robert J. (1993) 'Journalistic strategies for reporting long-term environmental issues: a case study of three Superfund sites', in Hansen, Anders (ed.) *The Mass Media and Environmental Issues*, Leicester: Leicester University Press, pp. 22–50.

*The Economist* (1996) 'Tiger or tortoise?' 27 April, pp. 27–9.

Gadgil, R. and Guha, R. (1992) *This Fissured Land: An Ecological History of India*, New Delhi: Oxford University Press.

Gerbner, G. (1990) 'Advancing on the path of righteousness, maybe', in Signorielli, N. and Morgan, M. (eds) *Cultivation Analysis: New Directions in Research*, Beverly Hills: Sage, pp. 249–62.

Gupta, C. and Sharma, M. (1996) 'Speaking in tongues', *Himal*, August.

Hall, S. *et al.* (1978) *Policing the Crisis: Mugging, the State, and Law and Order*, London: Macmillan.

Khubchandani, Lachman M. (1983) *Plural Languages, Plural Cultures: Communication, Identity and Sociopolitical Change in Contemporary India*, Hawaii: East–West Centre, University of Hawaii Press.

Kumar, Keval J. (ed.) (1993) 'Mass media and the environment: critical perspectives', *Vritta Vidya*, Special Issue, Pune: Dept of Communication and Journalism, University of Poona.

—— (1994) *Mass Communication: a Critical Analysis*, Bombay: Vipul Prakashan,.

Malik, Saroj (1989) 'Television and rural India', *Media, Culture and Society*, 11: 459–84.

Naik, J.P. (1963) *Selections from Educational Records of the Government of India*, vol. 2: *Development of University Education, 1860–1887*. New Delhi: National Archives of India.

Reddi, Usha (1989) 'Media and culture in Indian society: conflict or cooperation?', *Media, Culture and Society*, 11: 395–413.

Said, Edward (1978) *Orientalism*, London: Routledge & Kegan Paul.

Shanahan, J. (1993) 'Television and the cultivation of environmental concern: 1988–1992', in Hansen, Anders (ed.) *The Mass Media and Environmental Issues*, Leicester: Leicester University Press, pp. 181–97.

Stasik, Danuta (1994) *Out of India: Image of the West in Hindi Literature*, New Delhi: Manhoar.

Tomalin, E. (1994) 'Bharat Mata: Indian environmental values', unpublished BA dissertation, Lancaster University.

Turner, Bryan S. (1994) *Orientalism, Postmodernism, and Globalism*, London: Routledge.

Vasudeva, Sunita and Chakravarty, Pradip (1989) 'The epistemology of Indian mass communication research', *Media, Culture and Society*, 11: 415–33.

Wynne, B. (1991) 'Knowledges in context', *Science, Technology and Human Values* 16(1): 111–21.

Zachariah, B. (1996) 'The rise of the development ideal in India, *c.*1930–1947: a few preliminary observations', paper presented to the British Association of South Asian Studies Annual Conference, University of Bradford.

# 2    The environmental context

## 2.1 INTRODUCTION

Although the title to this chapter could refer in principle to any kind of environment – intellectual, financial, spiritual – the ESRC's Global Environmental Change programme which funded this research was quite clear that the environment meant the planetary physical environment – and that global meant either changes which were global such as climatic change – or changes that were local but repeated globally because of prevailing socio-economic conditions – such as deforestation.

The idea of a separate physical environment superficially seems simple but is not on closer inspection. As air-conditioning spreads around the world, so does legionnaires' disease, a lethal environmental hazard – but not quite part of the intended ESRC research programme. However, in this chapter we are going to risk the inevitable raised eyebrows by attempting to say something about significant characteristics of the physical environment in Britain and India: not to demonstrate that they are separate and not part of the planetary system but to demonstrate those differences which will lead to difference in perception and understanding of each other's predicament.

## 2.2 THE BRITISH ENVIRONMENT

The period of the most recent Ice Age (there have been other earlier ones in earth history) is known as the Pleistocene, which, roughly speaking, is synonymous with the Quaternary era of geology and covers the last 1.5 million years of earth history. During this time the ice has advanced and retreated several times; in Europe the understanding is that there have been many major advances, perhaps as many as 17, with durations of over 100,000 years, and interglacial warm periods in between when conditions in Europe have been as mild as now or milder. These periods have lasted approximately 60,000 years.

In northern Britain the valley glaciers finally melted only 15,000 years ago. At the height of the last glaciation sea levels may have been as much as 150 metres or more below current levels, but as the ice retreated and as sea levels recovered, the North Sea basin was flooded, and then, but only about 6,000 years ago, the English Channel was breached and Britain became an island. The removal of the overburden of ice (the ice sheets of north England may have been 2 kilometres thick) has meant that the northern parts of Scandinavia and Britain have been rising again like corks in water, relieved of weight. Scotland is still recovering at a rate of 4 mm a year. At the same time the south of England has been sinking – in absolute terms as well as being inundated by the global recovery of sea levels in the postglacial period. The southern coasts are being continually attacked and eroded. Medieval villages such as Dunwich in Suffolk have been lost to the sea. The fight to keep a small road open to a coastguard watch station on Flamborough Head has been given up and the headland is now accepted as an island. When it happened three years ago, ITV had a small tail-piece item on the 10 p.m. News, and completely without foundation attributed the severance in a small sound-bite to 'global warming'.

Contemporary Britain, like the majority of developed countries, is a mid-latitude country within the temperate climatic zone. It is actually a high mid-latitude country (50–60° North), looking over a major ocean at the western fringe of the European continent, beneath the predominantly westerly circulation of the polar front, where cold winds from the North Pole do battle with warmer winds from the subtropical high pressure belt at about 30° North. The result is that, although there is seasonality in the climate, seasonal temperature changes are much lower than for an inland 'continental' climate, there is considerable variation in temperature and rainfall on a daily basis within each season, annual variations in rainfall totals are relatively low, and there is a high expectation at any time of the year of at least moderate if not strong westerly winds passing over the country. Since the country is a small island, few rivers have a long passage to the sea – and only a few have major conurbations in their upstream courses. The more densely settled parts of the country are underlain by massive sedimentary strata, many of them permeable, thus forming very large groundwater aquifers. It has abundant coal reserves (we are not arguing about the political economy of coal, merely making a geological statement), significant oil and gas reserves, and some of the most attractive sites for renewable energy generation by winds, tides and waves – if the technology advances fast.

Possibly because it is hard to imagine a country less easily able to pollute itself, the British have had a more cavalier attitude towards pollution until the last few years than adjacent countries in Western Europe. In the 1950s during some of those few consecutive days in winter when stable conditions prevailed and there were temperature inversions over London, the smog from domestic coal fires and coal-fired power stations built within the city developed into 'pea-soupers' which killed a few thousand persons each winter. These conditions never prevailed long, the winds and rain returned, and the worst got (and still gets) shifted out over the North Sea and dissipated over Europe. Nevertheless, domestic soft-coal burning was prohibited in urban areas and smokeless coal fuels devised, the thermal electric power stations were replaced by massive new ones outside the cities, which they supplied through the grid, and those that were coal-fired were given very tall chimneys, so that dispersal of noxious gases took place over a longer range. (The empty hulk of the Battersea power station in London pops up as the subject of one of the news stories captured in our sample.) The air quality in cities was consequently restored (until the growth in motor traffic re-enveloped them in a new kind of air pollution). Given the ample and steady supply of water all the year round (the most a water shortage usually amounts to is a brief ban on the use of hosepipes to water domestic gardens with high-quality purified water during dry periods – although the 1995 hot dry summer may change people's expectations), all properties are connected to water-flushed sewage, and most of that could, from the point of view of human health, be dumped into a river pointed at the sea. (We are not claiming the beaches then remain healthy for bathers.)

Given a modicum of sense in their agricultural management, the soils of lowland Britain are remarkably robust. The thick clays of East Anglia have sustained intensive cereal farming for centuries, and have absorbed more fertilisers, herbicides and pesticides than modern sense seems to suggest is sustainable, but yet without proven significant damage. The change of heart has therefore not occurred because there is proof of the irreversible damage to these soils, but because nitrogen, thought to stem from this agriculture, has seeped into ground-water supplies, and because of growing public interest in wildlife diversity – of birds, and butterflies in particular. The chalk grasslands are fragile if ploughed up, and the fact that many have been is a sign of irresponsible policies and irresponsible responses. The upland soils used for sheep farming and evergreen plantations are much more sensitive to acid deposition, though the way in which acid deposition, the vegetation and the soils interact is far from simple, and there is a growing realisation that acid rain is not a trivial problem. The amount of 'original' and biodiverse forest cover is almost infinitesimally small. The trees of lowland Britain are also thought to be beginning to suffer in the summer months from excessive low-level ozone, odourless and invisible, which forms as a result of photochemical action on the automotive exhaust clouds of the big conurbations, and which drifts very quickly into rural areas. This seems to cause the trees to go into a premature autumn – pretty in its colours but inhibiting to the growth of the trees.

Partly because of mounting pressure within the European Community, partly because of growing public awareness, and because walking in the countryside is one of the nation's greatest pastimes all the year round, many of these local problems are being tackled, if slowly. The biggest stumbling-block would appear to be the reluctance of the public to forgo its love affair with motor-vehicles, despite a growing epidemic of asthma, particularly amongst the young, thought, but not proved, to be the result of exhaust fumes. Since the location and physical layout of so many shopping areas and entertainment complexes now assumes that their customers have cars, it is not easy to switch away from car-dependency.

Britain has dragged its feet over international treaties on environmental issues, though it did sign up to the conventions proposed at the United Nations Conference on Environment and Development at Rio de Janeiro in 1992. Paradoxically, it may be one of the few countries that can meet its European obligation to limit $CO_2$ emissions because of an accidental outcome of government policy undertaken for other reasons. The privatisation of the electricity industry has been accompanied by a radical shift away from coal towards gas-fired stations – with a substantial reduction in $CO_2$ output. ($CO_2$ is thought of as the major greenhouse gas but unburnt natural gas is a much greater greenhouse forcer. Depending on the leakage rate in distribution systems, the switch to gas may reduce the targeted $CO_2$ emissions but actually lead to an increase in greenhouse warming potential.)

In our view, Britain is still a fairly green and pleasant land, not yet, except in a very few parts, irreversibly damaged. There are strong conservationist movements, and different sites are categorised by environmental value. At the large scale National Parks seek to preserve visual amenity more than anything else – they are anti-'development'. At the small scale Sites of Special Scientific Interest theoretically protect rare habitats with rare plants and animals. Many of these sites are the result of centuries of human occupation – a thousand years of sheep grazing the chalk, of cattle grazing water meadows – so conservation is not necessarily of 'the natural'.

## 2.3 INDIA

By common usage, India is referred to as a subcontinent (which technically includes Pakistan, Nepal, Bhutan and Bangladesh). Modern plate tectonic theory explains why it is a subcontinent. Most of the Southern land masses (that is south of the line which runs from the Persian Gulf through the Mediterranean to the Caribbean) once formed one large land mass known as Gondwanaland. About 200 MA (million years ago) Gondwanaland broke up, parts drifting off to Antarctica, a part which is Australia, the major part remaining as Africa, but another chunk, what is now the Deccan block of peninsular India, drifted north across the Arabian seas. About 80 MA it first bumped into the southern flank of Asia. The impact of this collision is clearly seen on a world map – a rim of massive wrinkled mountains to the west, north and east of India. The process continues: the Deccan block is still pushing into Asia, at the rate of 6 cm a year, and the mountains are still going up at the rate of 6 cm a year. This is the arena of the strongest of inter-plate collisions on earth.

India is most easily divided into three macro-geological zones: the young mountains of the north, the highest and most extensive on earth; the old Deccan block, this errant fragment of Africa; and the massive river flood plains, the biggest and deepest alluvial plains on earth. These components are locked into a causal chain together with the climate, which is extreme. The most significant feature of the climate is the extreme seasonality of the rainfall. For much of India this is restricted to a period of say the three or four months of the monsoon. The monsoon is part of the annual movement of the inter-tropical zone of convergence, the same phenomenon that gives the Amazon and the Congo their rainfall, but in those areas the northward and southward movement of the rainy season is less dramatic. The massive Himalayas are high enough to inhibit the northward movement of the high-altitude subtropical jet stream in summer, until a sudden switch occurs some time in late June. The monsoon winds that then sweep across India are 6,000 metres deep, compared with the 2,000 metres of the East Asian monsoon, and the relative humidity is higher. When they collide with the Meghalaya block north of Bangladesh, they give Cherrapunji the highest rainfall in the world. And for much of the mountain front of India, rainfall is immense. Given this, and the height of the still lifting mountains, and given the often comparatively soft geological strata, the natural rate of erosion is the highest on earth. The rivers debouch onto the plains, laying the massive alluvial deposits, and finally in Bangladesh, where the Ganges and Brahmaputra are confluent, the rivers form the world's largest and most active delta. At peak flood the Brahmaputra in Bangladesh may be 40 km wide – wider than the English Channel.

The wettest parts of India are in general the western coastal areas of the Deccan, the lower Ganges valley and the Brahmaputra valley (Assam and surrounding mountainous states). Only in the extreme southwest in Kerala is there a short enough dry season for the area to be called 'tropical moist'. In the other 'wet' parts the dry season extends between seven and nine months, and in the summer months of April to June, as temperatures climb to 40°C or as high as 45°C, the country is desiccated. In the central Deccan and the upper parts of the Ganges, rainfall totals are lower, and inter-annual variability greater. Most of these areas are drought-prone, and without irrigation agriculture is an extremely risky business. Finally, in Rajasthan and parts of Gujarat on the Pakistan border, there is a real desert, the Thar desert.

The variation in river discharge is therefore also extreme. In the Ganges basin many of the south-bank rivers, massive rivers in the wet season, dry up completely in the summer. The north-bank rivers show almost equally great extremes of discharge but they do not dry up in the hot season because of glacier and snow melt in the mountains. Thus neither the Ganges nor the Yamuna should cease to flow.

For perhaps nine months of the year, away from the coasts, there is very little wind. The weather is stable and predictable in a way which would warm any cricket lover's heart, and outdoor badminton courts are a frequent sight. In north India, Delhi for example, in the winter months of December and January, local temperature inversion of the sort that gives London a smog for a few days may persist for the whole two months.

Bombay and Madras, two port cities established by the British, may be coastal. Calcutta looks coastal, but is in fact as far inland as Birmingham, up a distributary of the Ganges delta which has been losing its river flow through natural causes over the last few centuries. All the other big cities are far, far inland, some like Delhi near a perennial water course, in this case the Yamuna, others like Hyderabad or fast-growing Bangalore are in the drought-prone central Deccan.

The big cities mix industry, power stations and housing in close proximity. They are overrun by two-stroke scooters and badly adjusted diesel trucks and buses. The use of soft coal, wood and even rubbish for cooking is common. For much of the year particulate matter simply lands again in the same area from which it came. During the winter inversions, not even the gases escape outwards. These cities have some of the world's worst air pollution – a level unimaginable till experienced. They are short of water – particularly, of course, in the hot summer. Less than half the households may be connected to a public water supply, and certainly less than half have flush sanitation. Where would the water come from to flush the other half, even if the equipment were installed? The newspapers daily run stories on the water wars between Haryana and Delhi and yet the urban areas continue to grow and grow. The abstraction rate of water from the Yamuna at Delhi has more than once reached the point where water is being drawn back upstream from what should be downstream sections – into which untreated effluent is poured. Urban rubbish collection systems are inadequate and in many areas rubbish creates an eyesore, is a health- and sometimes a fire-hazard and an olfactory insult. But, as Indians are fond of pointing out, they produce far less per capita than westerners; proportionately much more of it is recycled (more of this later) and more of it is biodegradable. It's just that to the westerners and many thinking Indians the small amount of plastic represented by billions of small, thin polythene bags seems to have managed to spread itself over every part of the country, town and country-side, to the apparent unconcern of the majority of the populace. It is not just a visual offence; the plastic litter gets stuck in the inadequate drains and blocks them, and is forming a non-permeable membrane in the topsoil. (Bangladesh suffers from the same problem and recently officially banned the manufacture of plastic bags, also encouraging the local consumption of jute and other fibres. But the law is not being enforced for 'the time being'.) The westerners, by contrast, are wealthy enough to devise ways of hiding their garbage – out of sight and out of mind – but also may have more of a cultural aversion to seeing litter around them. At least in respect of air pollution it might be thought that rural areas are cleaner than urban areas, and for men they are. For women it is less true, since rural kitchens rarely have chimneys, and cooking fuel – often dung – is very smoky. Even now there are many villages with only meagre and contaminated supplies of surface water available in the summer months, increasingly liable to industrial pollution as small-scale industry spreads, although with increasing electrification and tubewells, many now have much more reliable and healthier supplies than before.

The following two extracts were placed adjacement to each other on the front page of the *Statesman*.

## The Statesman
Delhi, 6 June 1995
### Khurana Threatens Fast

The Delhi Chief Minister, Mr Madan Lal Khurana,[1] on Monday threatened to sit on fast on June 7th to get more water from the Jamuna even as the city announced a 20 per cent reduction in the water supply so as to avoid any crisis in the coming days. [The dispute is between Haryana, Uttar Pradesh and Delhi.]

### Medha Patkar[2] arrested near Rishikesh
### TEHRI (GARHWAL), 5 June

The Narmada Bachao Andolan leader, Ms Medha Patkar, was arrested early today at Narendranagar, 16 km from Rishikesh, when she was coming to see the noted environmentalist, Mr Sunderlal Bahuguna,[3] who is on a fast to protest against the resumption of work at Tehri dam, reports UNI.[4]

People immediately protested against Ms Patkar's arrest and blocked traffic on Gangotri–Badrinath road for two hours.

Mr Bahuguna, who has been on a fast for the past 27 days, refused to give up his fast ...

In a Press release it said it was ironic that the noted environmentalist [Patkar] had been arrested on World Environment Day ...

*Notes:*
1  Chief Minister Khurana and the two other states' Chief Ministers reached an agreement which made the lead item on DD's evening TV news broadcast on 7 June. Delhi was to get an extra 100 cusec of water from each of the other two states.
2  Medha Patkar will be referred to many times in Chapter 4.
3  Another paper reported Bahuguna to be weak, but lucid.
4  The Tehri Dam is thought by some to be at extreme seismic risk – but if it were built and if it were safe it would solve many water disputes in north India for some time to come.
5  The government announced a further committee of inquiry into the project, permitting Bahuguna to end his fast, but the building continues.

As befits a subcontinent with such a diversity of environments and such an idiosyncratic history – 60 million years an island in the Arabian Sea – the natural biodiversity is very high. Much of it would have been covered with forests, mostly tropical dry deciduous forest, i.e. the trees lose their leaves in the hot dry season, not the cold winter months as in Europe. But India's colonial history and recent urbanisation and population explosion have seen to it that many of these resources have been reduced to rump status. Perhaps a little more than 10 per cent of India remains forested, and only some of that in any kind of pristine condition. Of course it is hard to get accurate data on forest cover because the topic is beset with definitional problems about density, crown-cover, age and variety of tree cover, and because the best of what is left remains intact mostly because it is inaccessible. But modern techniques of remote sensing are helping, and the government takes the issue seriously, without external prompting by the world community. In a country where perhaps half the population still relies on renewable energy for cooking – cow dung and wood being the commonest such fuels – forestry is not a matter which is left to an English-style Forestry Commission (now renamed 'Forest Enterprise') with the simple caveat that the public should enjoy rights of access for Sunday walks with the dog. Society and trees interact at every level. In the Bengal delta, there is virtually no remaining natural tree cover, yet each hamlet is surrounded by, buried within and shaded by trees and woody species like bamboo. Each species may have several uses: for boat building, for furniture, for resins to proof fishing nets, for fruit, for fodder for animals, for agricultural implements, and for lopping for fuel. This is a productive and in some sense biodiverse arboreal landscape. But it is not the natural biodiversity, and it is certainly less diverse than the original forest. Is the idea of the hamlet surrounded by carefully planted trees not unlike the rural idyll of green and pleasant England?

The other great fear that inspires modern environmentalist apprehensions of rural India's fate is the fear of massive soil erosion and soil impoverishment through exhaustion and/or incorrect use of chemicals. We will not enter here the debate about the high mountains (see Chapman and Thompson, 1995, for a fuller account of that), since natural rates of erosion there are high anyway, and actually the proportion of India's population in these areas is comparatively small. We are therefore concerned here more with the gentler slopes of rain-fed agriculture in the Deccan and the Gangetic plains, and with the terraced wet paddy lands (which may be irrigated or rain-fed depending on local conditions) of the same two regions. The stories about soil impoverishment through excessive use of fertilisers mostly emanate from the most intensively farmed area of India – the Punjab. The problem is that the truth of the matter is not known, the evidence anecdotal, repeated and generalised until it is a self-evident truth. As far as we are aware there has been no state-wide survey which carefully links the empirical with the theoretical: this is an example of the failure of the thinkers and the doers to work together. Since tropical and subtropical soils have a lower organic content than temperate soils, recycle nutrients faster and have higher rates of bacterial activity, it is quite possible that some kinds of damage occur faster than with equivalent doses on, say, English clay soils. But this remains conjectural at the moment. Soil erosion is often exemplified by the ravines and gullies of the badlands of the river Chambal and its basin, some hundreds of kilometres south of Delhi and in the northwest of the Deccan. This, however, happens to be a late Pleistocene loess dump – wind-blown fine sedimentary material. The largest example of such a loess dump is in China, often described by geomorphologists as the world's largest waste tip, an area of massive erosion and catastrophic slumping. These deposits are way out of equilibrium with current conditions, and it is in their nature to erode and gully massively. This is nature working long-term, and man has little influence over the rates observed. Erosion on degraded forest land is quite clearly and quite often a problem, which in places is checked by reforestation. Massive erosion on good agricultural land is not common, however – it is too precious, and farmers are too careful. But then one can look at the formidable extent of brick pits surrounding urban India. These use local earth and are never dug deep, because there is only a low level of mechanisation, and because of monsoon flooding. Typically only the top 2 to 4 metres of earth are dug away. They are therefore very extensive – a kind of dug environment mirroring the built environment. But even here on close inspection the situation is not as obvious as first appears. Given the depth of alluvial soil in many places, the mineral content of the revealed surface usually returned to agriculture is quite often better than the well-used soil that was removed. It does, however, have a lower organic content. Further, it is not unusual for the pit user to want to enhance the value of the land by digging the pits so as to enhance the possibility of irrigation. The situation is not wholly a bad one.

Indian agriculture has increased its yields through a combination of factors – new techniques, inputs, increasing the number of crops per year – but all of them mostly work only in combination with irrigation. Many are the sources of irrigation water, but the two dominant ones are the large surface canal systems which lace many of the river plains, fed from river barrages, with the rivers themselves often fed from dams in the hills. The other major source is tubewells, which have exploded in number all over India in the last two decades. In much of the hard-rock Deccan, these are the only possible sources of water, but the Deccan rocks have variable and usually small aquifers. In many areas the pumping is lowering the water table. In the Ganges plains the situation is far more complex. Until a few years ago most of the alarm – and even now a great deal of alarm – was expressed because the leaky canal systems were causing a rise in water tables, leading to constant evaporation from damp soils, which turns them saline. On the other hand, one reads in the press – although there are few academic studies of the causes – that tube-wells even here are reducing water tables to the point where irrigation is no longer possible. Farmers like tubewells because they get water when they want it, not when or if the canal system gives it to them. In the technical literature much is now written about 'conjunctive use' of surface and ground-water, but in practice there is little done since the two systems are under different control. The ground-water reserves of the Ganges plains ought to be adequate for much more extensive exploitation than has so far occurred – but pumpsets use either electricity or diesel power.

In the cities the hot summer (reaching 45° by day and staying above 30° by night) is insufferable, unless a person can work indoors in an air-conditioned space and sleep in an air-conditioned room. The power demand peaks in the summer, at the same time that the farmers' demands for pumping power are maximised (preparing seedbeds and fields for the coming monsoon). The government has so far, though, paid even less attention to the standards of building insulation to keep heat out than the British have, despite their poor track record, to keep heat in. The demands for power increase annually at a staggering rate, and the policy is to try and satisfy those demands from big dams and from thermal power stations burning India's large supplies of somewhat dirty coal rather than by stemming demand. Was demand management part of Europe's 'development'? The combination of power demands by Bombay and of irrigation and drinking water demands by Gujarat are the essential driving force behind the contentious Narmada Valley Development Project that has become a *cause célèbre* around the world because of its impact on some of India's remaining forested areas with both a rich wildlife and an indigenous tribal population. It will feature in the news analysis from time to time.

There is a thesis that a country pollutes its environment badly while it increases its national income per head, until that figure reaches about $5,000 per capita per year. At that point part of the wealth created is used to start cleaning up. The thesis is based in historical comparisons and on the concerns and activities of those countries, like Taiwan, which are currently crossing the threshold and doing just that. If a country is poorer, then jobs, incomes, shelter and food have much higher priority. In India as elsewhere, there is a lot of evidence that the common man has just those priorities – he is prepared to work in a polluting factory if it gives him an income to feed his family, and rarely is such a worker interested in getting the place shut down. The leaders and orchestrators of India's environmental movement (though not necessarily the majority of its members) are mostly from the more secure, wealthier and educated classes. It is interesting to note, however, that their wealth on average is not high if one uses standard international monetary comparisons. Yet part of one's true income is psychic – the standing and esteem within the community and comparison with one's peers. Add this in, and their income is much higher.

We have written far more, yet still only sketchily, on India's environment than on Britain's. India is so much larger, so much more extreme, so much more varied, so much more contradictory, so much less researched, that one should write at as much length as possible, yet even then it would be impossible to avoid over-simplification and broad generalisation.

## REFERENCES

Chambers, R. (1988) *Managing Canal Irrigation*, New Delhi: Oxford and IBH Publishing.
Chapman, G.P. and Thompson, M. (eds) (1995) *Water and the Quest for Sustainable Development in the Ganges Valley*, London: Mansell.
CSE (Centre for Science and Environment) (1985) *Second Citizens Report on the State of India's Environment*, Delhi: CSE.

# 3 The view from the newsrooms of the UK and international agencies

## 3.1 INTRODUCTION

> News editors around the world yawned when the World Climate Conference in Berlin drew to a close.
>
> Paul Brown, *Guardian* Environment Correspondent, April 1995[1]

This chapter is the first that reports directly on our own research findings. As we indicated in the Foreword, we first break into the circle of society–media–society by interviewing the editors and journalists who are responsible for press and broadcasting output. The purpose of the chapter is to illuminate the media production process in the UK and the international agencies, and in particular to examine the issue of how journalists decide to privilege one story category over another – in other words to attempt to throw new light on news values. Chapter 4 will consider these issues in India.

Paul Brown's comment above is significant because past studies of the media production process have in general looked at relations between reporters and their sources as opposed to relations between reporters and their editorial controllers. As Michael Schudson (1989: 149), in his review of the literature of the sociology of news production, has observed: 'most research has focused on the gathering of news rather than its writing, rewriting and "play" in the press ... this is particularly unfortunate when research suggests that it is in the "play" of a story that real influence comes.'

This project builds on the work of Alison Anderson (1991, 1993), whose work provides important insights into the relationship between environmental reporters and their sources. The research reported here moves her investigation on by looking at the relationships between reporters and their editors and between editors and the wider social and political context within which they are operating. It is based on in-depth interviews with more than 30 reporters and editors in those news organisations that play a substantial role in the international news traffic flowing from the West to India. Hence, in addition to covering broadsheet newspapers, radio, television and the newsagencies in the UK, the research has also involved observation of and interviews with reporters and editors at Cable News Network in Atlanta, the Associated Press in New York and the Asian Broadcasting Union in Kuala Lumpur. The conclusions are also based on observation in the newsrooms of the interviewees, a postal survey of editors of all Britain's national newspapers and finally Ivor Gaber's more than twenty years' experience as both a reporter and editor on regional and national newspapers, and on network radio and television in the UK.[2]

## 3.2 COMPETITION IN JOURNALISM

The journalistic process is commonly assumed to be one in which editors, or more usually an editor, send reporters out on assignments and these assignments then materialise into written or broadcast stories. In fact journalists, even editors, have far less freedom of action, in terms of the day-to-day news production process, than they or their audience tend to assume. They are part of a newsgathering and news distribution mechanism whose very structure usually predetermines the type and manner of coverage. Journalists are very deeply integrated into this culture. They know precisely what the process requires. An editor of Gaber's acquaintance, on being told by a union official that if he introduced a certain change he would not have 'a happy newsroom', replied, 'I do not want a happy newsroom'. This is because newsrooms are intensely competitive environments. There is competition between the different newspapers, radio and television stations, between the different sections of the paper or broadcasting station (such as features, news and sport) and within news between the various areas such as politics, business, crime, etc. Even within the same news areas there is competition among the individual journalists to try and gain the greatest amount of column inches or minutes and seconds of air-time.

One of the main functions of this competitive environment is to ensure that the newsgathering machines are generating more stories than their newspaper, TV or radio programme can accommodate. This is because, in the event of stories not materialising, substitutes must be available. On an average day, when enough stories do materialise, competition for space and time is intense. Apart from this being a sensible strategy, in terms of having stories available in reserve, it is also supposed to ensure quality – this is based on the assumption that within this competitive jungle journalists will strive that much harder to ensure that their stories are of the highest quality and, in a reverse of Gresham's Law, the good will drive out the bad. But in such an atmosphere the amount of time and space given to relatively new subject areas, such as the environment, comes under intense pressure; consequently the nurturing that new subject areas require is often neglected.

Nigel Wade, Foreign Editor of the *Daily Telegraph*, noted how, in this competitive milieu, he was successful in fighting: 'to keep the Earth Summit in Rio off the foreign pages'.[3] But Wade's comment also points up another important aspect of the way newsrooms are organised: they are highly segmented, and this is an additional reason why it is particularly difficult for relatively new subject areas to fight their way into the system (see Tuchman, 1978: 149). Wade explained:

> one of the greatest practical difficulties in the paper is to find a proper home for copy about environmental issues ... a London conference on hardwood forests in Sarawak, for example – is it a story about Sarawak or is it a story about environmental issues?

## 3.3 THE EDITORIAL PROCESS

Broadly speaking, news organisations are divided between input and output operations. The input, or newsgathering process, is headed by the news editor and his or her assistants; working to them are the reporters and correspondents and in the broadcast media, producers andresearchers. The output side of a newspaper consists of what is known as the 'back-bench', the Chief Sub-Editor, the Night Editor, the designers and the sub-editors. In television and radio the output team consists of programme editors, producers, scriptwriters and sub-editors. For the purposes of this chapter senior members of the input and output teams are referred to as 'editors'.

They form an important part of this study for, despite their lack of specialist interest in this field, they play a decisive role in determining whether or not stories about the environment get printed or broadcast. Meetings and conversations between the editors themselves and between editors and reporters are the crucial moments in any news operation; it is here the negotiations take place which result in space in the paper or time on the programme. Reporters know the system and recognise the importance of finding stories that will interest their editors rather than the public, an observation first made by Herbert Gans (1979), but echoed by many of our respondents. Alex Kirby, BBC Radio's Environment Correspondent noted:

> What makes me hold off stories is not thinking that the public may not be interested as thinking that producers and editors may not be interested ... So there is an awful lot of what I would consider environmental stories that I don't bother with, not because I don't think they are interesting but because I don't think they are going to get anywhere.

Paul Brown, Environment Correspondent of the *Guardian*, was even more scathing:

> Because newsdesks are as short-sighted as politicians and news is immediate, sales are day-to-day, newsdesks can't see further than the end of their noses. That day's news-list (the News Editor's daily list of stories) is all that matters.

However, within the journalistic process environmental reporters have a particular difficulty, which is that, contrary to the Anglo-American norm of supposed journalistic neutrality, it is often assumed that environmental correspondents are, if not campaigners themselves, then sympathetic to the environmentalists' cause, an assumption made explicitly by Lowe and Morrison (1984: 81) who wrote: 'On the evidence of our interviews with specialist journalists working in the environmental field, they are sympathetic, often strongly so, to the cause of environmental protection.' And they went on to argue that because the environment was not seen as a site of party political contest, editors were happy to collude in the supposed partisanship of environmental reporters.

Support for the Lowe and Morrison position came from Ian Jack, who at the time of this research was editor of the *Independent on Sunday*. He said:

> I used to think that every reporter should be a kind of empty vessel, willing to go along somewhere and say on the one hand he says this and on the other hand he says that. But in a way it is an impossible thing. I mean human beings aren't like that. Defence correspondents are interested in armies and get on well with officers, you would be daft to assume that our defence correspondent was in some way impartial between pacifism and militarism. He is clearly not. And the same goes for our environmental correspondent, he is not neutral about environmental protection.

Yet this view was very much *not* the consensus encountered by this research project. Editors, and perhaps surprisingly environmental correspondents, vigorously resisted the notion that because, in the words of Lowe and Morrison only 'the truly demented would in principle demand more air, water and noise pollution' (1984) they should be less than neutral in this particular area. Bill Ahearn, for example, Vice-President and Executive Editor of Associated Press, New York, was emphatic: 'We do not crusade ... we're not joining in anyone's movement.'

Garron Baines, Deputy Editor of Channel 4 News, a daily news programme targeted at a more up-market audience, found anything that threatened traditional notions of journalistic impartiality worrying:

You're into slippery slopes territory here. Once you actually start agreeing on something as laudable, as improving the environment and actually effectively campaigning for that, then as a broadcaster you are moving into territories that I feel very uncomfortable with.

Alex Kirby of the BBC was in no doubt where he stood on the issue:

I am not a campaigner. If I were a campaigner I would go and apply for a job with Greenpeace or Friends of the Earth or someone. I am a journalist and it is my job simply to try and tell people more about their world to help to expand their understanding of it. The conclusions they draw have to be their conclusions.

Justin Jones, who was ITN's Environment Correspondent and then left to set up his own agency to provide specialist coverage of environmental issues, remained firmly committed to notions of journalistic balance and impartiality:

It's not my business to ram the environment down the public's throat in a prejudicial way. It's not our job. The Government can do that or Greenpeace can do that or Friends of the Earth can do that. That's their job. That's what they do.

Geoffrey Lean, one of the doyens of the Environment Correspondents, who at the time we spoke to him was working for the Sunday newspaper the *Observer*, took a more considered view. He argued that it was difficult to divide the work of the environmental journalist from that of the environmental educator. In answer to the question as to whether the environmental correspondent should help change public attitudes, he replied:

Yes, but not as a prime aim. One's prime aim is to report the truth ... if one goes into this kind of job with a fixed view of particular issues and you try and bend people to your view, you're not being a journalist, you ought to be a campaigner with Friends of the Earth or someone like that.

In subsequent correspondence he went on to explain how he saw a distinction between the way reporters such as himself, working on a longer time-scale, and those involved in on-the-spot reporting, could come to have a slightly different perspective on the balance between reportage and comment.

There is a continuum from the immediate news reporter (whether broadcasting or agency), through the daily newspaper journalist, the weekly or Sunday journalist, to the columnist or documentary-maker as to how much one is expected to come to a judgement about what one is describing. As a Sunday journalist I am quite a long way along this continuum ... and on the kind of paper I have always worked for, one stands or falls by how sound and fair one's judgement turns out to be. Unlike broadcast news programme or news agencies, newspapers, particularly Sunday ones, do campaign, if a correspondent, his editor and senior colleagues agree an issue is sufficiently important, and again there is a penalty for misjudgment.

This, he explicitly argues, does not mean alignment with any campaigning group. Lean contends that:

There is, of course, an enormous difference between running the occasional newspaper campaign on the environment – after going through the process of reaching a consensus – and being a campaigner for Friends of the Earth or Greenpeace. I find in practice one has to go into each story with a perennially open mind.

This discussion of how journalists see their role in society is an important aspect of understanding the news production process; equally important is understanding the journalists' notion of news values and their relationship to the environment as a distinctive news category. It is these values, as perceived by both reporters and editors, that profoundly affect the way environmental stories are selected and then covered. For in the sort of rigid structures that exist within most newsrooms, journalists, anxious to see as much use as possible made of their material, are obliged to try and fit their stories into the existing concepts of 'news'.

Media academics have devoted much time and effort to analysing and defining 'news' – for journalists it is less problematic. James Wilkinson, BBC TV's Science Editor, for example, offered this definition: 'it has to be new, it has to be interesting, it has to be true and it has to be important.'

Nigel Wade, Foreign Editor of the *Daily Telegraph*, put some flesh and blood onto these bare bones:

> I would look for universality and also I'd look for people, – identifiable goodies and baddies, particularly baddies who are always interesting. An element of conflict is important, we want to see battles, people locked in combat in some way just like a Greek drama. Somebody is for, somebody is against. They're fighting it out in some way, whether in a committee room or in a demonstration in the street or whatever. Also transparency, readers have to be able to get to grips with the story.

David Lloyd, Channel 4's Senior Commissioning Editor for Current Affairs, took on the argument that journalists resolutely sought out 'bad news':

> They're about reporting the tensions between interest groups and society at large, individuals and institutions, individuals and individuals. The tension is the flashpoints. That's why news is normally bad news because most things land you with an irresolvable set of conditions.

Turning to the definition of news as specifically applied to stories about the environment, Lloyd felt that first and foremost he would give priority to: 'novelty. Secondly, there must be a scale, in other words it must have an effect and an importance outside of a tiny community. And thirdly there must hover over or through it public policy questions.'

Armed with this sort of intelligence environmental reporters seek to narrow their search for stories to those that will fit into these existing news values. Paul Raeburn, who was until recently the Science Editor of Associated Press in New York and therefore responsible for that agency's world-wide environmental coverage, explained how he saw the process:

> I hate to say it but what can be a hard seller (to the newsdesk) are some of these stories about problems far away. If I wanted to do a story about a pollution problem of some sort in India I would have to make a case either that it was a kind of problem that was happening in a variety of places so there was general concern or there was a unique human interest angle, that, for example, this one uniquely affected children of a particular religious sect. It would have to be something novel or unusual about it for me to sell it because otherwise the reaction would be 'so, its another polluted river'.

Richard Tait, now ITN's Editor-in-Chief but at the time of interview the editor of ITN's Channel 4 News, reflected a similar view, but from the other side of the editorial table:

If you're looking at an environmental story the single most important criterion is the scale of the problem which is either believed or proved to exist, and that either means that the actual case you are looking at is a very serious case or it is an example of a wider problem which has serious implications.

Much of this discussion of news values draws upon the work of two American media scholars, Dorothy Graber (1988) and Gaye Tuchman (1978). Graber developed the notion that audiences make sense of information from the media by organising the diverse material that is thrown at them into mental maps or schemata – journalists, it can be argued, use similar techniques to arrive at decisions as to whether something is or is not 'newsworthy'. Ten years earlier Gaye Tuchman, drawing on the work of Erving Goffman (1974), developed the notion of 'news frames' to describe the way journalists processed new information: 'An occurrence is transformed into an event, and an event is transformed into a news story' (1978: 192).

An example of this came from Sue Inglish, Foreign (now Associate) Editor of Channel 4 News, who found an investigation into pollution from a secret former Soviet nuclear installation editorially irresistible:

> Chariabinsk for example was a brilliant picture story, wonderful pictures inside a secret city where nuclear weapons are being produced. No other Western journalist had ever been there. We had stunning pictures which we cut with pictures of the local children playing on the banks of a highly radioactive river; nobody had warned them, nobody had ever told them that this river was polluted. And then you work from that to children dying of leukaemia in the local hospital and doctors saying we don't know why but there is a hell of a high rate of child cancer out there.

Inglish's response identifies elements that can be seen as crucial to journalists' notions of what constitutes news values. The story has a strong visual element, there is exclusivity (at least within the UK), there are 'goodies and baddies', the victims are children, the heroes are doctors, there is the presence of cancer – an illness which has acquired a particularly threatening resonance for western audiences – and there is, underlying it all, a political theme about the alleged criminal incompetence of the former Soviet authorities.

As a result of the continuing predominance of these traditional news values, coverage of environmental issues tends to be erratic – if the story doesn't fall into a predetermined set of news criteria, then its chances of receiving media exposure are slim. One dangerous consequence of this phenomenon is that the environmental reporter, anxious as ever in the competitive environment of a newsroom to gain maximum coverage, can find him or herself colluding with the sort of schema outlined above, even though, as experts in the field, they are aware that, as approximations of the truth, such schemata are inadequate. Hence, we often witness the spectacle of environmental stories being presented in terms of an impending armageddon which in fact never comes to pass – coverage of the possible impact of Operation Desert Storm on the flora and fauna of the Persian Gulf is a case in point.

Obviously the media do respond to disasters or potential disasters, such as the Gulf War aftermath, but more importantly the question that has to be addressed is why media interest in the environment is so cyclical. As Anders Hansen (1993: xv) observes: 'The ups and downs of public and political concern about the environment are a poor indicator of the state and nature of environmental degradation.'

Compare, for example, the way that other subjects move in and out of the popular agenda – the economy or politics for example. The interest, or lack of it, in these areas might appear to be equally cyclical but coverage does bear some passing relationship to external events – economic developments are always taking place but depending on the state of the British economy these events attract greater or lesser interest. Similarly, the political process is continuous – but media interest will reflect certain external realities such as the prospects of a party leader being replaced, a government being toppled or the proximity of elections.

The environment, however, is different. Environmental degradation is, it is fair to assume, a continuous process: global warming, the growth in the hole in the ozone layer or the increase in atmospheric pollution, to take three obvious examples, are facts of life. Yet media interest in these topics has demonstrated dramatic shifts – at one moment intense, followed by perhaps years of the subject in the media doldrums.[4] So if external reality is not causing these dramatic shifts in media interest, what is?

## 3.4 AGENDA SETTING AND ENVIRONMENTAL ISSUES

At the core of this question is the notion of agenda-setting, a much discussed concept in the literature of media studies. It is one of a number of devices that help us to understand how the news production process relates to society at large. But in seeking to explain the presence or absence of environmental issues in the news media, the concept of a single news agenda is of only limited utility. We argue, based on Ivor Gaber's own professional experience, particularly as an input editor, that it is more useful to characterise the agenda-setting (or building process) as multi-layered, a process of three competing yet intertwined agendas – the media's, the public's and the politicians' – which together form what could be described as the 'national' agenda.

However, within these three broad categories there are important distinctions to be made. First, the phrase the 'media agenda' conceals as much as it reveals. The agendas of the press, television and radio differ, sometimes radically. And within these media there are also great differences. It would be fallacious to argue, for example, that the agendas of the tabloid press and the broadsheets represent anything that could usefully be described as a consensus. Similarly, it would be difficult to see a vast overlap of common interest between, say, BBC2's late night current affairs programme Newsnight and ITV's morning news magazine Good Morning Television.

The 'public agenda' is a concept perhaps most identified with the work of the German social theorist Jurgen Habermas (1989), who developed the notion of the 'public sphere' as an autonomous set of ideas and values circulating outside the mainstream political and economic institutions, reflecting the concerns and interests of the general public (although critics would argue, only certain strata of the general public). These are themes to which we will return in the concluding chapter.

Why should it be the case, then, that only at very specific moments does the issue of the environment suddenly come to prominence in the mass media and then just as suddenly disappear from view? One possible explanation lies in the activities of the environmental pressure groups and scientists who, as Anderson (1993) and others (Cracknell, 1993; Hansen, 1993; and Friedman, 1991) have demonstrated, are key sources for environmental journalists. However, it can be argued that in fact sources, important as they are, are only a sub-set of the public agenda since, whilst they clearly do have membership of, and are active participants in, the public sphere, they are not members of, and are not active participants in, either the politicians' or the media's internal spheres of activity (although that is not to deny their demonstrable abilities to influence and impinge upon these areas).

However, the major reason for the appearance and disappearance of the environment as a media issue is the predominance of the third of the agendas referred to above – the political one. For, based on Gaber's experience as a broadcast news editor at Westminster, it is clear that politicians, or rather the political class (which includes senior civil servants, lobbyists, media advisers, etc.), play a crucial role in determining the media's agenda, the public's agenda and hence the national agenda. This is an assertion that can never be 'proved' as such and certainly it is in both the politicians' and the media's interests to deny it, but the evidence from respondents to this project lends some credence to the arguments originally advanced by Hall *et al.* (1978) about the significance of 'primary definers', although with the particular gloss that as far as this study is concerned politicians stand in an apparently privileged position, compared to the myriad of other potential 'primary definers' in this field.[5]

The history of environmental coverage in this country illuminates this point. The most recent period of sustained media interest in environmental issues began in the second half of 1988 and through the subsequent two years (see n. 4). As the general level of knowledge around issues such as global warming and the hole in the ozone layer rose, two specific issues sharply focused media attention. First, there was the mysterious 'seal plague' in the North Sea in the summer of 1988 that was dramatically brought to public attention by a campaign spearheaded by the *Daily Mail* (Anderson, 1993). This was followed in September by an incident involving a freighter, the *Karen B*, that was trying to find a port in the UK where it could unload its cargo of toxic waste. After a great deal of public pressure the government announced that it was refusing the *Karen B* permission to dock and forced it to go in search of another country willing to accept its cargo.

According to the 'folk memory' of environment reporters, these two issues, particularly because they were given such prominence by newspapers that had a long record of supporting the Conservative Government, played a key role in forcing Mrs Thatcher to turn her mind to environmental issues, which she did in a speech to the Royal Society in London in September 1988. Amanda Brown, the first and last Environment Correspondent for the UK's national news agency, the Press Association, observed: 'what really did it as far as greenery is concerned is Mrs. Thatcher going public at the Royal Society ... unquestionably she made it her issue and put it, on the political agenda'. But the transformation, as one of our interviewees put it, of 'the iron lady into the green goddess' was, as the memoirs of Lady Thatcher (1995) reveal, a more complicated process. On the one hand her interest in environmental issues predated the seals campaign and the *Karen B* by a year. In 1987 she claimed to have taken measures to respond to growing public (and her own personal) concern about the environment by reorganising government funding of scientific research. And second, on an explicitly political level, she was troubled that what she saw as legitimate concern about the environment was being used 'to attack capitalism, growth and industry' (1995: 640). In other words she was seeking to reclaim the issue for the Conservatives. It is also worth noting in passing, that the Royal Society speech, now seen as a defining moment, passed by barely noticed. Thatcher amusingly recalls the problems caused by lack of television coverage of the event:

It broke new political ground. But it is an extraordinary commentary on the lack of media interest in the subject that, contrary to my expectations, the television did not even bother to send film crews to cover the occasion. In fact, I had been relying on the television lights to enable me to read my script in the gloom of the Fishmonger's Hall, where it was to be delivered; in the event, candelabra had to be passed up along the table to allow me to do so.

(1995: 640)

Following the positive reception given to Mrs Thatcher's speech in September 1988, she returned to the environmental theme a month later in her speech to the annual Conservative Party conference. The effect of those two speeches was dramatic in terms of their impact on the amount of space environmental coverage received in the national media and on the timing of the appointment of a whole new raft of environmental correspondents. It was all a far cry from the situation recently described by one of the veterans of environmental reporting Geoffrey Lean (*Independent on Sunday* on 23 April 1995), who wrote: 'For many years the subject (the environment) went so out of fashion that I would meet the entire press corps in the shaving mirror each morning.'

For a year this politically induced shift of interest towards green issues sustained the media's newfound environmentalism. However, the elections for the European Parliament in June 1989 saw the hitherto virtually unknown Green Party capture 2.3 million votes, representing almost 15 per cent of the poll. At this point, perhaps frightened by the genie that they had released from the bottle, the main parties sought to remove environmental issues from the political agenda, with the media following suit over the next twelve months. The number of environmental correspondents on the non-broadsheet press disappeared virtually overnight. In May 1989 there were 12 dedicated environment correspondents working on the national press; within two years the only remaining correspondents were on the broadsheet newspapers, even Independent Television News (ITN) decided it no longer needed an environmental specialist.

Andrew Veitch, Environment Correspondent at Channel 4 News, described the changing political environment as follows:

> There was the realisation in Conservative-supporting newspapers that you could report the environment massively as long as it was perceived in the ranks of the Conservative Party that the environment was important ... When green issues came to be seen as a threat to financial interests, then the enthusiasm declined suddenly. In other words, as soon as the environment correspondents dug in and started looking for the real causes of things, then Conservative editors became rather less keen on running these stories.

Michael McCarthy, who was Environment Correspondent for *The Times*, and who found himself out of a job when his newspaper ceased having a dedicated correspondent, perhaps surprisingly defended the decision to cut back: 'I think an environment correspondent for lots of papers has just been a totally unnecessary luxury because it's not justifiable in terms of the numbers of stories that they would produce which would be considered of sufficient public interest.'

However, the notion that either Conservative-leaning newspapers took fright when environmental issues seemed to threaten basic Tory values or that the mainstream politicians feared the electoral impact of the Green Party are not sufficient explanations in themselves. What in fact finally drove environmental issues off both the politicians' and the media's agendas was the onset of the recession of the late 1980s and early 1990s. Amanda Brown of the Press Association rejects the notion that either the media or Mrs Thatcher simply lost interest in the issue:

We didn't lose interest. What happened was there was a change of Prime Minister [John Major] and the economy took over. Mrs Thatcher led interest in the environment. I think she was genuinely interested, as a scientist. ... The trouble was that the recession came along and she went and the economic agenda just took over.

This is a point made strongly by many of our interviewees – that the ebb and flow of interest in green issues is directly related to the state of the economy. Michael McCarthy of *The Times* recalled a conversation with a senior Conservative:

In September 1989, the height of environmental concern, I interviewed Michael Heseltine who was then 'the prince over the water out of office' and he said that he believed that if prosperity disappeared environmental concern would disappear with it. And he has proved absolutely right. It was a subject that played no significant role in the 1992 General Election, there were no votes in it and I think that environmental concern is a distant issue, it's about things that are happening to future generations and they may be happening a long way away or perhaps in a distant future. It's not an immediate concern.

Amanda Brown argues that the environment came to be seen as a

luxury item ... that people can afford, they like to talk about recycling and doing green things when there is no threat of the old man losing his job. But I am afraid that when the economy is in a muddle it just goes down the pecking order of events.

The theme of the environment as a 'luxury item', as part of the 1980s 'yuppy culture', was also taken up by Grant Mansfield, who edits BBC TV's environmental series *Nature*:

during the eighties the economy was booming and people had quite a lot of money to burn and I think that, if one wants to be slightly cynical about it, people driving round in their GTI's also wanted to have their Greenpeace stickers on their windscreens.

However, one of the most interesting points raised by our respondents was the notion that the onset of recession not only pushed interest in environmental issues down the public's agenda (a notion that is not sustained by an examination of polling data for the period – see note 5 and Worcester, 1993) but also made those in key editorial positions ponder a little more deeply the complexity of the relationship between environmental concerns and economic development. Alex Kirby, the BBC Radio Environment Correspondent, observed:

I am fond of saying, very boringly, that there is no such thing as an environmental free lunch and I think that realisation is coming home to more and more of us now ... for every environmental solution there is another problem.

He was supported by his colleague Michael Buerk, one of BBC TV's main news presenters, who has had a long-standing interest in both environmental and development issues:

the deteriorating economic situation has led the audience to believe, and I think the journalists themselves to believe, that a lot of the decisions that appeared to be clear cut in times of prosperity are no longer clear cut. That, for example, there are trade-offs between a protected environment and jobs. Significantly in this recession middle class people in middle management jobs have been losing their jobs and environmental concern has been predominant with such people. ... They too have been feeling vulnerable about their jobs. And I think that people have come to realise that there are fine judgements to be made between wealth creation and job creation on the one hand and protection of the environment on the other.

## 3.5 ENVIRONMENTAL ISSUES ARE TOO COMPLEX

This perceived complexity has tended to 'scare off' journalists from environmental stories. For one of the basic adages of journalism, as Nigel Wade of the *Daily Telegraph* noted earlier, is the notion that at its heart is 'conflict' and the job of the journalist is teasing out and exposing that conflict. Superficially many environmental stories may appear to be classic 'goodies against baddies' – an ideal journalistic scenario. But Mark Damazer, who is Head of BBC Television News, recognises the dangers of being seduced by such simplistic approaches. Environmental stories have, he says:

to be stripped of this one-sided goodies and baddies style argument, which is not to say that anybody who is a mass polluter ought to be applauded or celebrated, but sometimes there are interesting reasons why somebody is polluting which, if you don't bother to explain to the audience, you're stripping the informative level out of the story and ridding it of its important dimensions which the audience needs to know.

An example of this was a Channel 4 News story about pollution in the river Stour near the English carpet-town of Kidderminster, which was being caused by the discharge of effluent from a local carpet factory. Sue Inglish of Channel 4 News explained:

On the face of it, it was a simple case of nailing the company for the damage they were doing to the river. However, when we looked a little deeper we discovered that the pollution was being caused because the factory was using yarn imported from India which contained a high concentration of pesticides and it was the washing of this yarn that was causing the pollution. So who do we point the finger at? The factory which was trying to maximise profits and keep employing workers in a town with a high unemployment level? The Indian farmers who were trying to work their poor quality soil as effectively as possible? Or the conservationists who were putting the state of a local river above the welfare of the workers in Kidderminster and the farmers in India?

But it is not just the imperatives of economic development set against the demands of conservationists that can blur the edges between heroes and villains; there can be, as Nigel Wade discovered, environmentalists on both sides of an argument. He quoted the case of a story about a planned bypass route around the English cathedral city of Salisbury which, after they had inadvertently presented it as a 'bad thing', led to an embarrassing climb-down:

We had talked to all the people who were against it. And for some reason the editing was lax and nobody noticed that the piece didn't quote at all the people who were for it and the next day we had a lot of equally concerned conservationists who actually lived in Salisbury and who were talking about taking the pressure off the old town and who said this by-pass is very badly needed. And so we later had to put their side. And this wasn't a case of the developer versus the conservationists. It was two groups of conservationists seeing the issue in different ways.

Specialist environmental journalists, and their editors, tend to view environmental stories as peculiarly complex. This is often cited as a major reason why environmental stories are frequently sidelined in the battle for space and time. One of the main reasons for this perception is that, in terms of normal journalistic time-frames, these stories tend to move at a glacial pace. Gans (1979) makes the point that stories that follow the daily time-frames of the news media have a much greater likelihood of being covered than those that move at a significantly faster or slower rate. The first time that journalists 'discovered' global warming or the hole in the ozone layer they were able, in the words of their trade, to 'splash' the story. They could for the second or even the third time, but a point soon came where they are unable to argue for the time on the bulletin or space in the paper on the basis of one more scientific paper adding to the weight of evidence. Justin Jones explained:

> news desks started to say 'oh no, not another story on global warming'. They saw it as a one-off story or a two-off story. They didn't see it now as they see other subjects. News editors see global warming as a subject, they don't see it as an issue.

Environmental stories are also complex because at their heart they are scientific stories and once the basic phenomenon has been described the journalist can only 'take the story on' if he or she is both able to understand the science at the heart of the problem and also explain it to a mass audience. Lawrence McGinty, ITN's Science Editor, summed up the problem:

> Two years ago you could do a story saying 'this was the problem'. Now people expect you to go a bit further and say 'well, ok, that's the problem. We all know about that. We know all about it from both sides. What's the wrinkle?' And the kind of second stage stories are actually technically and politically quite complicated and are not easily done on TV. Whilst, for instance, you could write an article analysing whether the recent tightening up of the Montreal Protocol was a good or bad development, on television it could be actually quite a complicated thing to explain.

## 3.6 PICTURING THE ENVIRONMENT

Television journalism, in particular, has a complicated relationship with environmental stories – on the one hand they are seen as particularly difficult to translate into visual terms, on the other hand they can produce some of television's most impactful news coverage. Amanda Brown of the Press Association, whose clients are newspapers, radio and television, observed the impact of the seal virus coverage in 1988 upon general perceptions about the importance of conservation issues. She was asked why she thought it was the seal virus in particular that seemed to draw the public's attention to the environmental agenda and replied: 'Well it's a good telly thing. I mean TV leads all these issues and you've got seals with big brown eyes sticking their heads out of the water, it was irresistible. . . . TV, it drives everything.'

But, to state the obvious, if dramatic pictures are missing, television journalists have real problems. Lawrence McGinty put the issue starkly: 'Can we show the threat, can we demonstrate that threat in pictures. If we can't, well, forget it.' This view was shared by his colleague Andrew Veitch from Channel 4 News for, despite the fact that his programme is aimed at a more up-market audience, for him, too, pictures remained paramount: 'If something is not picturable [sic] then we wouldn't cover it.' ITN's former environment correspondent Justin Jones faced a similar dilemma:

> pictorially it's a tricky one to show global warming because obviously they're showing something of the future. You cannot show what it is. So after you've done your fancy graphics and you've shown a few bits of East Anglia that have been battered by rising sea levels, there's not much more you can do before you have to use the same pictures again with a slightly different script.

From the perspective of an international news agency, where pictures have an even greater significance, Dave Modrowski, News Editor of Worldwide Television News, was frank about the problem:

> In our editorial conference someone might say that we ought to be doing something about global warming for instance. We all say 'Yes we ought' then someone else says 'How the hell are we going to illustrate it? How do we show a hole in the ozone layer? How do we get a camera to Antarctica to show the thinnest area of the ozone layer and so on?' We try but it's difficult, it's very difficult.

However, there are occasions when powerful pictures can combine with a strong story-line to promote a story that, on traditional news values alone, would probably never see the light of day. Sue Inglish of Channel 4 News gave the example of a Channel 4 report about the drying of Lake Chad in Africa – drying because the four countries that bordered Lake Chad were unable to agree upon an equitable way of distributing the waters that fed the Lake:

> it was a brilliant picture story . . . but in order to give it the edge that we needed we actually had to look at why the politics of the four African countries involved prevented them from collaborating with their neighbours on any kind of water conservation work.

But there is another way in which the 'picture power' of environmental stories can work to ensure coverage, and that is because in some political cultures environmental stories are seen as 'uncontroversial'. Humayun Choudhury, who runs the Asiavision News Exchange in Kuala Lumpur, believed that environmental stories were harmless 'bulletin-fillers': 'They are the safest bet . . . highly non-controversial, non-political . . . a good reporter can do a pure environmentalist story without it having too many political ramifications elsewhere. That's what they try to do mostly without getting into any controversy.'

Even where pictures are available, news editors and producers tend to be attracted to environmental stories that have exotic settings, despite the fact that there is no shortage of compelling environmental stories available within the domestic sphere. Sue Inglish of Channel 4 News:

> Now, one of the problems of environment stories is, quite a lot of them tend to be in places which are quite hard to get to and are quite expensive and might take quite a few weeks to film, either because the travelling is difficult or because it is difficult to get people to talk to you or whatever. So you have to make a number of calculations about what it's going to cost you.

## 3.7 IS ENVIRONMENTAL REPORTING WORTH THE COST?

Environmental journalists, fighting to get space on bulletins or on the page, recognise the fact that the cost of their stories is the single biggest factor inhibiting the amount of coverage they can achieve. Amanda Brown of the Press Association:

> Everyone knows about killing whales. That story always makes it and if it's in the UK we'll cover it but for example this year the story was in Japan and it was going to cost PA a lot of money to send me, perhaps up to £3,000, so we just picked up the story from Reuters.

A point echoed by Sue Inglish's environment correspondent Andrew Veitch who in answer to the question 'What would you say are the major constraints on environmental issues getting more coverage?' replied:

> Money. It's simply money, to go and film. Obviously if I want to go, as I am doing, and film in the middle of Russia next week, its thousands and thousands and thousands of pounds and that is a big consideration.

Cost considerations are significant as far as domestic news producers are concerned but for those journalists working for organisations that seek to provide an international news service, they can have a significantly distorting effect, as Melissa Ballard, the Environment Correspondent of Cable News Network (CNN), makes clear in answer to the question as to the balance between American and international environmental coverage. 'It's heavily domestic because of costs. We would do more international stories if we could afford to. It's certainly a factor that makes us go for a story as against another one.' Even when asked what the major pressures affecting her selection of stories were, Ballard found it impossible to escape the cost argument: 'Make it interesting to the audience and the producers ... They're the only pressures. The traditional ones and oh yeah money. Can we afford to do it?'

For those involved in the more direct activity of earning their living on the basis of their ability to 'read' a market then the choices become even more stark. John Sutton, Managing Editor at Reuters TV, the world's largest television news agency, put it thus: 'Well, there is no point in sending out something to clients that may be seen as very worthy but is so dull that no broadcaster is going to bother airing it.'

Sutton's American counterpart addresses the issue in a slightly different way. Bill Ahearn, a Vice-President and Executive Director at Associated Press (AP), claimed that his organisation believed that AP should be seeking to lead, rather than follow, its clients' news priorities:

> I don't think the Associated Press should wait for members to initiate the request that we cover topics of global interest. I think it's for the AP to initiate that action; for the AP to decide these are the priorities based on what we know is happening in society, what people are reading, what they care about.

The former Science Editor at AP, Paul Raeburn, shared this view by claiming that if they were not providing their clients with what they required, they would soon become aware of it: 'I think that's a reasonable assumption that we are pretty close in meeting their requirements. If we were too far off the mark we would be getting feedback on it.'

## 3.8 CULTIVATING THE DEVELOPING WORLD MARKET FOR NEWS

Among editors working with international news agencies, a notion of universal news values, which do not challenge their own domestic news values even when catering for audiences coming from very different cultures, appears common. Bill Ahearn, for example, questioned whether AP's news agenda was 'satisfactory' for a Third World audience, responded:

> Well, I think I have to ask you what do you mean by 'satisfactory for a Third World audience'? What do you think the needs of the Third World audience are versus any other audience? Do you think they have specific needs that are different from those of another audience?

Pat Meeney, who covered environmental issues for the London-based Reuters news agency, believes that story-selection was a relatively simple market decision – largely dependent on whether the country involved was a Reuters customer or not:

> You also think about who your audience is. My audience is different from PA's audience or from the Times audience. For example, if Britain came out with a policy criticising or supporting Bangladesh cutting down more rain forests, PA wouldn't care about that because they don't have any people in Bangladesh, whereas I have got to think about who is going to read this, who my audience is.

Meeney's opposite number at Associated Press, Paul Raeburn, had a different approach. Although his material was supposed to have a world-wide audience, he in fact recognised that he was writing from an American perspective for an American audience. He left it to AP's World Desk, the sub-editors responsible for distributing AP material outside the United States, to give his copy a more global stance:

> The World Desk would have to go through my copy and get out all the specifically local material. And if I write a story that includes statistics from US environmental protection agencies then they'll substitute global statistics from the UN or something.

Raeburn accepts the fact that, when writing a story not based in the US, he faces particular difficulties:

> I find it quite difficult when I am doing stories that involve an issue like say the one about compensation of genetic resources to find knowledgeable people in far away places who I can find on the telephone and then be able to have a conversation where I really get the information I am looking for. So there is a concern there that the stories are not going to be balanced. Even if you try, they're still going to be tipped toward a developed world perspective more than you want them to be.

He gave as an example a recent story he had covered about American companies prospecting for oil in Ecuador; it was based on a piece that had earlier appeared in the *New Yorker* magazine:

> I wrote a short story here and talked to the Natural Resource Defence Council and the oil company. So I got from two sides what was going on. It was just a short story of no great consequence. But the key point in the article was the impact this exploration was going to have on some of the indigenous groups that live in the area. And there is no way I can talk to them unless I go down

there which, in terms of the time, the cost and the significance of the story, could not be justified.

Nor was the problem necessarily eased by Raeburn leaving his desk in New York and going to where the stories were actually taking place. He described how, at an international conference, cultural differences can inhibit the work of even the most conscientious reporter:

I was very concerned throughout that conference that I wasn't getting enough from the developing countries that were there because many of the important discussions went on in closed sessions. American participants are accustomed to being accosted as soon as they walk out of the door and they'll tell you what's been going on because they feel they have an obligation to talk to the press, but many developing country participants that I talked to just simply refused to talk, feeling that they should protect the confidentiality of the meeting. So therefore it was very difficult to get their points of view on some of those stories.

Another aspect of the gap between journalists in the developed world and potential Third World audiences was highlighted by Melissa Ballard of CNN. She recognised that audiences in some societies were less adept at 'reading' her television pictures than her domestic audience. She realised that conventions of American (and British) television news, largely derived as they were via the newsreels from Hollywood, are based on the willingness of the audience to collude with the film-maker in accepting time and space compressions, in other words to accept the legitimacy of the editing process. Some audiences found this problematic, as Ballard explained by way of example. She recounted the difficulties an audience, unfamiliar with American television news, would have in trying to 'read' a story about the effects of a tanker oil-spill on sea birds:

Where did that oil come from, they'd want to know. It's not obvious how it was spilt from a ship. You have to show cause and effect. If you're dealing with an audience that is not visually literate, you're going to have to show the ship, then show the oil inside the ship, then the hole in the ship, the oil seeping out onto the water, the oil hitting the animal, the animal hitting the beach and so on. They're not used to supplying the information in between the edits. American audiences are used to television news. They're used to seeing a ship, oil around the ship and then cut to the picture of an oil-affected animal on the beach and they're used to that going into their brain without even thinking 'oh, this animal was affected by the oil spill'.

However, the problem goes further because, as Ballard herself admits, many of her reports are based on knowledge and assumptions which would be reasonable to make for an American audience but are less easy to justify for audiences elsewhere:

You have to assume a lot. There is no way to do a news story about a development in the ozone hole without assumptions such as what an atmosphere is. You're assuming that these chemicals are floating around in the air and if you are trying to reach an uneducated scientifically illiterate audience, there is no way to compensate for that in news. They're going have to be reached somehow else. Education, that's the only way.

The issue of how different societies interpret differently the same media texts is the subject matter of Chapter 5, which investigates and analyses how audiences in

India and Britain interrogate and resolve identical texts in sometimes very different ways. This chapter has sought to demonstrate how journalists have only very limited autonomy in defining, for themselves, what aspects of the environmental story should receive prominence, and it concludes with this bleak characterisation of how former ITN correspondent Justin Jones came to see his role as an environmental correspondent:

> The news is not true. News is what the row is about and maybe what the consequences are and what's likely to happen. You really rarely have the time to find the truth. All this thing about journalistic truth is fine in documentaries, but in news, as far the truth is concerned, you just haven't got the time.

## NOTES

1   The *Guardian* newspaper, 20 April 1995.
2   The British tabloid press have not featured in this chapter (other than as respondents to a postal questionnaire) for two reasons. First, because they now devote very little space to environmental issues, for example see Longman and Lacey (1994) for preliminary findings of a survey of British newspapers' coverage of the Earth Summit which revealed how this major conference was well-covered by the broadsheet newspapers but 'simply did not occur in the tabloid press' (p. 51). At the time of writing there were no environmental correspondents working for UK tabloid newspapers, and second, because, like the regional media (which are also excluded from this study), they have very little involvement in the flow of international news between the UK and India.
3   Unless otherwise stated all designations are of the position held by the respondent at the time of interview in 1993.
4   This is borne out by an indicative research exercise undertaken for this project. A word-search using three key environmental phrases – 'global warming', 'ozone layer' and 'pollution' – was undertaken in five UK national newspapers (*The Times*, the *Guardian*, the *Daily Telegraph*, *The Independent* and *Today*) covering the period April 1987 to March 1995. It revealed a very distinctive pattern. Coverage between 1987 and the end of 1988 was minimal. It then began to climb, peaking in the middle of 1990 and, apart from a brief upturn at the time of the Earth Summit in 1992, fell back, not quite to its 1987 levels but to something like half the level it achieved early in 1990. Some of the reasons behind this trend are explored in Anderson and Gaber (1993a).
5   The debate about the significance of 'primary definers', i.e. those in positions of authority in society, being able to determine the news priorities of the mass media has raged on through the past 17 years. It began with Hall *et al.* (1978) and has been responded to in a variety of ways, perhaps the most significant being Schlesinger (1990) and very much echoed by Anderson (1993) in her work specifically on the environment.

## REFERENCES

Anderson, A. (1991) 'Source strategies and the communication of environmental affairs', *Media Culture and Society* 13(34): 459–76.
—— (1993) *The production of environmental news: a study of source–media relations*, unpublished Ph.D. thesis, University of Plymouth.
Anderson, A. and Gaber, I. (1993a) 'The yellowing of the Greens', *British Journalism Review* 4(2): 49–53.
—— (1993b) 'The road from Rio: the causes of environmental "antisappointment"' *Intermedia* 21(6): 27–9.
Clayton, A. *et al.* (1993) 'Change and continuity in the reporting of science and

technology: a study of *The Times* and the *Guardian*', *Public Understanding of Science* 2: 225–34.

Cracknell, J. (1993) 'Issue arenas, pressure groups and environmental agendas', in Hansen, A. (ed.) *The Mass Media and Environmental Issues*, Leicester: Leicester University Press, pp. 3–21.

Curran, J. and Gurevitch, M. (1991) *Mass Media and Society*, London: Edward Arnold.

Friedman, S. (1991) 'Two decades of the environmental beat', in LaMay, C. and Dennis, E. (eds) *Media and the Environment*, Washington, DC: Island Press.

Gaber, I. (1993) 'A cold shoulder for the environment', *British Journalism Review* 4(4).

Gallup (1993) 'The environment – research undertaken for the South-North Centre for International Environmental Policy Studies', unpublished.

Gans, H. (1979) *Deciding What's News: A Study of CBS Evening News, NBC Nightly News and Time*, New York: Random House.

Goffman, E. (1974) *Frame Analysis*, Philadelphia: University of Pennsylvania Press.

Graber, D. (1988) *Processing the News: How People Tame the Information Tide*, New York: Longman.

Habermas, J. (1989) *The Structural Transformation of the Public Sphere*, Cambridge: Polity Press.

Hall, S., Crichter, C., Jefferson, T. and Roberts, B. (1978) *Policing the Crisis*, London: Macmillan.

Hansen, A. (ed.) (1993) *The Mass Media and Environmental Issues*, Leicester: Leicester University Press.

—— (1993) 'Greenpeace and the press coverage of environmental issues', in Hansen, A. (ed.) *The Mass Media and Environmental Issues*, Leicester: Leicester University Press, pp. 150–78.

Henley Centre (1993) *Media Futures* 93–94 London: Henley Centre.

Longman, D. and Lacey, C. (1994) 'Despatches from the doze-zone', *British Journalism Review* 5(2): 49–53.

Lowe, P. and Morrison, D. (1984) 'Bad news or good news: environmental politics and the mass media', *Sociological Review* 32: 75–90.

Schlesinger, P. (1990) 'Rethinking the sociology of journalism: source strategies and the limits of media centrism', in Ferguson, M. (ed.) *Public Communication: The New Imperatives – Future Directions of Media Research*, London: Sage.

Schudson, M. (1989) 'The sociology of news production', *Media Culture and Society* 11: 262–82.

—— (1991) 'The sociology of news production revisited', in Curran, J. and Gurevitch, M. (eds) *Mass Media and Society*, London: Edward Arnold.

Soloski, J. (1989) 'News reporting and professionalism: some constraints on the reporting of news', *Media, Culture and Society* 11: 207–28.

Strentz, H. (1989) *News Reporters and News Sources: Accomplices in Shaping and Misshaping the News*, Ames: Iowa State University Press.

Tuchman, G. (1978) *Making News: A Study in the Construction of Reality*, New York: The Free Press.

Worcester, R. (1993) 'Are newsrooms bored with greenery?', *British Journalism Review*, 4(3): 24–6.

# 4 The view from the newsrooms of India

## 4.1 INTRODUCTION

This chapter will discuss the conclusions we have drawn from our analysis of the transcripts of interviews with Indian media personnel, conducted in Bombay, Pune and Delhi in August 1994. Before we do so however, it is best to amplify the brief description of the media given in Chapter 1, and in particular to highlight the question of language. The State TV network Doordarshan (known as DD) broadcasts in 26 specified languages and a further number of unspecified languages, as shown in Table 4.1.1, out of the between 500 and 1,000 languages (depending on the definitions used) spoken in India. The unspecified languages used by DD mostly relate to tribal groups, many of whose tongues are not Indo-European but Indo-Chinese in origin. Quite a number of these languages have never been systematically attached to an alphabet, so it is impossible to be literate in them. There are many more languages in which there are no broadcasts at all.[1] Even those shown in this table may only be used for a few hours' transmission at specific times during the week. Obviously the language issue is equally significant for the press.

## 4.2 LANGUAGE AND BROADCASTING

Assam is the state with the widest range of language broadcasts. The figure for its neighbour Arunachal Pradesh is unknown, since it is given only as 'local languages' – and of these there are many. These two states are in one of the two major areas of India which are currently 'disturbed', meaning that there is an armed insurgency. The whole of this disturbed area includes Nagaland, Tripura, Manipur, and sometimes Meghalaya as well. (Jammu and Kashmir, adjacent to Pakistan in the northwest of India, is the other area of insurgency.) Many of the local population of the northeast are tribals, some of whom have been converted to Christianity by missionaries or have remained animist. Their disputes are not only about incorporation within India, but also between themselves. As a government-owned enterprise, Doordarshan has to play its part in the modernisation and integration of these areas, so it is not surprising to see the effort put into broadcasting here.

Geographically, the most widely used language is Hindi, followed by English, and then Urdu. This latter fact reflects two things: one, the scattered nature of the Muslim population, mostly in urban centres; and two, secular India's need to incorporate people of different communities.

Overall the number of TV sets is actually quite small and heavily biased towards urban areas. We do not have the urban–rural breakdown of TV sets by state, but Doordarshan (1994) gives the national figure for rural penetration. For the population of 627 million rural inhabitants, there are 12 million households with TV sets, reaching an estimated 60 million family viewers. This figure is augmented by 'other viewers', some of whom have access to community sets, and thus a national figure is given of 110 million rural viewers – let us say roughly 15 per cent of the rural audience. It is further stated that in only four states do more than 15 per cent of households in rural areas have TV sets – Punjab, Haryana, Kerala and Maharashtra. All this puts the recent clamour about satellite TV and foreign penetration into perspective. It is a real phenomenon, but very much limited to urban areas, and particularly the major metropolises.

We do not have equivalent data for radio. AIR is, however, much more widespread and has many more local stations than TV. Kumar (1994: 140) gives the number of AIR stations as 10 in 1947, 90 in 1986 and 205 in 1990. He gives the 'reach' of radio as 64 per cent in urban areas, and 46 per cent in rural areas in 1989 – although by 1995 it was estimated that there were 111 m. receivers in the country and that transmitters could theoretically cover 97 per cent of the population. AIR puts out 89 national news bulletins in 19 languages each day, 134 regional bulletins in 64 languages and dialects, plus an external service in 7 Indian and 18 non-Indian languages (All India Radio, 1995).

## 4.3 LANGUAGE AND THE PRESS

The newspaper industry is in private hands and does not have to satisfy all the minority groups which DD and AIR attempt to reach. Table 4.3.2 shows that on average half of India's population is literate – but literacy is biased towards males and towards urban areas. The Indian Newspaper Society (INS) survey on which Tables 4.3.1a, b and c are based lists publications in 16 languages. English used to be the lead language, but now the greatest number of dailies is published in Hindi (146) followed by English (54), the language of 2 per cent of the population. In Maharashtra there are 62 daily papers published in 6 languages, and in Uttar Pradesh, 53 papers in 4 languages. In terms of total circulation, Hindi again comes first, but English is not so far behind. The number of publications also reflects the number of places in which papers are published. It is far from the case that all papers are published in the state capital. In Uttar Pradesh, for example, the 53 papers are published from 22 towns.

In the UK it is difficult to run a competitive daily newspaper with a circulation of less than 400,000 – but this table shows average figures only reaching 115,000 in Tamil (the major language of Tamil Nadu), and 119,000 for Malayalam (the major language of the adjacent state of Kerala). Averages are of course misleading – but taking the cut-off point of 400,000 there are two papers in Gujarati published from Gujarat, two papers in Malayalam from Kerala, one paper from Tamil Nadu in English, and one paper in Bengali from Bengal which cross this threshold.

Although the figures for average circulation are small by western standards, it is difficult to interpret them for a few 'national' papers – those few major papers which are published in different editions from several centres, covering different states. The INS data count these as separate newspapers, but in Table 4.3.3 we aggregate the figures for two leading papers, *The Times of India* (English) and the *Navbharat Times* (Hindi).

In general this kind of exercise will increase the scale of a few papers, and of the English papers more than the non-English papers. But even then the circulation figures are still not very large.

The papers are 'small' in another way too. Partly because of government rationing of newsprint in the past, and partly because these papers have not grown in the

*Table 4.1.1* Languages used by Doordarshan TV transmissions by states and union territories, 1994

| | Andhra Pradesh | Arunachal Pradesh | Assam | Bihar | Goa | Gujarat | Haryana | Himachal Pradesh | Jammu and Kashmir | Karnataka | Kerala | Madhya Pradesh | Maharashtra | Manipur |
|---|---|---|---|---|---|---|---|---|---|---|---|---|---|---|
| TV households in '000s 1994 | 3580 | 17 | 692 | 2379 | 91 | 2652 | 946 | 116 | 123 | 2500 | 1409 | 2558 | 5828 | 66 |
| 1991 population in millions | 87 | 1 | 23 | 86 | 1 | 41 | 16 | 5 | 8 | 45 | 29 | 66 | 79 | 2 |
| % urban | 27 | 13 | 11 | 13 | 41 | 35 | 25 | 9 | 24 | 31 | 26 | 23 | 39 | 28 |
| % literate, urban | 66 | 72 | 79 | 68 | 80 | 77 | 74 | 84 | na | 74 | 92 | 71 | 79 | 71 |
| % literate, rural | 36 | 37 | 49 | 34 | 72 | 53 | 50 | 62 | na | 48 | 89 | 36 | 56 | 56 |
| Assamese | | | x | | | | | | | | | | | |
| Bengali | | | x | | | | | | | | | | | |
| Bhojpuri | | | | | | | | | | | | | | |
| Dogri | | | | | | | | | x | | | | | |
| English | x | x | x | | | | | | | | x | x | | |
| Garo | | | | | | | | | | | | | | |
| Gorkhali | | | | | | | | | | | | | | |
| Gujarati | | | | | | x | | | | | | | x | |
| Himachali | | | | | | | | x | | | | | | |
| Hindi | x | | x | x | | x | x | | | | x | x | x | |
| Jaintia | | | | | | | | | | | | | | |
| Kannada | | | | | | | | | | x | | | | |
| Kashmiri | | | | | | | | | x | | | | | |
| Khasi | | | | | | | | | | | | | | |
| Konkani | | | | | x | | | | | | | | | |
| Local languages | | x | x | x | | | | | | | | | | |
| Malayalam | | | | | | | | | | | x | | | |
| Manipuri | | | x | | | | | | | | | | | x |
| Marathi | | | | | | | | | | | | | x | |
| Nepali | | | x | | | | | | | | | | | |
| Oriya | | | | | | | | | | | | | | |
| Punjabi | | | | | | | | | | | | | | |
| Sindhi | | | | | | x | | | | | | | | |
| Tamil | | | | | | | | | | x | | | | |
| Telugu | x | | | | | | | | | | | | | |
| Urdu | x | | | x | | x | | | x | x | | | x | |
| Total | 4 | 2 | 7 | 3 | 1 | 4 | 1 | 1 | 3 | 3 | 3 | 2 | 4 | 1 |

*Table 4.1.1*   Continued

| | Meghalaya | Orissa | Punjab | Rajasthan | Tamil Nadu | Tripura | Uttar Pradesh | West Bengal | Delhi¹ | Total | Mizoram | Nagaland | Sikkim |
|---|---|---|---|---|---|---|---|---|---|---|---|---|---|
| TV households in '000s 1994 | 34 | 828 | 1401 | 1415 | 3734 | 46 | 4338 | 3740 | 1726 | | | | |
| 1991 population in millions | 2 | 32 | 20 | 44 | 1 | 3 | 139 | 68 | 9 | | 1 | 1 | 1 |
| % urban | 19 | 13 | 30 | 23 | 9 | 15 | 20 | 27 | 90 | | 46 | 17 | 9 |
| % literate, urban | 82 | 72 | 72 | 65 | 81 | 83 | 61 | 75 | 76 | | 93 | 83 | 81 |
| % literate, rural | 41 | 45 | 53 | 30 | 54 | 56 | 37 | 51 | 67 | | 72 | 57 | 54 |
| | | | | | | | | | | | language details not available for the rest of the table below | | |
| Assamese | | | | | | | | | | 1 | | | |
| Bengali | | | | | | | x | x | | 3 | | | |
| Bhojpuri | | | | | | | x | | | 1 | | | |
| Dogri | | | x | | | | | | | 2 | | | |
| English | x | | | | x | | | x | x | 10 | | | |
| Garo | x | | | | | | | | | 1 | | | |
| Gorkhali | | | | | | | | x | | 1 | | | |
| Gujarati | | | | | | | | | | 2 | | | |
| Himachali | | | x | | | | | | | 1 | | | |
| Hindi | | x | x | x | | | x | x | x | 14 | | | |
| Jaintia | x | | | | | | | | | 1 | | | |
| Kannada | | | | | x | | | | | 2 | | | |
| Kashmiri | | | | | | | | | | 1 | | | |
| Khasi | x | | | | | | | | | 1 | | | |
| Konkani | | | | | | | | | | 1 | | | |
| Local languages | | | | | | | | | | 3 | | | |
| Malayalam | | | | | x | | | | | 2 | | | |
| Manipuri | | | | | | | | | | 2 | | | |
| Marathi | | | | | | | | | | 1 | | | |
| Nepali | | | | | | | | | | 1 | | | |
| Oriya | | x | | | | | | | | 1 | | | |
| Punjabi | | | x | | | | | | | 1 | | | |
| Sindhi | | | | x | | | | | | 2 | | | |
| Tamil | | | | | x | | | | | 2 | | | |
| Telugu | | | | | x | | | | | 2 | | | |
| Urdu | | | | | | | x | x | | 8 | | | |
| Total | 4 | 2 | 4 | 2 | 5 | 1 | 3 | 5 | 2 | | | | |

*Note:*  1 The source also says 'All Indian Languages'

*Source:* 'Doordarshan – 1994: facts and figures', Audience Research Unit, Doordarshan, New Delhi, 1994.

*Table 4.3.1a*  Number of daily newspapers by state of publication and language, 1993

| | Andhra Pradesh | Assam | Bihar | Goa | Gujarat | Haryana | Himachal Pradesh | Jammu and Kashmir | Karnataka | Kerala | Madhya Pradesh | Maharashtra | Manipur | Meghalaya | Orissa | Punjab | Rajasthan | Tamil Nadu | Tripura | Uttar Pradesh | West Bengal | Chandigarh | Delhi | Total |
|---|---|---|---|---|---|---|---|---|---|---|---|---|---|---|---|---|---|---|---|---|---|---|---|---|
| Assamese | | 5 | | | | | | | | | | | | | | | | | | | | | | 5 |
| Bengali | | 3 | | | | | | | | | | | | | | | | | 4 | | 9 | | | 16 |
| English | 1 | 4 | 3 | 2 | 1 | | | 2 | 2 | | 2 | 13 | 1 | 2 | 1 | | 1 | 2 | | 5 | 5 | 2 | 5 | 54 |
| Gujarati | | | | | 18 | | | | | | | 6 | | | | | | | | | | | | 24 |
| Hindi | | 2 | 17 | | | 2 | 1 | | | | 30 | 7 | | | 1 | 3 | 23 | | | 45 | 5 | 1 | 9 | 146 |
| Kannada | | | | | | | | | 9 | | | | | | | | | | | | | | | 9 |
| Konkani | | | | 1 | | | | | | | | | | | | | | | | | | | | 1 |
| Malayalam | | | | | | | | | | 13 | | | | | | | | | | | | | | 13 |
| Marathi | | | | 2 | | | | | 2 | | | 33 | | | | | | | | | | | | 37 |
| Nepali | | | | | | | | | | | | | | | | | | | | | 1 | | | 1 |
| Oriya | | | | | | | | | | | | | | | 9 | | | | | | | | | 9 |
| Punjabi | | | | | | | | | | | | | | | | 12 | | | | 1 | 1 | 1 | 1 | 16 |
| Sindhi | | | | | | | | | | | | 1 | | | | | 1 | | | | | | | 2 |
| Tamil | | | | | | | | | 1 | | | | | | | | | 9 | | | | | | 10 |
| Telugu | 8 | | | | | | | | | | | | | | | | | | | | | | | 8 |
| Urdu | 3 | | | | | | | 1 | 1 | | 1 | 2 | | | | 3 | | | | 2 | 2 | | 3 | 18 |
| Total | 12 | 14 | 20 | 5 | 19 | 2 | 1 | 3 | 15 | 13 | 33 | 62 | 1 | 2 | 11 | 18 | 25 | 11 | 4 | 53 | 23 | 4 | 18 | |

*Source:*  Calculated from *INS Press Handbook 1993*, New Delhi: Indian Newspaper Society, 1993.

*Table 4.3.1b* Total circulation of daily newspapers in '000s by state of publication and language, 1993

| | Andhra Pradesh | Assam | Bihar | Goa | Gujarat | Haryana | Himachal Pradesh | Jammu and Kashmir | Karnataka | Kerala | Madhya Pradesh | Maharashtra | Manipur | Meghalaya | Orissa | Punjab | Rajasthan | Tamil Nadu | Tripura | Uttar Pradesh | West Bengal | Chandigarh | Delhi | Total |
|---|---|---|---|---|---|---|---|---|---|---|---|---|---|---|---|---|---|---|---|---|---|---|---|---|
| Assamese | | 162 | | | | | | | | | | | | | | | | | | | | | | 162 |
| Bengali | | 69 | | | | | | | | | | | | | | | | | 107 | | 1068 | | | 1244 |
| English | 74 | 97 | 75 | 33 | 17 | | | 116 | 169 | | 27 | 844 | 10 | 39 | 21 | | 3 | 730 | | 146 | 344 | 194 | 461 | 3400 |
| Gujarati | | | | | 1602 | | | | | | | 269 | | | | | | | | | | | | 1871 |
| Hindi | | 52 | 703 | | | 46 | 22 | | | | 885 | 317 | | | 21 | 395 | 1117 | | | 1598 | 208 | 48 | 578 | 5990 |
| Kannada | | | | | | | | | 574 | | | | | | | | | | | | | | | 574 |
| Konkani | | | | 2 | | | | | | | | | | | | | | | | | | | | 2 |
| Malayalam | | | | | | | | | | 1545 | | | | | | | | | | | | | | 1545 |
| Marathi | | | | 23 | | | | | 54 | | | 1662 | | | | | | | | | | | | 1739 |
| Nepali | | | | | | | | | | | | | | | | | | | | | 42 | | | 42 |
| Oriya | | | | | | | | | | | | | | | 604 | | | | | | | | | 604 |
| Punjabi | | | | | | | | | | | | | | | | 458 | | | | 23 | 24 | 63 | 16 | 584 |
| Sindhi | | | | | | | | | | | | 3 | | | | | 38 | | | | | | | 41 |
| Tamil | | | | | | | | | 18 | | | | | | | | | 1032 | | | | | | 1050 |
| Telugu | 622 | | | | | | | | | | | | | | | | | | | | | | | 622 |
| Urdu | 56 | | | | | | | 22 | 1 | | 11 | 45 | | | | 84 | | | | 27 | 37 | | 47 | 330 |
| Total | 752 | 380 | 778 | 58 | 1619 | 46 | 22 | 138 | 816 | 1545 | 923 | 3140 | 10 | 39 | 646 | 937 | 1158 | 1762 | 107 | 1794 | 1723 | 305 | 1102 | |

*Source:* Calculated from *INS Press Handbook 1993*, New Delhi: Indian Newspaper Society, 1993.

Table 4.3.1c  Average circulation of daily newspapers in '000s by state of publication and language, 1993

| | Andhra Pradesh | Assam | Bihar | Goa | Gujarat | Haryana | Himachal Pradesh | Jammu and Kashmir | Karnataka | Kerala | Madhya Pradesh | Maharashtra | Manipur | Meghalaya | Orissa | Punjab | Rajasthan | Tamil Nadu | Tripura | Uttar Pradesh | West Bengal | Chandigarh | Delhi | Average for all states |
|---|---|---|---|---|---|---|---|---|---|---|---|---|---|---|---|---|---|---|---|---|---|---|---|---|
| Assamese | | 32 | | | | | | | | | | | | | | | | | | | | | | 32 |
| Bengali | | 23 | | | | | | | | | | | | | | | | | 27 | | 119 | | | 78 |
| English | 74 | 24 | 25 | 17 | 17 | 23 | 22 | 58 | 85 | | 14 | 65 | 10 | 20 | 21 | | | 3 | | 29 | 69 | 97 | 92 | 63 |
| Gujarati | | | | | 89 | | | | | | | 45 | | | | | | | | | | | | 78 |
| Hindi | | 26 | 41 | | | | | | | | 30 | 45 | | | | 132 | 49 | | | 36 | 42 | 48 | 64 | 41 |
| Kannada | | | | | | | | | 64 | | | | | | | | | | | | | | | 64 |
| Konkani | | | | 2 | | | | | | | | | | | | | | | | | | | | 2 |
| Malayalam | | | | | | | | | | 119 | | | | | | | | | | | | | | 119 |
| Marathi | | | | 12 | | | | | 27 | | | 50 | | | | | | | | | | | | 47 |
| Nepali | | | | | | | | | | | | | | | | | | | | | 42 | | | 42 |
| Oriya | | | | | | | | | | | | | | | 67 | | | | | | | | | 67 |
| Punjabi | | | | | | | | | | | | | | | | 38 | | | | 23 | 24 | 63 | 16 | 37 |
| Sindhi | | | | | | | | | | | | | | | | | 38 | | | | | | | 21 |
| Tamil | | | | | | | | | 18 | | | 3 | | | | | | 365 | | | | | | 105 |
| Telugu | 78 | | | | | | | | | | | | | | | | | 115 | | | | | | 78 |
| Urdu | 19 | | | | | | | 22 | 1 | | 11 | 23 | | | | 28 | | | | 14 | 19 | | 16 | 18 |
| Average for all languages | 63 | 27 | 39 | 12 | 85 | 23 | 22 | 46 | 54 | 119 | 28 | 51 | 10 | 20 | 59 | 52 | 46 | 160 | 27 | 34 | 75 | 76 | 61 | |

Source:  Calculated from INS Press Handbook 1993, New Delhi: Indian Newspaper Society, 1993.

*Table 4.3.2* Population and literacy in India, 1991

| | Population (millions) | Literacy % | | |
|---|---|---|---|---|
| | | Total | Male | Female |
| All India | 846.3 | 52.2 | 64.1 | 39.3 |
| Urban | 217.6 | 73.1 | 81.1 | 64.1 |
| Rural | 628.7 | 44.7 | 57.9 | 30.1 |

*Source:* Census of India, 1991.

*Table 4.3.3* Regional circulation of some major papers

| | Circulation in '000's | | |
|---|---|---|---|
| | Times of India (English) | Express (English) | Navbharat Times (Hindi) |
| Bombay | 348 | 134 | 105 |
| Chandigarh | | 36 | |
| Madras | | 260 | |
| New Delhi | 162 | 74 | 213 |
| Ahmedabad (Gujarat) | 44 | | |
| Patna (Bihar) | 30 | | 49 |
| Lucknow (Uttar Pradesh) | 26 | | 29 |
| Jaipur (Rajasthan) | 31 | | |
| Total | 641 | 504 | 396 |

*Note:* Local variants of main city editions of the *Express*, such as the Pune edition, have their figures subsumed under relevant city, e.g. in this case Bombay.
*Source:* Calculated from *INS Press Handbook*, 1993.

way that western ones have, with new sections such as Reviews or Personal Money, the number of pages is actually small. For many it may be only 10 to 12. Even the big English papers often have only 16–20 pages. Paradoxically, journalists claim that this means that (a) there is too much competition for space, and therefore environmental stories, for example, will face fierce competition; and (b) the papers are small because they lack the resources to fill more space. In this as in other things there is also a difference between the English press and broadcasting, and the vernacular language equivalents. The language press and broadcasting stations claimed they were always much more strapped for all kinds of resources and infra-structure than the 'English' or central institutions and therefore found it harder to pursue independently any particular environmental issue, even if they wanted to.

In the previous chapter Ivor Gaber's researches showed that there were not enough resources to send teams around the world getting good footage of envi-ronmental problems. It is ironic that with so many more resources in Britain, environmental issues should still be squeezed for budgetary reasons. The fact that this occurs in both India and Britain at very different levels of funding suggests that perhaps the budget may be an excuse – the environment is simply not a high-priority item.

## 4.4 FRAGMENTATION OF THE MASS MEDIA IN INDIA

Overall, therefore, we have an image of a very fragmented media industry, spread over many centres, using many languages, and on average catering to local needs. The State maintains an all-India service through DD and AIR. But even then the 'national' picture is fragmented because of the many local centres within those services. The national television news is broadcast in English, Hindi, Sanskrit, Urdu and some other languages. But even in Bombay there is no national service of news in Marathi, and that language is used for the regional news broadcasts. For national news the viewer must understand either Hindi or English. Satellite channels of DD and private companies (such as Asianet, Sun TV, Jain TV, BITV) offer news in English, Hindi and some regional languages but their penetration is still mostly limited to more urban and upper income families. This is not just for economic reasons: the language used – sometimes a mixture of Hindi and English – is decidedly not local and vernacular.

No newspaper comes anywhere near a nation-wide circulation. The papers with the widest circulation are English-language papers, but even these are dominant within regional domains. *The Times of India* reaches out to other parts of north and west India from Bombay. The *Hindu* reaches into many parts of southern India from Madras. The *Statesman* is powerful in Calcutta and has a small circulation in Delhi. A few examples of the Hindi press are beginning to attain a more powerful stature, but in the north only. There are all-India services such as the Press Trust of India (PTI) and the United News of India (UNI).

What accounts for this fragmentation? Perhaps it is not so fragmented if one remembers the scale of India – as big as Europe and with three times the population. But the perception is not wholly wrong: the sheer number of small cities from which publications flow is enough to show an industry which has not 'rationalised'. In many ways this fragmentation is a statement about the way that the modernisation of communications – both telecommunications and transport – has been limited to major metropolitan areas. Outside these areas telephone and fax services, though improving, are by international standards abysmal and expensive. There are so-called 'express' train services between major towns, but they are slow by international standards, and the majority of trains are extremely slow. There is nothing which remotely resembles a high-speed train, certainly not a French-style TGV. Train journeys between north and south frequently take between two and three days. Road transport has penetrated deep into rural areas since Independence – it is one of the most significant transformations of rural India – but average speeds on poorly surfaced and narrow roads which are shared by buses, lorries and bullock carts, private cars, motor-scooters and bicycles, are, inevitably, slow. New toll highways have just entered the scene, but it will be decades before any kind of national network is built.

In essence, daily newspapers cannot be serviced at a distance. They are therefore, like dairies which cannot transport milk far, or like vegetable wholesalers whose produce cannot be kept fresh, essentially local affairs. As was pointed out in Chapter 1, even papers like *The Times of India* run editions which peddle increasingly outdated news the further from Bombay the edition is produced. For this reason only the front page bears the day, month and year. The inner pages omit the day and have simply the month and year. It is perhaps a consequence of this that there are weekly and fortnightly news magazines of a very high standard which are marketed nationally. *India Today* is the best known, a fortnightly starting in English in 1975 and currently with a circulation of 348,000, but now published in Hindi (starting 1986: circulation now 251,000), Malayalam (1989, 117,000), Tamil

(1989, 145,000), and Telugu (1992, 145,000). *India Today* has an Environment section, which publishes one or two stories per issue, near the back of its 180 pages. *Sunday* is a weekly in English only (1975, 76,000). *Frontline*, a news magazine from the publishers of the daily the *Hindu*, is yet another magazine that provides good coverage of environmental issues in the Indian and the global context. The same publisher brings out annually *The Hindu Survey of the Environment*.

## 4.5 ANALYSIS OF THE INTERVIEW TRANSCRIPTS

### Introduction

With these background comments in mind, we will look at the analysis of the interviews with representatives of newspapers and broadcasting. The interviews were obviously angled towards environmental matters but, quite properly more general issues were touched on too. The most important of these is the issue of language. The interviews were transcribed from tape into a Word Document; this was further manipulated using an Excel spreadsheet to code different parts of the interviews with key words for different themes. The spreadsheet allowed the ordering of material by these key words.

We have therefore identified some major themes in order to organise this material. Most of the views expressed, however, can be related to more than one, if not several, themes and the themes themselves are interrelated. A key background feature is the distinction between urban and rural areas – but that in turn might be some sort of surrogate for the distinction between the better and the less well educated, between the centres where English language users may be found and the vernacular hinterland (or *mofussil*, meaning the 'interior', or 'the sticks'), and most importantly, between the areas which in many ways have a degree of development and affluence and those that do not. Even the very concept of environment – which is as vague and abstract an idea as one could conjure up – seems to have differential reality, being more real in urban areas, where 'civic' problems are seen to be synonymous with environmental ones.

### Language and the audience

We start with the intended audience the publishers are trying to reach. Any audience has, of course, both specific and general components. In Britain Radio 4's Today is a news magazine listened to avidly by the professional classes. But it is more than that: for many it is the main political agenda-setter for the day – something noted above in Chapter 3.

In India political matters have for long dominated the press and given an urban and metropolitan bias to much of it. It has often seen itself as more addressed to the political classes than to anyone else. But that is changing: Arvind Das (Delhi news desk, the *Statesman*, English language, Calcutta and Delhi):

> our target is not only politicians; it's the common people also. ... That's why we publish lots of stories of common, human interest also. Earlier this paper used to be eighty percent news, used to be politics based. ... But now – our emphasis is on features and different kind of human interest stories, like population problem, then environment, ecology; lots of other things.

Although Das uses the phrase 'the common man', this description cannot really be applied to the English-reading audience. The élitist nature of the targeting for the weekly and fortnightly type of magazine is illustrated by Ranvir Nayar, a specialist correspondent working for *Sunday*, interested in the environment amongst other things: 'Our target audience is, let's say, the middle class readers in all the major cities. . . . And that's where we sell the most. And a few copies do go to the rural areas, but that's very few.' The journal has no plans to follow *India Today* with language editions. (It is common practice in India to refer to 'the language press', for any non-English language publication).

> But I think even an English language journal like ours can sell very well if we reach out to the people. There is 850 million people of which 2 percent understand English. Even that makes it a very big audience.

A specialist correspondent of the *Express* (English), Usha Rai, noticed a selectivity within the English readership:

> I find that in India, let me tell you the *Express* is not an élite paper like *The Times* [of India]. The *Express* . . . has a very large readership. It's a middle-class readership . . . lower middle-class. I am saying it's not the cream of the English reading public.

She writes a weekly feature page on development issues, gender issues *and* environmental issues amongst others, which, however, is not targeted at the general public so much:

> The development page for example is read by a lot of NGOs and the others who can then translate this into the languages that they want to explain things to the people.

> Well, I know that the development page for example is kept by most NGOs; they clip it and keep it.

> I've heard of people making television programmes on non-conventional energy, on environment and development issues. They also keep looking on that page for information. People in the Rural Development Ministry like that page. But I don't know if the general public likes a page like that. I don't think the young – the younger people would be interested in reading it.

Interestingly, Das also thought that the English language was less élitist than it used to be – perhaps the reason why he felt he could say 'the common man'. But the common man seems uncommonly metropolitan:

> CHAPMAN:   How far out of Delhi would you have to go before you wouldn't find any English paper?
> ARUN KUMAR (wire service, Press Trust of India):   Just about 30, 35 kilometres. That's all.
> CHAPMAN:   And then you won't find an English paper?
> ARUN KUMAR:   It would be difficult to get an English newspaper during the day. Suppose you have a regular newsagent who – he has two, three, four regular readers of English newspapers; he can deliver it at your home. But if you go to the news stand you wouldn't find it.

At the all-India level in fact English is the dominant working language of the media. We continue with Arun Kumar, who put it like this:

> Yes, all stories do come in English. As far as PTI is concerned we collect news in English and we distribute news in English. We have a sister service called BHASHA, which stands for language Hindi.

We have close to five hundred newspaper subscribers, not all of them . . . I mean only a small percentage of them is English newspapers. The rest of them publish in their own languages so they translate it into their own language and publish it.

BHASHA, as I said, it's only for Hindi language. . . . So, like even if I send a BHASHA story say to Kerala, they would not – I mean they would not be able to understand that.

This immediately raises a question about the translation of different technical terms from English to the language media, and specifically with reference to environmental terms (even the word itself). This was tested out several times:

CHAPMAN:   Can you tell me; is it difficult to talk in Marathi, to write in Marathi about greenhouse gas, CFC, ozone, global warming, climatic change?
ARVIND GOKHALE (Editor of *Kesari*, a Marathi daily paper, Pune):   In fact we have published a number of articles about this. We use the English terms . . Because other colleagues we spoke to had the same view. We don't – yes, we don't find any parallel words in Marathi.
CHAPMAN:   But what about greenhouse gas? (*Arvind Gokhale laughs.*) You could translate that?
ARVIND GOKHALE:   We put it as 'greenhouse gas' and in brackets we can translate it in Marathi.
CHAPMAN:   So, you use the phonetic value of Marathi to transliterate exactly 'greenhouse gas'?
KEVAL KUMAR:   But sometimes you use your own 'coined' terms.
ARVIND GOKHALE:   Yeah, yeah. Then put it into brackets there is the English term.
PRAKASH KARDALAY (Editor, Pune Edition, Indian Express, English): Marathi cannot really express technical things so well, my observation is that it is not very adequate, and the latest trend is to retain the English words.

A similar line was followed with Man Mohan Gupta, of the Delhi desk of *Dainik Jagran*, a Hindi daily paper based in Bhopal, Madhya Pradesh.

CHAPMAN:   For example, if one was to talk about a greenhouse gas or global warming or ozone; is that difficult to translate into Hindi or don't you carry these stories?
MAN MOHAN GUPTA:   Then the words which are difficult to translate into Hindi are used [untranslated] as it is in order that the people understand the meaning of these words now.
CHAPMAN:   So in other words you're saying that where there is a technical vocabulary needed you normally use the technical vocabulary that exists.
MAN MOHAN GUPTA:   Exactly.

Mukul, a specialist development and environment correspondent for one of the most influential Hindi papers, the *Navbharat Times*, answered the same questions as follows:

It is a very difficult and challenging situation before language journalist. Hindi or Marathi have to evolve a new language, have to take up this new challenge, even if it is very difficult. So, we have to find out the idioms and the words and the sentences to express these issues.

Vishram Dhole (Pune Edition, *Indian Express*, English):

It happens because of time pressure, we have to send the translated text quickly to the composing section: so what we did, where we translate into Marathi, we go not just sense-wise, but also in construction. The Marathi construction has been greatly influenced by English construction – for example the subject–object–verb relations and especially the passive construction. Ordinarily in Marathi we did not have much of passive construction, but since in English it is common to report it in third person passive, we have accepted it to such a level, that passive construction has dominated the largest chunk of Marathi we read in newspaper nowadays. You can definitely point out the difference between current literary Marathi and current journalistic Marathi, because there are two totally different ways of construction.

Prakash Kardalay (Editor, Pune Edition, *Indian Express*, English):

Our local man complained he had too much translation to do, but Marathi is a regional language and after all most of the information is bound to come through English, and therefore for better coverage they have to do that, and they shouldn't complain, otherwise the windows to the world will be closed.

Prem Vaidya, ex-*Films Division*, Government of India, had this to say:

Because we have such an amount of stock in our – especially I could talk authoritatively on Hindi language, where you have a good amount of words and if it is simplified and if it is done in a proper manner I am sure it will work out.

I find at present, at this moment of talking, there is no substitute for English.

But he also saw an element of professional status and money at work:

You see, what happens in India particularly: the reporters who are working in another language other than English they are paid very poorly. There is no attraction. Even as a newspaper reporter. If you consider the English reporter is getting much higher salary than a language man.

Usha Rai (Specialist Reporter, the *Express*, English language) also admitted to the difficulties involved:

I suppose it would be the same as trying to explain a difficult subject to the reader in a simple manner. It's the same kind of problem. If you translate 'greenhouse gases'; it's very difficult for me to explain greenhouse gases to my readers in English too. You have to find the correct expressions too and then give some kind of explanations to what it is. And it's by repeated use of those expressions, of those translations in Hindi that people will begin to understand. I mean the Hindi word for typewriter for example is – it's such a long labourious word that – it's so much simpler to call it a 'typewriter' and even in Hindi people would understand it.

CHAPMAN:   They've seen a typewriter. They haven't seen greenhouse gases.
USHA RAI:   But it is possible to do it. And there are a lot of very committed journalists in the regional languages who are writing on these issues and who will be able to.

We are not so sanguine about the difficulties involved. Journalists in India – perhaps because they are relatively few and they have to cover too much ground, perhaps for other reasons – commit more major errors of understanding even in English, let alone translation, than their peers in the UK. Some of these errors proceed from the common perception that India is separate from the rest of the world, that its problems and the rest of the world's problems are set apart from each other. We return to Ranvir Nayar, specialist correspondent of *Sunday*:

> But, yes, I do agree that it is a problem created largely by the West and it has to be solved by their efforts. They have to cut down their current levels. They have to cut down their consumption levels and we should also ensure that our consumption doesn't reach that level, so that we are not, you know, we aren't ready for any footsteps of the West, creating our *own little ozone holes here* [emphasis added].

An article in the *Pioneer* (Lucknow and New Delhi, English) in August 1994 said that India certainly had its problems but by and large was lucky. Most of its bio-systems are still intact – unlike in the western industrialised countries, where everything is in a terrible mess and where 'they' are suffering terribly from greenhouse effects. Just as for the British public, the problems of the environment happen somewhere else, a long way away.

All of this raises quite clearly the issue touched upon in the first chapter and debated by educators in the nineteenth century. At that time the British decided that higher education had to be in English, waiting until the vernacular languages caught up. But the current general models of language development suggest that the vernacular languages are trying to catch a moving target – and what could illustrate this better than 'greenhouse gas' and 'sustainable development'? To the extent that vernacular languages are coping, they are doing so by learning new vocabulary whose linguistic roots are not instinctively obvious and by adopting new grammatical constructions and styles.

## The urban–rural divide and nascent internationalisation

At the apex of society, the new metropolitan India is being specifically catered for. An emerging new readership has been given an intriguing label by the managers of *The Independent*, in Bombay, a label which perhaps the *European* in Europe has not yet identified, but which it too hankers after.

Bachi Karkaria and Bharatkumar Raut (*The Independent*, modelled on the UK *Independent*, located in *The Times of India* building, English) identify it as follows:

> KARKARIA: The concept behind the *The Independent* and the *Metropolitan* [weekend magazine edition] is what we call a post-nationalist concept. This means we are appealing to readers who do not have to approve nationalism, and who are not intensely concerned about the nitty gritty of events in the hinterland of India. This really is an urbane readership which is interested in the developments and happenings in their immediate surroundings, that is the city and beyond that the world. And that is really largely a Bombay reader that we have in mind as far as these two papers are concerned.
>
> RAUT: Because they see a lot of TV, the world is coming closer. People in Bombay particularly the niche for which we are catering, they see BBC, they see CNN, so their concerns are not limited only to the city or to other states in India. They are equally interested in what is happening in Washington, what is happening in New York: their economic interests are involved. So we try to cater for those needs also. Either you talk of Bombay or you talk of the world cities, so that is the priority.

[When asked in detail what 'post-nationalism' meant, there was some degree of laughter, about a new term looking for a meaning, and no-one knew what it was.]
KARKARIA:   (*Laughter*) I don't think anyone understands the word even in *The Times of India* building. We were first a little dismissive about this term, we thought this is one more marketing phrase, dreamt up by some marketing genius sitting in the upper echelons of the *T o I* administration. But we did put it into effect, partly because we had no choice, but having started as scoffing at it, I think we have stayed to pray.

The term applies to a set of attitudes within a group of Bombayites (and in other major cities) that assumes that India now exists and has coherence, that their part of this India has developed, that they are now aware of the outside world, and are keen to communicate with it. But it would be false to suppose that this means that they are international in the sense of being supranational in outlook, part of an undifferentiated world cosmopolitan class. Amongst these people will be found some of the fiercest critics of western hegemony, and although there may be a general awareness of international forces, in practice, issues at home make the news.

CHAPMAN:   Is there an international dimension to the environmental stories you would pick up on?
KARKARIA:   You know we have so much that we need to address in our own country . . . I don't think we would worry so much about global warming because we need to talk so much about deforestation in our own backyard. So to that extent, we concentrate much more on our own issues, which is as it should be, because that is how the reader will identify with it, rather than think about something which is remote from his sphere of influence.

I don't think any Indian paper is looking at such a broad level. . . . When they start dealing with it over here, they start dealing with it from the Indian perspective. And I think this is true about most Indian papers. Darryl D'Monte [President of the International Union of Environmental Journalists and former Editor of *The Times of India*], even he realised that he needs to talk about issues which are close to the Indian people, because if he does talk about global warming, people are not going to get very excited about it, and are not going to get so concerned about it. A dead dog on the street outside is more important than a flood in China. It's really the basic principle of news.[2]
DINA VAKIL (Editor, *The Times of India*, Bombay, English):   All issues related to the environment have become fashionable in the last few years. In other words environment is a sexy subject area, and judging by the response of my own young reporters that is a fair assessment. . . . This is an area, urban development generally, environment issues generally, these are areas which seem to have high appeal for young journalists.
RAUT:   All environmental interests are essentially civic interests. You cannot split them. Because if there is a sewage problem, if there is a garbage problem, if this is to become reader-friendly instead of giving them big names like environmental issues call them civic issues – people will be more interested in that.
DINA VAKIL:   A lot of urban readers who have increasingly become conscious with growing affluence of the quality of life are paying greater attention to environmental issues. They see environment issues generally in a positive light, and they look positively on the environment in the environment and development debate.

PRAKASH KARDALAY (Resident Editor, Pune Edition, *Indian Express*): To us the environment means the urban environment, the deterioration of the conditions in this urban environment, the violation of the FSI rules [Floor Space Index], we also have zoning, the development control rules which are being flouted, unauthorized constructions occur. Since I am catering to the English speaker who is basically educated and urban, and highest of his priority is his immediate neighbourhood.

They [language papers] are picking up from us all our civic issues. The local vernacular dailies are picking up from us. I am picking up stories from the *mofussil* [the 'interior', 'the sticks'] from them. The average level of education makes a lot of difference.

Within India the English press does seem more concerned about environmental issues than the language press, and it is true that it has more resources, but it seems more likely that the values of the audience and readership are different, and correspondingly the attitude of the editors and journalists. The greatest poverty and backwardness (measured by literacy and levels of education) is found in rural India – which is not to deny that urban India does not also contain large elements of both. But much of urban India has left rural India 'behind'. When Chapman was first working in India in 1970 in Bihar, urban people knew their native village and usually returned there for marriages and for the women to give birth. In 1980 when he was working in a rural area in south Bihar, a local college student asked if he could accompany the group to the field. When the car left the black top road for a dirt track, he wound up the window and shrank down in his seat, saying he had 'never been so interior before'. In 1995 there are middle-class urban Indians who have never been in a village.

Sonora Jha Nambiar is the Editor of *Changes*, an English-language quarterly put out by the British-based charity, Action Aid. She was a successful journalist, who found she was not satisfied by snap reporting on murders and political intrigue. She wanted to work more in depth on rural India.

I mean, for my own example; I come from Bihar which is probably the most backward state in India. And the last time I went there was when I was two. And I have grown up as a Bombay brat, like I say.

So, for example, my brother doesn't know now what I am doing [writing on development] he keeps asking me 'what are you going down to the village for, I mean what is happening in the village that could possibly interest a newspaper person?'

But where translating of *Changes* is concerned; we are not looking at that now because we are looking at the urban Indian middle-class which is an Indian – and English speaking, an English reading middle-class. ... Since they are the mainstream we are trying to sensitise them and they are the ones who are really most divorced from these issues and most divorced from reality of the larger part of India. India is developing at such a fast rate and you know, it's becoming very achievement orientated and you have this thing called the global citizen and things like that. Now, we are looking outward. We are looking at what's happening in the world. So we are hardly able to look back at what's happening in another part of our own world, you know.

But maybe, after you have read all your technical magazines and all your information and everything else; if you could possibly spend ten minutes just getting to know out of curiosity what is happening in someone's life somewhere else. It could just be a story, it could just be like reading a short story, you know. But that – that just reminds you of another India that is existing. And you are part of one India which is moving in this fast pace and there is another much larger India which is, you know, staying behind.

The remoteness of rural India is also reflected in these remarks by Bipin Chowgule who works for DD's Children's Educational TV in Pune:

Because, you know, the language differs from place to place. We use the standard Marathi language which is mostly spoken in Pune. But whatever language is spoken in villages is different; whatever concepts and whatever associations we have with the things are a bit different from whatever they have.

But the environment as you have defined it; no, we don't make any films on that concept. Somewhere the topic might come but we never directly mention these things because our targeted group is very limited. Up to eleven years of age the child doesn't know who Chief Minister is or what the problem with the Bombay beach is, he hasn't even seen a beach. He hasn't – most of the children in Maharashtra haven't seen the sea or photos of coastal areas. Many of them haven't seen a train. . . . We introduce them to the natural environment as created by the nature.

In the previous chapter the inability of the audience in the Third World to understand some of the conventions of western broadcasting was cited as a reason for failing to communicate internationally. Here we have heard from someone facing the same kind of problem within India itself. Not everyone thought these attempts successful. Usha Page of AIR Pune/Ahmednagar said of TV documentaries:

The one main reason is television programmes – the [greater] percentage of television programmes I found is in English. Actually the programmes should have been broadcast in regional language. So they should reach to the cross-section of people and to the grassroots. But they are not reaching. There are some programmes on awareness, conservation, environment; they are broadcast in English and for affluent audience only. It is in English and the level of the programme is rather. . . . It is for intellectuals. It is not for the common viewer.

Keval Kumar teaches on a Master's course in Communication Studies at the University of Poona, which attracts students from all over the country. As an impromptu test of the technical language translation problem in August 1994 students in the class were paired according to their native language; then one of each pair was sent to an adjacent room. The first of each pair was asked to translate 13 terms from English into the native tongue, and these translations were then taken to the adjacent room, where the other member of each pair translated the words back. The results are shown in Table 4.5.1. Note the following: there are a considerable number of blanks, according to the students mostly because they were no longer fluent in their native tongue; that drought is interpreted in terms of its human consequence; that 'development' once comes back as 'grain' – this is reminiscent of the Bengali word *dhan* which means both rice and wealth (like the hippie 'bread'); that there is no long-term notion of 'climatic change' only the idea of seasonal or annual change.

*Table 4.5.1* Double-translation meaning-consistency test

| English | Hindi | Hindi | Marathi | Marathi | Marathi | Assamese | Assamese |
|---|---|---|---|---|---|---|---|
| **drought** | drought | falcon/barren | | famine | famine | arid | famine |
| **environment** | environment | change | environment | environment | environment | environment | environment |
| **development** | progress | promotion | improvement | development | grain | development | |
| **fertiliser** | fertiliser | chemistry | manure | fertiliser | letters | fertiliser | fertiliser |
| **nitrogen** | | | nitrogen | nitrogen gas | | bio-gas | |
| **pollution** | pollution | pollution | pollution | pollution | pollution | pollution | pollution |
| **acid rain** | rain | rain | acid rain | acid rain | | acid rain | |
| **two-stroke engine** | two-stroke engine (via phonetic transliteration) | | two-stroke engine (via phonetic transliteration) | | | | |
| **television** | tele-communication | television | television | television | television | television (Door-darshan) | |
| **mass media** | | plebs | advertisement | communication skills | | mass media | |
| **erosion** | | | | | | eroded | |
| **communication** | transmission | | dialogue | | contact | contacts | |
| **climatic change** | change of seasons | | climatic changes | weather changes | change in temperature | change of weather | |

*Note:* The English word on the left has been translated by one person into the target language given by the column heading, and by another person back into English (Pune, MA students in Mass Communication (English medium), August 1994).

**The scope of coverage**

Although Sonora Nambiar sees the new urbanite as looking outward, the general press and broadcasting certainly does not. Explaining the low level of international coverage, Arvind Das (Press Trust of India) had this to say:

> See, you forget, because we don't have that much resources. We don't have so many correspondents about. Of course we have ties with foreign newspapers and if there are any stories – or we have ties with agencies. We get stories, we do publish. Otherwise we don't, because we don't have that much resources. And we have at the moment about eleven foreign correspondents and twenty part-time foreign correspondents.

Specifically with regard to environmental stories he also said:

> Something which happens close to you gets higher priority rather than something that happens in a remote part of the world. So, if there is some environment related story, say coming from a part of Bangladesh or Pakistan, probably that finds high prominence or would be given more attention, even by my own desk rather than say something coming from some part of Africa or somewhere.

Mrs Bimla Bhalla, Editor of DD's national news service observed:

> For the international footage the wire agencies give us the news but for pictures we subscribe to two or three international television news agencies. That way we are better off as far as the international footage is concerned. Whereas at the national level we are a little restricted at the moment, because our network is very limited.

> CHAPMAN:   So at the national level, essentially you use a lot of verbal stories and then some of your library footage to provide the graphic images to go with it. You also implied [earlier] in what you said that you cover local and national stories much more than international environmental stories.
> BHIMLA BHALLA:   You see, basically we try to relate it to the viewership. If there is an international event as I mentioned earlier, news is related to events and developments; then we try to relate it to the national or the regional level also. So, to that extent, if statistics are given, if there are [international] statistics given at a conference, we try to also bring in the regional, the national statistics for that. People can understand the enormity of the problem nearer home, rather than as an abstract situation.

In fact, the international agencies are not much help for a country such as India in getting much international environmental coverage anyway, as Gaber outlined in the previous chapter. It is echoed by the comment of a TV newscaster from Bombay (who wished to remain anonymous): 'Well, let's be frank on this. I think Visnews (now Reuters TV) doesn't really concentrate very much on environmental issues, does it?'

The restricted interests of the language papers come out clearly in these comments by Arvind Gokhale (Marathi daily, Pune):

> But we are definitely interested in the local issues and local environment and local problems, because the selling area is all over Maharashtra, not outside Maharashtra. And that is particularly because the regional readers are there, regional language readers are there.

Even then, at the local level many of the smaller papers find it hard to cover environmental stories. Man Mohan Gupta, *Dainik Jagran* (Bhopal, Hindi): 'Environmental we do weekly. Reporting is there basically on trees, forests and on pollution. But pollution is only seen in smoke and all that. Nobody talks about the noise as such which is also a very important aspect.' His comment on noise is not idle: in much of Calcutta noise levels exceed WHO limits for 24 hours a day, seven days a week.

CHAPMAN:  So government is a big source of stories. That would be in Delhi. What about in Bhopal? Where would the stories come from there?

MAN MOHAN GUPTA:  All these regional publications bring the news down from capitals, state capitals – from Delhi as such and all that. So all my publication is basically from Bhopal and Madhya Pradesh. We do get stories from Delhi and send it to Bhopal. And then state government also doing something, you know about this.

CHAPMAN:  So you don't pick up many stories from inside Madhya Pradesh, from your reporters inside Madhya Pradesh on environmental issues?

MAN MOHAN GUPTA:  We do, but very few.

Mukul, the specialist correspondent of the much larger *Navbharat Times* (Hindi) had this to say:

Yeah, actually there are so many examples, I can briefly touch upon two examples. Suppose there is a very great environmental problem in the Ganges region, in northern India, Bihar and Uttar Pradesh, and in this area there is a lot of erosion and a lot of environmental degradation and a lot of movements and so many activities. [This refers to political agitation and banditry, common where people are dispossessed.]

But in English newspapers, either published from Delhi or Bombay or Calcutta you rarely find a coverage on the problem of the Ganges belt, or the problem of Bihar region and that is also huge region. But if you go to Bihar or Uttar Pradesh and if you survey the newspapers of Bihar and Uttar Pradesh in Hindi or regional language you can find so many reports, so many coverage, so many news items related with these problems of the Ganges region. Whether it can be environmental degradation, criminalisation, or land erosion.

Mukul was celebrated as someone with both the resources and the motivation to seek things out. The committed investigative journalist of this sort is rare in India. The way to the top is still via reporting politics and business.

MUKUL:  So far as my thing is concerned, I rarely use PTI as my source of information. Because I cover this Ministry of Environment also I have to use the source of ministry; I have to cover the policy issues; I have to cover the press conferences and the various issues coming out from ministry in Delhi. But all the time in all the years my priority is always to cover the grass root issues; to go out from Delhi; to do fieldwork, to go to the Ganges region; to go to Narmada. So I try to make a balance between government policies and between grass root initiatives and movements and environmental issues emerging out of the field.

CHAPMAN:  But you can't cover the whole of India. Do you use stringers much to help you?

MUKUL:  Yeah, I would like to use the source of stringers much in our newspaper, because so many times the coverage by PTI and UNI I am sorry to say it's very dry and stale and not catering to the needs of, or the demands of our readers. Even then we miss some issues because we are very few in our newspaper to cover all these *developmental issues* [emphasis added]. One special correspondent is very insufficient. [Note, in passing, that Mukul slides from environment to 'all these development issues'.]

So there are some NGOs which are – which I can say roughly are very élite and not caring for language press, not doing any work in regional languages, or sometimes they don't bother about the language press. But there are so many important NGOs who are dealing with grass root initiatives and movements and all those things and they are very sensitive to language press also. But in some cases if some – these types of NGOs do some discrimination then we also don't care for this type of discrimination.

CHAPMAN:  I am not trying to fish around anywhere, but just out of curiosity, would you say that WWF and CSE[3] were élitist in this way or do they provide you with quite a lot of material?

MUKUL:  You are right; I can't say that WWF and CSE – I am sorry to say that so many times they do their work only in English, only in English; their press releases, their press conferences.

I personally speaking want to use all these materials in language press also, because I think that they are also useful . . . But sometimes its difficult to disseminate all this English material in Hindi or regional press.

Not only is covering India an impossible task, but there seems little motivation either to look beyond India to international stories. Mukul again:

There is very little of it which is about global issues or about issues in other countries. It's so many times – it's quite some time difficult to focus upon or write about these global issues, because our editors or bureau chief so many times caution us about the readability of and acceptability of these issues in terms of our readership.

Ranvir Nayar, the correspondent for *Sunday*, an élitist English language weekly, was even more blunt.

CHAPMAN:  Do you think that issues like global warming or ozone holes are issues which concern your readers?

RANVIR NAYAR:  We did have a few articles on them at the time of the Rio conference. But, no I don't think it's the top of the priority for Indian readers, even if they are environmentally aware. They don't see – I mean maybe they are mistaken – but they don't see it as India's problem. They see it as a problem which has originated in the West, it has to be tackled by the West. That's where it is. So, they don't see it as an Indian problem. They are more concerned about forests and land use and with industrial pollution, with water pollution, air pollution. They haven't really gone up to the ozone holes.

Usha Rai of the Indian Express gave similar views:

But I don't think my readers are really interested in Global Environmental Facility or the Montreal Protocol to the same extent as he would be in what is happening in a village where people on their own are constructing a dam to store water. . . . I think – I really do think we should be much more focused in reporting on national issues and local issues rather than international issues on the environment.

This kind of hostility to western viewpoints is in some measure a legacy of the Rio conference, where quite clearly many thinking Indians felt that the Third World in general was being held to account for the West's mistakes. Usha Rai:

I think this entire thing comes because the NGOs at the time of the – before the Rio Summit; there was a very strong feeling which was basically created by the NGOs that there was an attempt by the North to dominate the countries of the South and to show that they are the ones who are causing all the pollution. And there were reports to that extent, you know, which Centre for Science and Environment, Anil Agarwal and a couple of them got hold of and they brought this to the public notice. So, the entire Rio thing was very much a North–South debate – who is responsible [for environmental problems]. Because based on that we were to get funding. Perceptions of how India was to perform would depend on what was expected of India; what was expected of the developing countries. So some of those perceptions have stayed on and it is not going to be easy to get rid of them. Like even the Global Environment Facility; the funding – I mean there is – we're still saying that we are not the ones who are polluting. It is the West that causes more pollution. It's not that – it's not happening here. Even, even the whole question of population; the West is going on saying 'it's your population that is causing the destruction of the forest'. And to some extent they are correct. But it is politically inexpedient to say this on an international platform. You sort it out on your own level. But when you make a noise, you don't want them to use that same handle to beat you. So you have to say that 'OK, we have a greater population but look at our consumption patterns'. You know. So it is because of that, that there is some amount of mis-perceptions or wrong perceptions in the general public I would say. Even among journalists. Maybe they don't see it as an international issue and they see it more as a national issue.

## The environment as a category

Since the emphasis is so clearly on India and its environmental problems, it is not surprising to hear views that are very different from those of the West about what has fuelled the level and type of environmental coverage historically. There is no doubt that the amount of writing is increasing, and that increasingly there is a type of story recognised as environmental – at least by the English press and perhaps by the broadcasters – and the recognition of the category may be a kind of foreign import. Mrs Bhimla Bhalla of DD National Evening News recognised an environmental category of story:

CHAPMAN: But do you yourself or does Doordarshan recognise a category called environmental story?
BHIMLA BHALLA: Yes, we do. But we relate it to whatever event or the development may be. It – we would identify it as an environment but not slotted in a specific segment or something. We just treat it as news.
CHAPMAN: I've noticed you have less coverage of environmental issues than for example the UK television news does. In UK television it's always the last item on the news or something like that. It comes in a category of its own. With Doordarshan I noticed that environmental stories can come sometimes right at the beginning, maybe even the first two stories. Do you have a comment on that?
BHIMLA BHALLA: See, we don't fragmentise the news. We relate it to whatever the developments may be. And if there's an important activity relating to

the environment, then we do give it and bring it up. Otherwise if it's something which is not so important then there are other major developments, then we bring it down. It's not a question of a fixed slot or any such thing.

Local (state-level) Doordarshan TV in Bombay definitely saw development as covering environment:

FEMALE ANON:   But we don't have a separate slot for environment. But what we do have, however, is a newsletter which is devoted to a particular district of the state every day. I think it's three or four minutes. The caption is about three minutes. And a lot of this does deal with matters relating to the soil and thus the environment. [This is confirmed in the content analysis in Chapter 5.]

DINA VAKIL (Editor, *The Times of India*, Bombay, English):   The news we carry is of the day, event related, and of course a lot of environment issues tend to be stories that really are talking about process not necessarily tied to an event, and for those kinds of stories there is even less space ... when you are fighting for space those stories have to get less space than they deserve.

This will be evident in the content analysis in the next chapter.

Few people thought that the UNCED conference in Rio had played any major role in environmental coverage in India, although Nayar of *Sunday* said they had covered a few global issues at the time. Prem Vaidya of the Films Division (GoI) offered his view that environmental coverage went back to the tragedies of 1965–66. In those years the country had its worst drought since Independence, particularly hard-hitting in the poorest and most backward state, Bihar. The country became a massive food importer and there were many deaths from famine. There was another bad drought in Gujarat in 1974 accompanied by food riots.

Right from '65. Because then – after the biggest famine took place in Ganges Plain of this century they realised that this is because of the devastation of the natural environment which we're not preserving. The water direction, the wind direction, the cyclic rains of the seasons which were all disturbed; then only they came for the realisation. And after that we had started making films. Even an animation on this environment. Since then we had become a little conscious of this.

In other words, it is not abstract things like General Circulation Models (of the world's climate, on supercomputers) which have stirred the government and some people to action; it is real problems in the country. One of the results of the 1965–66 famine was a harder and faster push for irrigation – with its attendant dams and tubewells – and the fertilisers and seeds of the Green Revolution. The result is that while India's population has doubled since 1965, its food output has increased even more, and the country at the moment seems in most years to have a slight surplus of food grains. The 'Indianness' of this experience was stressed by Darryl D'Monte, formerly Editor of *The Times of India*, a committed environmentalist and president of the new International Association of Environmental Journalists.

Climatic change ... this is not a problem India is going to worry about, because we have already experienced that. We've been suffering droughts for a long time. We've been suffering famines. We know what change is all about. Nobody can threaten us with any greater change than we've already had.

**Development (language) versus environment (English) and the Narmada Dam**

The Deccan block is tilted, so that the western side is higher than the east. Although the central plateaux are bordered on both sides by ghats (hills/escarpments), the Western Ghats are more abrupt and higher. Thus on average the watershed between the Arabian Sea and the Bay of Bengal is located along the top of the Western Ghats, very close to the Arabian Sea. Nearly all the major rivers of peninsular India therefore drain eastwards – and they are of course highly seasonal in their discharge. They have all been dammed at various points to retain the wet-season flow. But where the Deccan crinkles into west–east ranges of ghats in its northern parts as one gets nearer the Ganges plains, there are two rivers which curiously drain from east to west in deep troughs between the hills – the Tapti and the Narmada. This ghat country is rugged and remote, and also in a drought-prone area. It is part of a belt of such country running across the northern Deccan, reaching almost to Bengal in the east. It contains much of what is left of India's non-Himalayan forest, and its associated wildlife and biodiversity, and it is also inhabited at a comparatively low density by tribal peoples, mostly farmers, who have nevertheless retained an understanding of the manifold uses of the forest for medicinal roots, fruits, wild silk, fodder, and for hunting. This is often called India's Tribal Belt. (There are other regions of tribals – the northeast being prominent.)

It is a curious fact of history that the Narmada has not been dammed before, except in a minor way on smaller tributaries. There is no specific reason why not – after all, history does not concern itself with proofs of why something has not happened. But there are many 'causes' about which one could speculate. British agricultural officers in the early part of this century thought about irrigation in some areas, but came to the conclusion that the black cotton soils, as they are known, became glutted with too much water, and that irrigation was unwise in many parts. The river could, of course, be dammed for the coastal plains, but then these are not very extensive; there is no large delta as with the east-flowing Kaveri (or Kavari or Cauvery) or Krishna-Godavari. In the days before Independence there was no electricity grid, so the idea of conducting power to Bombay did not arise. Since Independence the 'ownership' of the river has been rearranged, because of the states reorganisation which took place from 1956 onwards. After the creation of first the new Madhya Pradesh and the new Bombay State, and then the division of Bombay State into Maharashtra and Gujarat, a seemingly endless number of disputes erupted between the three over their share of the waters. These disputes were finally settled in 1968. By then India's competence in and confidence in large-scale projects was peaking, the electricity grid was being installed and industry in metropolitan cities expanding fast. In addition it was realised that this was the nearest water source to the truly drought-prone but quite heavily settled areas of southern Gujarat. Although water from the Narmada could not flow directly by gravity to this region, with a little help from lift pumps at the beginning and the right engineering, it could indeed do so.

The Narmada Valley Project as finally conceived is massive. It calls for two huge major dams on the main river, and hundreds of minor dams. The lower of the two big dams, known as Sardar Sarovar Sagar, is the one that is causing most controversy, the one referred to if anyone says simply 'the Narmada project' or the Narmada Dam. This is slightly confusing, since the upper dam is officially known as Narmada Sagar (*sagar* means 'ocean'). The project will produce power, although the actual amount is in doubt because of the need to use some of it to pump water. It will provide water for many urban areas in and near the catchment and will irrigate substantial areas of Gujarat, if all goes well, and provide drinking water too, improving the lives of up to 40 million people in 130 towns and 2,000 villages. It will also drown some of India's remaining wildlife reserves, destroy some botanically interesting forest, and uproot between 40,000 and 1,000,000 – not just people – but tribal people, the archetypes of the disempowered and downtrodden. It is also hugely expensive. Hence another player enters the scene, the World Bank.

Projects of this size take decades to mature, and in that time politics and technologies change. Work on the main dam at Sardar Sarovar began in the 1960s, when a few families were displaced, and promised compensation and resettlement. According to some authors this never happened, and it has never happened properly elsewhere either. They argue there is no spare land for resettlement, and that corrupt officials will run off with most of the compensation money.

Indian environmental politics seems anchored around strong personalities – campaigners who lead public demonstrations as Gandhi did, one eye fixed on the media, and continually threatening to fast until death. In this case a woman named Medha Patkar has campaigned ceaselessly, along with other well-known people, against the dams, in favour of the tribals, the forest and the wildlife. Her campaign has lasted long enough to see the World Bank call for a new environmental impact assessment, and then pull out of financing the project, mostly on the basis that the government has not shown it can manage the resettlement schemes. The Bank is now in any case more interested in financing structural adjustment loans as part of the policy of pushing liberalisation. Meantime, the dam is slowly, slowly being built. A compromise has been suggested. Lower the final height of the main dam by 50 feet, and then there will be a small reduction in power generated and water stored, but a proportionately larger reduction in the flooded area.

The campaign against the dam is world-wide, mostly because such schemes have become anathema in the eyes of many thinking and all unthinking environmentalists. The story clearly illustrates the tension between development and environment, and the inseparability of the two.

ARVIND DAS (the *Statesman*, English language):   See, generally we try to be objective. But in some instances we have to take a stand. Like that Gujarat thing, that Narmada dam. The thing is that we have never supported this kind of development programme. Because all the financing bodies they have always said that the rehabilitation programme is not foolproof. Looking at it, we are not against dam policy but what we are saying is that the scheme, the planning should be more appropriate. But now almost every week we carry those stories like Narmada, Tehri. [The latter is a dam currently half constructed on the Ganges in the Himalayas which is equally controversial and in an earthquake risk zone – see below.]

We are giving extensive coverage. Then two years back there was an earthquake in Uttar Karshi [near Tehri], then Latur [central Deccan and unexpected, in a low-risk zone].

During that event – these things were covered. The whole debate again came up. The new development process and the reporting of the development process and the politics and quality of this kind of large dam construction. See, what is happening? That Uttar Karshi [the earthquake] I'll tell you – naturally it was not a man-made disaster. But in very close to Uttar Karshi there is a place called Tehri, where the Tehri Dam construction is going on and there is a big popular movement against that. And the major argument against this dam is that this area, this particular area, is prone to earthquakes, and if you construct a big dam there and get a big earthquake the dam will collapse. And nearby cities will be destroyed.

AYAZ MEMON (Newspaper publisher, English and Urdu, Bombay): Like for instance the Narmada project where Medha Patkar has now become, you know, through consistent and constant exposure and write-ups and all that. So the issue has become familiar, people have become familiar, whether they are Hindi readers, Marathi readers. But I think that – this is my – I may be very wrong; I think the perception in the vernacular press is that she is perhaps in the wrong. That the country needs, you know, just development, that is the more important issue rather than relocating tribals or dislocating them.

RAUT (*The Independent*, Bombay, English): When I see the coverage on the Narmada, I have a different opinion on this. There is a clear divide between English newspapers and language newspapers. All English newspapers are generally unanimous against anti-Narmada agitation, while the language press in Gujarat or Maharashtra are very pro-dam. So that shows a complete divide in that. In my opinion, both the sides are not fair, because the English people are not giving enough coverage to pro-Narmada, while the language people are just not printing anything about what Medha Patkar and her people are saying. Both are having a monologue.

BHIMLA BHALLA (DD TV National News Editor, Delhi): Well, we see the Narmada story. We have handled it from various points of view. Sometimes it becomes political news when Medha Patkar went on a fast and there were lots of agitation connected with the dam. Otherwise we try and do it as a development story about specific areas how it is going to affect the lives of the people, what would be the impact, what would be the benefits. It depends what is the background [event] on which the particular news story is based on.

ARVIND GOKHALE (Editor, Marathi paper, Pune): As you might be knowing there is a lot of criticism against this Narmada Dam project. Because some people see it as progress; that we need the water for urban purposes; for Gujarat; for farming and for power and so on. ... I personally wrote about that project a number of times. But I am personally pro-project.

MUKUL (*Navbharat Times*, Hindi, Delhi): Broadly, if you survey the English newspapers published from the capital they are covering this issue either on an environmental basis, or on a policy issue. But if you go to the state of Gujarat where this project is based in big way there, in the Gujarat newspaper, you will find a total different type of coverage that is very pro-dam. I am not commenting upon the aspect of that which is right or wrong. But the Gujarat newspapers are in some way trying to fulfil the aspirations of Gujarati people and their need for the water. So, this type of difference of focus in terms of coverage of Narmada project in Gujarati and language newspapers, and English newspapers.

PREM VAIDYA (ex-documentaries, Films Division, Government of India): And the latest example is that Narmada Dam project, Sardar Sarovar and all that which is again having a political tone. You see, people living just by the side of Narmada river are very close to it, they don't get water, drinking water in the months of summer. And the river flows next to them, very near to them. This is just because we have not thought of preserving water in the needy days and we have just neglected. Now when the Government of Maharashtra, the Government of Gujarat, Government of Madhya Pradesh thought of it; now there is again some plan that it is going to ruin. The ruin means, you see the people who inhabit that area ought to be given proper accommodation somewhere else also with land, but that is what is not being done. ...

Nobody objects to building a dam which will provide enough water to all the three states or three provinces. The problem – the real problem is habitation of those people who were living in this area where the dam water will cover up. If the proper arrangement is made for them, then there is no problem of the dam or the dam water. The real problem is the implementation of the decision taken by the government. ...

This particular dam is really useful – you see the deep dams are well planned and if they are given some sort of an age [meaning if they are long-life projects] that it will be of benefit for the coming two hundred years, then it should be implemented. So, whilst you're giving age to any programme, and see the benefits to the people, then only it should be done.

RANVIR NAYAR (Correspondent for *Sunday*, English news weekly, Bombay): That's perhaps the question that is facing this government today, because on the one hand they realise that there are far too many poor people. We still have 40 per cent of the population living below the poverty line. And on the other hand they know that we can't stretch our resources too far. We are already reaching the limits so to speak.

We cannot spend all our resources and then – I mean, what are we left with? So we have to protect the environment, even if it means that a few more people are left without development for some more time, the so-called development.

Ultimately I think if you follow the traditional Indian way of living, if you do not go in for the so-called 'western consumerism' which means more power, more water, more of everything; if you live the way Indians are used to living there is no real dearth of resources. Let's say Bombay consumes perhaps more power than say, two, three states in the country, so, if Bombay's power consumption is brought down and if that was shared among the states we'll have enough for everybody. So we need to cut down on this consumerism. It is taking hold of this middle-class – 260 or 280 million people.

USHA RAI (the *Indian Express*): I was initially, you know, sort of fully supportive of the movement [anti-dam] and I do think they are making a point when they say we cannot have these large projects which displace so many people. There is no proper rehabilitation of these displaced people.

But I think what has happened in the case of the Narmada movement is that it's gone on too long. And it's not fair to people, I think you should allow people to settle [the dispute] in one way.

I think Medha and the others have to come down a little and make some adjustments. ... Come to a sort of compromise because I think it is not fair to people – nor is it fair to the country to allow a project to drag on and on like this and not allow issues to be resolved.

As a matter of fact the fear is that if it carries on any longer, then the movement will lose its public sympathy as it has already. So, it is not a question of is it development versus environment. Definitely there is – it's a question of – it is a development question and you do need power, you do need.

So, I think we – OK, since so much money has been invested, since the people of Gujarat feel so strongly about it; at least I think after you [anti-dam leaders] made your point that we have to look at other forms of development which are friendly to everybody.

## Development versus the environment

There is a fairly strong suggestion in the above quotations, and indeed something of a correlation, that the English press is more pro-environment and the language press more pro-development. But Arun Kumar of the PTI thought that the reason was not language *per se* – which to some extent is a surrogate for the representation of class – but that it was a matter of geographical propinquity: the language press is somehow closer to the areas where *development* occurs, the metropolitan cities further removed. Arun Kumar:

No. I wouldn't say that this kind of – this attitude is language related. I wouldn't say that the English papers are more negative or the language papers are more positive relating to those contentions. I think the basic point would be the proximity of the region for instance. It's not the language which is the deciding factor or which changes the perspective. But the place from where a newspaper is operating; its subscribers, its readership and – I think the regional press is more interested in local issues anywhere.

The environment versus development debate is, of course, also played out within any one language group, because human action is impossible without some consequential impact on the environment. As the BBC's Alex Kirby said, quoted in the previous chapter, 'for every solution, there's another problem'. This worried Usha Page of AIR Ahmednagar (Maharashtra):

CHAPMAN: Do you talk about sustainable development in your farming programmes?
USHA PAGE: Actually, we are broadcasting programmes on both. Particularly I was also baffled by two things. When, on one hand we are telling people to use fertilisers and on the other hand we are also broadcasting and telling people about that Japanese organic farming. We have reduced the percentage of programmes on fertilisers. Yes. Talking about DDT also. We have totally stopped now. We are not telling people to use these things. We are not encouraging them to use more and more fertilisers. Actually in my opinion we should start telling people to use less fertilisers. But being a government media we are still telling people to use fertilisers. Actually we are broadcasting advertisement also for various fertilisers.

Press correspondents on environmental matters never once claimed that their job was to be 'neutral' or 'objective' reporters – in the way that, for example, Kirby of the BBC hoped to be. Those who came closest to his view were, similarly, the broadcasters. Female Anon (Marathi Regional TV News, DD, Bombay):

I think as far as, you know, this question about development versus environmental politics; I think any major development project involves a little bit, you know, of all the three. So our stories would actually reflect a little bit of all the three issues, statements by government officials, bureaucrats or ministers on the issue, statements by people who are agitating, well I shouldn't say agitating, who are fighting for the project and against the project. And, in fact, Maharashtra has a very rich tradition of voluntary work.

**Correspondents and sources**

As far as we can ascertain, no paper and no broadcasting channel has an 'Environment Correspondent'. Several papers do, however, have science and/or development correspondents, who may also cover some environmental matters. There are also dedicated freelance journalists and dedicated campaigners who also write. But even then, for 900 million people, not many.

CHAPMAN:   My last question is; you mentioned Patkar and you could have mentioned Bahuguna. You mentioned Maneka Gandhi and you mentioned Darryl D'Monte and so on. These are people that keep cropping up and I've met quite a few of them. It seems to me that maybe there is a small group of people, both journalists and activists, who are dominating the agenda, setting the agenda. But a very small group?

AYAZ MEMON:   Very small group. And I think the lurking problem, the very immediate problem is that it could become a, you know, it could become very clannish or clubbish. You form a small club and you keep doing. . . . And what the real pitfall there is, I don't know if it's happening or not, I am not closely associated with any other person except maybe with Darryl who I know well and who writes for us, is that you might end up doing things just for pure group approval. The 'you know I write because five colleagues will read and then maybe they'll react and respond and maybe some NGOs abroad will volunteer some, you know, assistance to go somewhere else and do a project'. And it becomes – it shifts from the problem to maybe a . . . to a fashion. Or just to, you know, I mean like it becomes part of an industry, a small industry in itself where you have to survive.

DINA VAKIL:   I would say, given my limited experience, I would say that the agenda as of now would seem to be set by activists, whether it is the Narmada Bachao Andolan or another movement elsewhere, they have tended to be the most articulate, they have good PR with the press, the various NGOs in this country have developed a good deal of political awareness and know how to get their point across, and are not just seen as confrontationist – so they, more than government, seem to be setting the agenda in terms of coverage of environment issues.

KARKARIA:   There is a very large corps of journalists who are interested in development issues, who identify themselves with a number of NGOs, they are activists in their own right apart from being journalists, and these people sensitise their editors about the need for developmental issues. These people, the do-gooders, are dismissed by the conventional political and traditional journalists, there are very strong dividing lines between the two.

But the actual push to promote environmental issues comes from much wider and more diverse groups. India, for example, has active green lawyers, and the kind of public-interest litigation which Britain lacks, which can be used very effectively against polluters. This obviously provides journalists with news stories which have surfaced in court – and indeed judgement-related environmental articles are not infrequent. The other agenda agents were identified by Nayar as follows:

CHAPMAN:   Now can you tell me; how is it that stories on the environment get into the press? Do reporters go out and find out about them? Do local groups agitate and bring them to you? Do NGOs tell you about them? Does the government give you a release?

RANVIR NAYAR:   It's everything except the last. The government comes last because it's always the government that's at the wrong end of the stick. So they never want to leak stories on the environment. It's very largely when the locals get together they start agitating. Like for instance I am in Bombay, I already will have very little clue as to what's happening in, let's say, in Gujarat or in, let's say, Kajoorli which is on the eastern fringes of Maharashtra and is 1,000 km away. But if the locals do get organised and they agitate on something then I am obviously curious and I want to find out more. That's where the NGOs come into play because the people know I get in touch with them and they let me know more about it and from there I build up the story. . . . On our own it would be very difficult to engage what's happening really.

When asked who was pushing the agenda in India Memon observed:

I think it's a very diffused profile. I think the NGOs are perhaps . . . providing the major thrust. And some editors are very concerned. There is a lot of mock concern and there is absolutely no concern in some areas and among some editors.

But then he added a twist about the changing international scenario, which shows India's fear of the outside world:

Because I think with the opening up of the economy what you might have is a tremendous inflow of capital, tremendous, you know, development. Technological and industrial development without any concern for the environment. I mean it'll be virtually like the rape of the country. Unless you examine the problem, scrutinise the problem thoroughly and immediately. I mean we've had Union Carbide some time back [the Bhopal tragedy of 1985 in which 4,000 people were gassed to death and many others severely disabled]. If it had happened in a developed country we wouldn't have heard the end of it for another century.

Not surprisingly, the government-employed broadcasters did not see the sources in the same way.

BHIMLA BHALLA: The main sources for Doordarshan news are the wire agencies. The national wire agencies. The Press Trust of India and the United News of India and the network of correspondents of All India Radio, because they have a fairly extensive network and as far as Doordarshan itself is concerned, we have a very limited number of reporters who do specific coverage, basically in the capitals of the states and in metropolitan places like Delhi, Bombay, Calcutta, Madras. For the international footage the wire agencies give us the news but for pictures we subscribe to two, three international television news agencies. That way we are better off as far as the international footage is concerned, whereas at the national level we are a little restricted at the moment, because our network is very limited.
CHAPMAN: How do you relate to the Ministry of Environment and Forests?
BHIMLA BHALLA: See, they are keeping very close ties with us about their activities, because they want us to publicise about it in our news bulletins, in our other programmes. So, they keep in close touch with us and keep telling us about what their plans are, about what their activities are.

According to the two anonymous interviewees from local TV news in Bombay, environmental news was event-driven, but they were so short of resources, actually they could do little.

FEMALE ANON: I would say they are very much event-driven. I mean if it is connected to an event like, if for example, the Narmada agitation.

MALE ANON:　Then in Goa an oil tanker was hit at the Agwar Fort. An oil slick was there, that time we covered it.

FEMALE ANON:　But, I mean I think this is – let's not be very inflexible about this. There are stories about the environment which are not event-driven, which are related to broader aspects of the environment. Especially – especially from Delhi. Because they have ... more correspondents, more cameramen. They're able to go out there. So, as far as Bombay is concerned; yes, event-driven because we don't have enough infrastructure to do a story which can deal with the broader aspects of environment, because that means a lot of effort, a lot of research. We don't have – let me be frank – we don't have the necessary infrastructure to do that kind of story.

CHAPMAN:　Do you have dedicated environmental correspondents in your local service?

FEMALE ANON:　We don't have an environmental correspondent. We don't have a correspondent, let alone an environment correspondent. (*Laughter*) That's what I mean. We don't have correspondents. We're functioning – I mean one of the three of us goes to, you know, cover some meeting or whatever is taking place, an event. We rely a lot on our stringers because the local station does not give us a camera unit or even – not even one camera unit exclusively for news. The stringers give us visuals and stories on video.

CHAPMAN:　So, how many stringers do you have – I mean approximately?

MALE ANON:　Eight to nine stringers.

CHAPMAN:　And is a lot of what you show dependent upon what the stringers think you might want to get?

FEMALE ANON:　Well, I think normally we do give them instructions about what kind of visuals we want depending on what kind of story it is. So, I think it depends more on what we think. But you are right. The stringer is the man on the spot so we have to depend on what he gives us.

MALE ANON:　But basically, our bulletins are dry bulletins. They do not visually reach.

Anyone who has watched the rather stilted nature of one of the newscasts would agree with the last point – nearly all of it is the presenter reading the items. Visuals are rare, poor, and often silent with some sporadic voice-over.

At local radio level, however, the attitude was different. Usha Page of AIR said of the Pune station and its development broadcasts on population:

Actually, we don't get much material from Delhi, our head office, from the directorate. The main source of our material and our expertise comes from the family planning association, here working in Poona. So, there are doctors working on the campaign – family planning campaign, there are experts we call them. Besides that, we call the producers and we ask them to write dialogues, to write dramas, to write small pieces, to arrange discussions.

CHAPMAN:　And for environmental stories do you get anything from outside, from Delhi or from anywhere else?

USHA PAGE:　Only some stories.

But, as I would again tell ... [the ] government is ... working at district places. We receive material from district places. If you ask me I will say by my experience that NGOs are the best source, they are more steady, they have up-to-date knowledge with them, they have up-to-date information and data with them, we are using NGO expertise.

CHAPMAN: Do they approach you and say 'this is a story you should have'?
USHA PAGE: Actually, it is a both-way process. At a place like Pune, yeah, they come to us, because Pune radio has – is an established institution and that's why people know and they are aware of the importance of radio. But it's the – it's not fifty-fifty – you may call it twenty or thirty from NGOs and fifty, sixty, seventy from us.'

Usha Page had moved recently from Pune AIR to the local station at Ahmednagar. At this level one might have expected a closer integration of radio with people – but it doesn't happen. Local radio is broadcast on FM – and British listeners who live in rural areas know how problematical that can be even with a good receiver. But as most receivers in India do not have FM anyway, local radio listeners are more likely to listen to Pune on AM than Ahmednagar on FM. Usha Page:

There are some environmentalists and – actually my experience of the local radio stations so far they are not very effective due to so many reasons. Maybe one of them is budget and all these things, then here all local stations they are FM stations. Their reach is also very limited. The audience is very small number. ... They have got their own radio and they have not FM facility in their old sets. And we have, up to some extent, failed to establish the need and to cater for the needs, the daily needs, of the local people and local radio due to many constraints. Then they are listening to radio, particularly to our farming and home programmes which are being broadcast from AIR Pune.

The freelance journalists concerned specifically with environmental matters build their own contacts. Usha Rai had printed a story about arsenic contamination of water in West Bengal, which had resulted in 2,000 deaths.

Sometimes I think one is tipped off and at other times one has to investigate – for example this story on arsenic, I mean I just happened to be sitting with one of the senior officials in the rural development ministry under which drinking water comes. And I was just sort of talking to him about various things, generally trying to find out information on various things that the ministry is doing. And then I got – you know he talked about arsenic, so I asked for more information. Then I went back to him twice, thrice 'til I got the information that I wanted. I suppose if you have been in the profession for thirty years, you know your senses get hardened to 'here is a story'.

## Trees

The relevant Indian ministry is called the Ministry of Environment and Forests. The concern with forests predates the latest western wave of environmentalism by a long way. After the 1960s droughts questions were raised about the degree of deforestation and its impacts on local climates, and there have been many tree-planting schemes, some on state forest land but many for local people who are encouraged to plant along canal banks and roadsides and on communal land. The number of saplings sold at the subsidised rate of Rs 1/- each over the last 20 years defies enumeration. Public campaigns about trees are as common as public campaigns about 'Two adults, two children' ('Hum Do, Hamare Do' literally translates as 'We two, our two').

BIPIN CHOWGULE: There might be a drama, there might be a puppet show, there might be a dance, folk song and dance; where trees are dancing personifying the – well, I mean the different aspects of environment. Trees saying (words not very clear) speaking to you and saying 'oh, don't contaminate me, I will die one day and if I die you will die'.

I have restricted myself to introducing what is environment and more specifically I introduce wildlife and the trees.

CHAPMAN:   Is environmental news mostly bad news?

BHIMLA BHALLA:   No, not always. But I think to make people aware of the basic issue, to a certain extent you give the picture as it is which generally is not very happy. If you take the case of forests in India; deforestation they do it to a large extent.

ANON (local TV News, Bombay):   We do plenty of tree plantations.

2ND ANON:   Coverage of tree plantations.

ARVIND GOKHALE:   As we know that once upon a time Poona city, that is Pune city we call it, itself was full of trees and it was surrounded by good forest and hilly area. So, some trees are fallen in the meantime or some trees were totally destroyed. So some trees were destroyed by the builders, some trees were destroyed by the telephone people or say civic authorities, civic administration. So, at that time we had to put a very strong – we had to . . .

CHAPMAN:   So you developed a strong protest movement in favour of the trees in your paper?

ARVIND GOKHALE:   Yes.

In 1974 we faced a grave situation in the food front and there was a big famine because there was no rain. . . . Then only we started our campaign against the felling of the trees and cutting of the trees. My paper started a vigorous campaign against these activities.'

MAN MOHAN GUPTA (*Dainik Jagran*, Hindi):   Environmental we do weekly. Reporting is there about basically on trees, forests and on pollution.

RAJEEV NAG (*Dainik Jagran*, Hindi):   You see, in a country like India for the last five, six or ten years environment has become a major issue for Third World countries. . . . And a lot of aspects on environment that is – one is the traffic environment, the tree environment, the industrial; other aspects also. And the stories that are really kind of grabbing attention today in the country are definitely Narmada, that's on the tops and there are issues on deforestation.

RANVIR NAYAR (*Sunday*):   And, yeah, the issues of national forests being ravaged, or the clash between tribals who live in forests, or the so-called clash between tribals and the forest, the animals in the forest, the ecosystem, these are kind of issues that are played out very often.

USHA RAI:   For example: an immediate concern I would say is about the ministry's policy to give degraded forest areas over to industries to re-grow the forest in these areas and to use that material – that material that they grow as bulk food and raw material for their industry.

If the 'environment' nationally in India has a concrete form, it is trees. Urban environments may have pot-holes, blocked drains, corrupt administration, air pollution – but nationally, deforestation and reforestation are seen as the major issues: not global warming, not biodiversity *per se* (although this is not ignored in more sophisticated circles), not ozone holes.

**4.6 CONCLUSIONS**

Taken together, this chapter and the previous chapter have revealed many similarities and many differences between the opinions of the UK/North journalists and
editors, and those in India.

If we start with the differences, the obvious and greatest one is the issue of
language. All of Ivor Gaber's and Alison Anderson's interviews in America, the
UK and Singapore were in English, and it was not necessary to comment on
the fact at any stage. It is true that we did not survey other editors working in,
say, France or Germany, but had we done so we surmise that we would not have
found so much apprehension about the use of technical terms about the environment, either in their English originals, or in well-understood equivalent terms in
other languages. Certainly we do not suspect that the editors and reporters
in France and Germany are significantly less well paid and less well resourced
because they are not using English as their prime medium. But there are clear
differences in status and resources between the English and language media in
India. This difference also reflected the inequalities between urban and rural areas,
between the dominant metropoli and the *mofussil*. It relates to the difference
between the new post-nationalist class and the fragmented local publics. The post-
nationalist class is open to ideas emanating from the West, but it is staunchly Indian
and will only interpret these ideas in an Indian context. This is an issue to which
we will return in more abstract form in the concluding chapter.

In the UK, and with the international agencies, there was little problem in
accepting that there could be an 'environmental' category. In India there is considerable difficulty in separating this out, and any attempts to do so continually reverted
to discussion of development issues or civic issues – in fact, there was more than
one suggestion that the idea of the environment was too abstract, it had to be made
relevant to the daily problems people faced. When the reader reaches the conclusions to Chapter 6 on the focus groups' responses in both the UK and India, this
ought to be borne in mind, since in both countries the discussion always drifts to
local experience, even though the UK groups appear to be able to accept some
abstract idea of the 'environment' somewhere far off.

There are also differences in the values which journalists bring to their work.
Although novelty and surprise are the common elements of news values, virtually
all of the North reporters say that they will not campaign and that they try and
stay neutral over environmental controversies. In India this is not the case, and
several of the interviewees explicitly said they wished to campaign on particular
issues, although those who campaigned on an environmental issue rarely did so for
its 'greenness' alone – the Chipko movement to protect trees in the hills and the
anti-Narmada Dam campaign are both tied to the issue of the protection of poor
people and their access to livelihood resources – embedding the environment within
the argument about what kind of development is possible and desirable. Most such
campaigners are associated with the English press rather than the vernacular.

In both North and South there is agreement that the environment is a difficult
topic to cover, because it is about processes, often very slow-moving processes, and
not so often about events. The causes and outcomes of the processes are often
complex and it is difficult to do justice to the story in the brief space or time allocated to it. In the North, the visual value of a story is important, and many stories
cannot run on TV or perhaps (this is surmise) so well in the press either, because
they do not generate attention-holding imagery. Those that do, that project sick
seals or sick children as victims, will inevitably lead stories away from large, abstract

issues of habitat health and biodiversity, to specific species and specific locations. This constraint does not operate in India in the same way, since the visual development of news in both press and broadcasting lags far behind the western equivalents.

In both the North and India there is agreement, despite the vast differences in resources available, that these are expensive topics to cover, and that they do not have a high priority. In both North and South, as one would expect, stories that are covered have to conform to the audiences' conventions of understanding. In both there are directly comparable statements about how statistical examples would have to be adapted to local circumstance. Within financial constraints, the North is prepared to pursue stories from distant places; in India, both because it has been and is more autarkic and introverted, as well as because of cost and resources, there is hardly any concern about environmental problems in distant places.

The rhythm of fluctuating environmental coverage shares some similarities in the North and South. In the UK an economic downturn will place these issues low on the political agenda, and for most in India the struggle to attain economic development will similarly place these issues low. When there is a trigger for an upturn there are great differences. In the UK the political 'primary definers' can have a significant input, and the scientists' doomsday predictions of global change have made an impact: in India it is local trauma such as drought and famine and large-scale displacement of people which are more likely to be triggers of heightened attention.

## NOTES

1 When Chapman was conducting fieldwork into indigenous technical knowledge in tribal areas of Bihar in 1975 (see Chapman, 1983), the field group held a taped interview with Ho people in their village. At the conclusion of the interview, the tape was played back to the villagers. They smiled and clapped. This, they said, was 'Ho radio' – the first they had ever heard. However, one old man who was blind gave the party a fright. He had a screaming fit and collapsed. He had not been able to see the machine, and thought that he must have died, as he heard his own voice coming back to him.

2 A few days after this interview the front page of the Pune *Express* carried a photograph of a dog biting a human corpse lying in the road near the hospital in central Pune. The caption censured development without compassion.

3 The World Wide Fund for Nature has a large office in Delhi; the Centre for Science and Environment is an NGO based in Delhi led by former journalist Anil Agarwal, which has achieved international fame for its Citizens' Report(s) on the State of India's Environment.

## REFERENCES

All India Radio (1995) *Facts and Figures*, New Delhi: Directorate General AIR.
Chapman, G.P. (1983) 'The folklore of the perceived environment in Bihar', *Environment and Planning* A, 15: 945–68.
Doordarshan (1994) *Facts and Figures*, Doordarshan, New Delhi: Audience Research Unit.
India Newspapers Society (1993) *Press Handbook*, New Delhi: I.N.S.
Kumar, Keval J. (1994) *Mass Communication: A Critical Analysis*, Bombay: Vipul Prakashan.

# 5 The messages through the media

## The content analysis of media output on the environment in India and the UK

### 5.1 INTRODUCTION

Although we have heard in the preceding two chapters from editors and broadcasters about their attitudes towards their coverage of the environment, these opinions do not give any clear idea of exactly how much coverage there is, or on what kinds of issues. Part of this project was therefore to monitor the output of the press, TV and radio in the UK and India. We collected data for content analysis for six sample weeks spread out over the course of one year, one in each of the following months: February, April, July, September, November 1993 and January 1994. The same six sample weeks were used both in the UK and in India, and for the three different data sets which we collected. The November week coincided with the public opinion surveys in India and Britain (Chapter 7).

The first data set was on environmental items in two newspapers in each country – one chosen to be slightly more upmarket, the other more popular. We have commented on the selection of the newspapers – *The Independent* and the *Mirror* in the UK and the *Express* (English language) and the *Navbharat Times* (Hindi) in India – in Chapter 1. In the case of the UK papers it is probably easier to distinguish the first paper as appealing more to the intellectual classes, and the latter paper to the lower-middle and working classes, than it is possible to make any such simple distinction in the case of the Indian papers. Here the issues of language and region are so much more prominent. The selection of the *Express* (the Pune Edition) is not too problematic – it is appealing to the middle range of English-language readers – the middle élite, if that is not too oxymoronic. The choice of the *Navbharat Times* was more problematical. It is published in Hindi – we needed an Indian-language paper – and the chosen edition is from Bombay, the nearest city to Pune (Keval Kumar's base) where it is published. To some extent this immediately reveals it is not a 'typical' vernacular paper – perhaps a better representative of that would have been something like the *Dainik Jagran* published in Madhya Pradesh, a Hindi-speaking area (see Chapter 4). Most of the papers in Bombay serve its predominantly Marathi and Gujarati-speaking population, and the important English-speaking élite. But there is also a large and growing community of migrants from North India who are Hindi-speaking and who provide yet another market. The Bombay edition of the *Navbharat Times* is published for them. We should also point out that, as it belongs to a large group (The Times of India Group) and it is published in four cities (the others being in the Hindi belt – see Table 4.3.2), the Bombay paper has quite wide resources to draw on. We could have chosen a more local and vernacular paper – but had we done so it is clear that, like a local evening newspaper in the UK, it would have been dominated by local news, and we would have had to sample several such papers to get any general feel of their environmental contents.

There is one way in which the two UK papers are clearly different from the two Indian papers. Not only are they bigger, but they are (to repeat a well-known newspaper advertising pun) prone to 'go to pieces'. This means that they develop new sections – Business, Living, etc., or Colour Supplements, or Magazines, etc., in which vehicles they are often drifting from news reporting to surrogates for specialist magazines on travel, literature, food and drink, photo-journalism, etc. Anyone who has watched Colour Supplements replace such magazines as the now defunct *Punch* in doctors' and dentists' waiting-rooms will know what this means. The practical impact of this on our investigations is that it becomes difficult to say how long a newspaper is: which sections should one count? Therefore, although we know how much coverage in column centimetres there is on 'environmental' items – we recorded it for the supplements, etc., too – it is not so easy to say that $x$ per cent of the total imprint of *The Independent* is devoted to the environment. The Indian papers are only just showing the first signs of this proliferation with the bigger ones producing perhaps a four-page supplement at the weekends.

The papers were examined by trained investigators, whose task it was to pick out items – stories, features, editorials, letters, adverts etc. – which were either directly proclaimed to be environmental or which nevertheless reflected environmental interaction in some way. The rest of the paper was not recorded, because the time and cost involved in doing so would have been too high.

This procedure inevitably has an element of subjectivity in it. There are some issues which one person might think of as environmental, and another person might not. Chapman and Gaber, for example, found themselves in disagreement over reporting on animal rights activists in the UK who were protesting against the export of live animals to France for slaughter. Gaber felt this was an environmental story – implicitly dealing with man–animal interaction, whereas Chapman thought it was not, since the use of the environment for the rearing of animals was not examined, only the question of whether meat should be moved alive or dead. Another example comes from the *Tribune*, an English-language paper published in Chandigarh, Punjab. All through the very dry hot season (May/June) of 1995 the paper ran a series of articles on difficult civic conditions, each day looking at a different village or town in Haryana or Punjab. These invariably listed problems with inadequate drinking-water supply, problems with electricity, and problems with garbage collection and disposal. Clearly the first and last are environmental, yet the paper did not designate the articles as 'Environment' – though it does have an environment feature from time to time – often about afforestation or about the health of fish stocks in ponds. Just before and during the wet season (the monsoon starts around the middle of July) the paper ran a series of articles on the state of preparedness of flood defences, and then, as the 1995 monsoon became very heavy, stories on flooding, about house and road collapses, and the subsequent loss of life. In both sets of articles if there is any connection implied with some overall journalistic or social category it is revealed by what almost seems to be a catch-phrase – 'The village is crying out for development'. This reinforced the point made in Chapter 4, that it is difficult for many journalists and broadcasters in India to think of an 'environmental category'.

Trained investigators were used to record the data, all of them using the same Coding Handbook, which is reproduced in Appendix I. The word 'coding' is rather misleading, since the data collectors were not asked to code contents. They were asked to *describe* a story or item, using free-format input. The elements of the story would be coded later, using the system outlined below. Data collectors were therefore able to err at all times on the side of including anything which might be

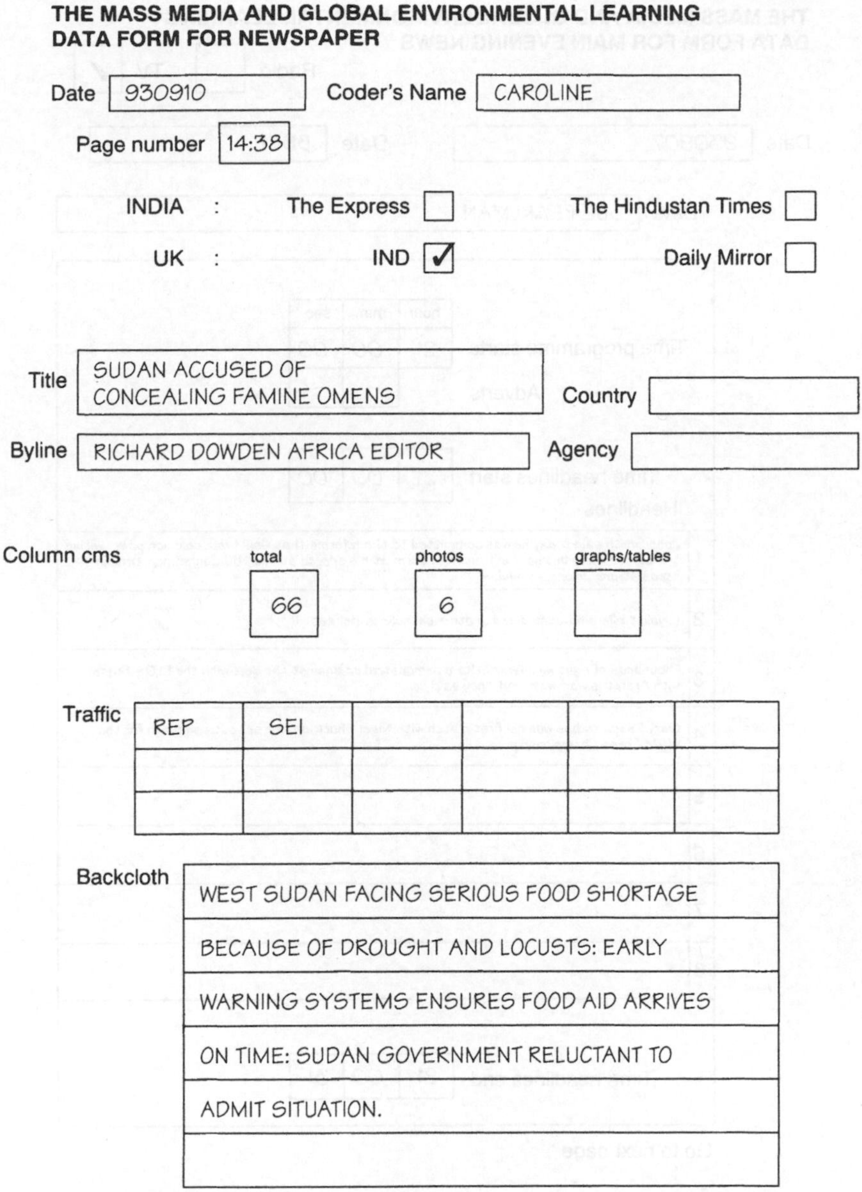

**THE MASS MEDIA AND GLOBAL ENVIRONMENTAL LEARNING**
**DATA FORM FOR NEWSPAPER**

Date  930910     Coder's Name  CAROLINE

Page number  14:38

INDIA  :     The Express ☐          The Hindustan Times ☐

UK  :        IND ☑                  Daily Mirror ☐

Title  SUDAN ACCUSED OF
       CONCEALING FAMINE OMENS        Country [        ]

Byline  RICHARD DOWDEN AFRICA EDITOR   Agency [        ]

Column cms        total        photos        graphs/tables

                  66           6

| Traffic | REP | SEI | | | | |
|---------|-----|-----|--|--|--|--|
| | | | | | | |
| | | | | | | |

Backcloth

WEST SUDAN FACING SERIOUS FOOD SHORTAGE

BECAUSE OF DROUGHT AND LOCUSTS: EARLY

WARNING SYSTEMS ENSURES FOOD AID ARRIVES

ON TIME: SUDAN GOVERNMENT RELUCTANT TO

ADMIT SITUATION.

*Figure 5.1.1*  Sample data sheet for newspapers

environmentally connected, leaving the task of rejecting or accepting and then coding a story to us as the data analysts. This was possible because all data were entered in free-format description in English – a small précis of what the story was about. A sample of the data sheets for the first data set, the newspapers, is shown in Fig. 5.1.1. This data set comprises 480 records, out of which 434 items were

**THE MASS MEDIA AND GLOBAL ENVIRONMENTAL LEARNING**
**DATA FORM FOR MAIN EVENING NEWS**

Radio ☐    TV ✓

Date [ 930907 ]    Date [ BBC1 ]

Date [ SUE PEARLMAN ]

|  | hour | min. | sec. |
|---|---|---|---|
| Time programme starts | 21 | 00 | 00 |
| Adverts | | – | – |
| Time headlines start | 21 | 00 | 00 |

Headlines

| | |
|---|---|
| 1 | John Smith said today he was committed to the reforms that would reduce union power within the Labour Party but he didn't mention them at his speech at the TUC Conference. Britain needs strong unions, he said. |
| 2 | Loyalist killers have murdered another catholic in Belfast |
| 3 | Thousands of right wing Israelis have demonstrated against the deal with the PLO – Peace with Arafat is war with God, they said |
| 4 | Gary Kasparov has won his first match with Nigel Short in their sell out campaign for the World Chess Championship. |
| 5 | |
| 6 | |
| 7 | |
| 8 | |

| Time headlines end | 21 | 00 | 31 |
|---|---|---|---|

Go to next page

*Figure 5.1.2*   Sample data sheet for broadcast news

finally accepted as 'environmental'. The way in which the data analysis then proceeded is discussed below.

The second data set was derived from recording the monitored contents of TV and radio news broadcasts in both countries. In this case the whole bulletin was recorded. This is a much simpler task than the equivalent complete recording of a

Items

| | | | |
|---|---|---|---|
| Time item starts | 21 | OO | 32 |

Source

Traffic

| NEW | REP | ANC | ROR | GRA | |
|---|---|---|---|---|---|
| | | | | | |

---

**Backcloth**   John Smith, Labour Leader went to TUC conference.

No change on one member, one vote proposal.

Strong pro union speech.

---

| | | | |
|---|---|---|---|
| Time item ends **or** Time edition ends | 21 | 05 | 21 |

---

| | | | |
|---|---|---|---|
| Time item starts | 21 | 05 | 22 |

Source

Traffic

| NEW | REP | ANC | ROR | | |
|---|---|---|---|---|---|
| | | | | | |

---

**Backcloth**   Loyalist gunman murdered hairdresser,

Mr Hughes in Belfast: Catholic Republican connections with

Mr Hughes denied.

---

| | | | |
|---|---|---|---|
| Time item ends **or** Time edition ends | 21 | 07 | 06 |

*Figure 5.1.2*   (Continued)

newspaper and there is therefore no element of subjectivity at this stage in deciding what is and what is not environmental. Essentially the same procedure for analysis is used for these data, although there are small variations in the way the computer programs worked because the data are not exactly the same in all respects. This can be seen by consulting the data sheet for broadcast news shown in Fig 5.1.2.

This data set resulted in 4,800 records, of which 4,200 are actual news items (the others cover the headlines and programme introduction).

The third data set is for TV and radio programmes more generally. For these data brief details about all *types* of programmes are recorded, but the actual *contents* of the programmes are not recorded unless they include environmental material. As was the case with the newspapers, we had to rely on the judgement of the data collectors, although again they were urged to err on the side of caution. This data set has 10,000 records, of which 900 give detailed descriptions of content for further analysis. This should not be taken to mean that all 900 were finally accepted as environmentally relevant, or that they represent 9 per cent of total broadcast time. The analysis below brings out the proper interpretation of the data.

For all three data sets an interactive computer program was written which reproduced on the monitor screen the data sheets shown in Figs. 5.1.1 and 5.1.2. Each box on the screen represented a free-format field – any field could be activated by use of a mouse. This allowed the data inputter to work comfortably and directly from the data sheets to produce the data files for analysis. The analysis itself was conducted using our own programs specially written for the task.

## 5.2 METHOD OF ANALYSIS

The methodology of data collection and analysis we have used is an updated version of that developed during an earlier project (the International Television Flows Project funded by the Social Science Council of the UK) described in detail in Gould, Johnson and Chapman (1984). This methodology owed a lot to ideas developed by Atkin and his Q-analysis (Atkin, 1974), which encourages, amongst others two particular approaches to the collection and analysis of data. The first is that numerical data – such as counting or frequency – are usually attached to some structure, either physical or conceptual, and that it is important to understand what that structure is to make sense of the data. The second is that the structure itself is usually defined in terms of qualities and combinations of qualities. This leads to the practical implementation of the methodology via set-theory. The fact that there can be combinations of qualities gives rise immediately to the idea of sets and sets of sets, which results in a well-defined hierarchy. The hierarchies are explicitly not partitional – they utilise cover-sets, which in practical terms will be shown below to mean that something does not have to belong to one and only one higher category. A 'flood', for example, can be a member of both 'climate' and 'water environment' categories.

The way in which structure and data relate needs to be explained a little further. The structure, known as backcloth, is thought of as somewhat static, and supporting the traffic which exists on it. A road system is a backcloth, and supports in this case the traffic composed of cars, lorries and buses. The backcloth 'roads' can be defined in terms of the structure of their interconnectedness, and the traffic can be described both in terms of its qualities – i.e. vehicle types, which can be arranged in different categories – and also in terms of quantities, such as numbers or speeds. Those wishing to know more about the implications of this for understanding data in the social sciences may care to consult Chapman (1996).

### The methodology applied to TV and radio programmes

We will use first the example of the data set on TV and Radio programmes to flesh out these ideas. The kind of data we required can be divided into three kinds:

1 Fixed descriptive data on such items as programme start, programme end, language, production company, channel, etc. This will enable us amongst other things to attach quantities such as total durations to the data structures.

2 The data on what is termed traffic (as explained above, a term derived from Q-analysis and retained here), which broadcasters would more usually refer to as programme type or form. These data are qualitative, using words such as 'Comedy' or 'Documentary' or 'Serial'.

3 The qualitative data on programme backcloth, or subject matter, such as 'books' or 'trees' or 'cooking'.

Most existing classification schemes for broadcasting data do not make these distinctions, and end in something of a mess. They may use categories such as Sport, or Light Entertainment or Culture or Music. What, then does one make of a quiz programme on sport? or a programme of ballet music?

The data collected on traffic are coded using short descriptive codes, such as SEL for Serial, CHA for Charity Appeal: for an explanation of these please see the Coder's Manual in Appendix I. The list of such traffic words was developed before the data collectors began their task, and the latter were allowed to invent new ones but discouraged from doing so unless they thought it absolutely necessary.

The backcloth data constitute the raw data on subject matter, and are free-format free-English descriptions by coders using simple phrases, divided by colons. Each short phrase represents a data item. For TV and radio programmes (but not for the broadcast news – see next section) these data are collected only for those items which are about the environment, environmental issues or problems. There are thousands of these data items, such as those displayed on the specimen forms shown here. We think of these data items as being at the bottom of a hierarchy, the hierarchy of backcloth denoted by B, and the bottom being denoted by the superscript 's' – thus $B^s$. The use of 's' comes from the idea of 'soup' – that we have a mass of stuff yet to be sorted out. These data items were then aggregated to a smaller number of words, at the hierarchical level named Level $B^1$, by Fraser, using a computer program displaying two matching windows on the screen. The purpose of this was to reduce the wide diversity of original descriptions to equivalent descriptions using a more restricted vocabulary. This resulted in a dictionary of about 500 $B^1$ words, as seen in the dictionary in Appendix II. From $B^1$ the computer procedures aggregate the record for each broadcast programme automatically to $B^2$, where there is a restricted backcloth of 26 words, for the highest level of subject-matter classification. The use of a computerised custom dictionary means that if a word – say 'trees' at $B^1$ – is once put into the $B^2$ category 'ecology', then any use of 'trees' anywhere will automatically be given the same higher level classification. More will be said on this in the analysis section below.

## Broadcast news

Dealing with broadcast news required the development of new forms, but following the same methodology, which had also been used recently by Chapman (1991) in an analysis of a single day of TV news world-wide. This also provides a further example of why one needs to use a carefully worked-out methodology such as ours. The then Editor of *InterMedia* had devised a form (which was not ultimately used) on which there were check-boxes of ten news categories, plus the inevitable 'Other', out of which coders were to tick one and only one for any particular news story. The ten categories included Politics, Economics, Crime and International. Chapman asked whether a story on British reservations about the European Union was 'International' or 'Politics' and a story on Colombian income from cocaine trafficking to Britain 'International', 'Crime' or even 'Economics'. The system described in this chapter was adopted instead.

First the outline sheets denote the programme fixed details and headlines. The subsequent sheets are used to record the backcloth of all news stories in free format and the length of each item. These stories therefore again have to be aggregated from the free-format descriptions to a more restricted higher level vocabulary – the mapping from $B^s$ to $B^1$. This again was done by Fraser using matching windows on the computer screen. The $B^1$ data were again automatically classified at $B^2$ using our own customised dictionaries. The higher level of data can then be used for analysis of the categories of stories. Subsequently and separately the Environmental stories were abstracted for analysis at a more detailed level.

### Newspapers

Finally, we mention the newspaper data. The selected newspapers were scrutinised during the sample weeks for any items of environmental interest. These were all recorded again using the same kind of methodology, but with specific forms. In this case column cm obviously replaces duration as the quantity element. Aggregation to $B^1$ is again manual, with aggregation to $B^2$ being automatic using the aggregations defined (by Chapman or Fraser) in the dictionaries.

There is one final point that needs to be stressed about the use of this system: our classifications are neither right nor wrong – they are simply what we use. But if anyone wanted different classifications, they could use the same data but change the customised dictionaries to their own liking, and produce different categories. Once the system is established, this is not a difficult task.

## 5.3 ANALYSIS

### The environmental content of the newspapers

From the newspapers, only the environmental stories (and letters, editorials, adverts, etc.) were recorded – other news stories, constituting the vast majority of coverage, were not included. Table 5.3.1 shows the number, total column length, and the page order of the stories, excluding the supplements of the UK papers. Treating the six weeks for the four papers as one data set, 55 per cent of the data set is accounted for by *The Independent*, 23 per cent by the *Express*, 15 per cent by the *Mirror*, and 8 per cent by the *Navbharat Times*. Thus *The Independent* in England has the greatest number of stories by far, but of course has on average more than twice the column cm that the *Express* (of India) does. Proportionately, the *Express* almost has the same percentage of space attributed to environmental issues. Proportionately, the *Mirror* carries less environmental coverage than the Hindi-language paper the *Navbharat Times*. The average length of story in *The Independent* and *Mirror* is the same at 50 column centimetres – which is quite large, although it is not clear whether this is larger than the overall average for all stories. The articles in the two Indian papers have shorter average length, at 40 col. cm (the *Express*) and 30 col. cm (the *Navbharat Times*), but it is still clear that these stories are comparatively quite long. This lends a little credence to the oft repeated remark of editors that actually these are quite complex stories. In terms of placing, none of the newspapers places many of its stories on the front page – in the case of

*Table 5.3.1* Page location of environmental items in the newspapers, main sections

| | India | | | | UK | | | |
|---|---|---|---|---|---|---|---|---|
| | Express | | Navbharat Times | | Independent | | Mirror | |
| Page number | Number of stories | Total col. cm | Number of stories | Total col. cm | Number of stories | Total col. cm | Number of stories | Total col. cm |
| 1 | 3 | 149 | 3 | 115 | 11 | 520 | 2 | 160 |
| 2 | 6 | 184 | | | 20 | 712 | 6 | 75 |
| 3 | 14 | 418 | 11 | 247 | 19 | 938 | 1 | 18 |
| 4 | | | 8 | 377 | 7 | 385 | 2 | 58 |
| 5 | 8 | 557 | 8 | 182 | 4 | 221 | 4 | 137 |
| 6 | 5 | 161 | 1 | 25 | 20 | 847 | 2 | 136 |
| 7 | 11 | 361 | 4 | 121 | 9 | 579 | 1 | 15 |
| 8 | 10 | 341 | 1 | 57 | 10 | 773 | | |
| 9 | 7 | 633 | 5 | 196 | 3 | 142 | 3 | 180 |
| 10 | 13 | 544 | 1 | 62 | 14 | 521 | 2 | 39 |
| 11 | 10 | 307 | 3 | 67 | 8 | 253 | 1 | 20 |
| 12 | 4 | 90 | 1 | 40 | 8 | 344 | | |
| 13 | 4 | 130 | | | 10 | 679 | 3 | 21 |
| 14 | 7 | 239 | 1 | 6 | 5 | 237 | 1 | 110 |
| 15 | 6 | 236 | | | 6 | 317 | 3 | 118 |
| 16 | | | | | 4 | 148 | | |
| 17 | 1 | 28 | | | 9 | 348 | | |
| 18 | | | | | 1 | 120 | 1 | 8 |
| 19 | 1 | 61 | | | 2 | 90 | | |
| 20 | | | | | 5 | 264 | 2 | 37 |
| 21 | | | | | 11 | 886 | 1 | 21 |
| 22 | | | | | 6 | 373 | 4 | 268 |
| 23 | | | | | 1 | 140 | 1 | 8 |
| 24 | | | | | 3 | 41 | 2 | 225 |
| 25 | | | | | 4 | 97 | 1 | 27 |
| 26 | | | | | 1 | 100 | 1 | 50 |
| 27 | | | | | 4 | 92 | 3 | 214 |
| 28 | | | | | 2 | 111 | 1 | 26 |
| 29 | | | | | 1 | 100 | | |
| 30 | | | | | | | 1 | 48 |
| 31 | | | | | | | 1 | 160 |
| 32 | | | | | 2 | 124 | 3 | 45 |
| 33 | | | | | 2 | 25 | | |
| 34 | | | | | 1 | 60 | 1 | 23 |
| 35 | | | | | | | | |
| 36 | | | | | 1 | 65 | 1 | 150 |
| 37 | | | | | | | | |
| 38 | | | | | 1 | 8 | | |
| 39 | | | | | | | | |
| 40 | | | | | | | 1 | 120 |
| Total for pages > 40 | | | | | 1 | 7 | 5 | 307 |
| Totals | 110 | 4439 | 47 | 1495 | 216 | 10667 | 61 | 2824 |
| Col. cm as % of total dataset | 22.9 | | 7.7 | | 54.9 | | 14.5 | |

*Note:* Six sample weeks, 1993/94.

*Table 5.3.2* Subject matter of environmental stories in newspapers: percentage of each paper's stories devoted to particular categories[1]

| | India | | UK | |
|---|---|---|---|---|
| | Express | Navbharat Times | Independent | Mirror |
| afr | 2.8 | | 3.2 | 15.1 |
| cis | 1.5 | 1.8 | 1.0 | 0.4 |
| int | 11.3 | 14.4 | 6.3 | 1.6 |
| nam | 3.6 | 7.2 | 16.7 | 4.5 |
| nme | | | 0.8 | 0.4 |
| poa | 0.2 | 1.3 | 0.7 | 1.0 |
| ind | 84.9 | 76.8 | 0.5 | |
| oas | 11.7 | 1.8 | 8.9 | 3.8 |
| unk | 0.7 | | 64.1 | 69.3 |
| oec | 3.6 | 5.9 | 14.1 | 5.7 |
| sca | 2.8 | | 4.0 | |
| | | | | |
| air | 6.6 | 3.9 | 7.4 | 2.0 |
| lan | 34.1 | 21.2 | 27.3 | 46.1 |
| wat | 22.2 | 33.5 | 15.1 | 14.3 |
| urb | 8.7 | 19.0 | 6.7 | 0.5 |
| rur | 7.4 | 2.9 | 4.7 | 7.6 |
| dis | 5.3 | 2.7 | 14.6 | 8.4 |
| mal | 45.8 | 48.5 | 30.2 | 25.0 |
| mit | 27.4 | 23.7 | 17.5 | 5.5 |
| plu | 20.5 | 44.1 | 18.8 | 14.9 |
| com | 11.7 | 3.2 | 19.3 | 5.5 |
| con | 32.0 | 22.8 | 21.0 | 39.0 |
| ecl | 46.0 | 40.4 | 31.6 | 36.2 |
| ecn | 40.4 | 37.3 | 35.7 | 53.9 |
| hea | 17.9 | 17.9 | 10.6 | 13.1 |
| hum | 18.3 | 24.9 | 9.9 | 19.6 |
| ins | 22.8 | 13.8 | 21.1 | 26.9 |
| dev | 18.1 | 11.3 | 2.8 | 1.7 |
| pol | 38.5 | 47.1 | 34.0 | 9.0 |
| res | 10.7 | 2.7 | 12.2 | 6.6 |
| tec | 7.4 | 5.7 | 16.9 | 5.0 |
| val | 26.5 | 35.5 | 33.6 | 39.0 |
| | | | | |
| afr ecl | 1.9 | | 1.4 | 10.1 |
| afr ecn | 1.0 | | 1.4 | 13.5 |
| afr hum | | | 0.9 | 8.4 |
| afr ins | 0.6 | | 0.4 | 5.0 |
| afr mal | 2.8 | | 2.3 | |
| afr val | | | 0.5 | 5.0 |
| int ind | 5.3 | 1.3 | | |
| int unk | 0.7 | | 3.5 | |
| int oec | | 2.9 | 1.9 | |
| int air | 2.5 | | 0.9 | |
| int lan | 1.1 | 2.9 | 1.3 | |
| int wat | 2.8 | 1.0 | 2.9 | |
| int mal | 5.2 | 8.5 | 2.4 | 1.6 |

*Table 5.3.2* Continued

| | India | | UK | |
|---|---|---|---|---|
| | *Express* | *Navbharat Times* | *Independent* | *Mirror* |
| int mit | 4.2 | 6.0 | 1.9 | 1.0 |
| int plu | 2.0 | 7.9 | 3.1 | |
| int con | 2.8 | 0.3 | 1.1 | |
| int ecl | 4.2 | 0.4 | 0.5 | 1.6 |
| int ecn | 5.7 | 5.8 | 2.4 | |
| int hum | 2.0 | 6.6 | 0.3 | |
| int ins | 7.2 | 6.5 | 2.8 | |
| int pol | 8.2 | 13.1 | 3.7 | |
| int tec | 1.8 | 1.2 | 3.7 | |
| int val | 2.6 | 4.2 | 1.2 | 1.6 |
| nam oas | 1.3 | | 2.5 | |
| nam unk | | | 5.7 | 0.8 |
| nam oec | | | 3.6 | |
| nam air | | | 2.3 | |
| nam rur | 1.3 | | 0.7 | 3.6 |
| nam dis | 1.3 | | 7.0 | |
| nam mal | 1.0 | 2.7 | 3.5 | |
| nam mit | | 3.1 | 2.1 | 0.8 |
| nam plu | | 2.7 | 4.9 | |
| nam com | | | 6.7 | |
| nam con | | | 2.4 | |
| nam ecl | 2.3 | 4.5 | 2.1 | 3.6 |
| nam ecn | 1.3 | | 3.4 | |
| nam ins | 1.0 | 0.4 | 3.4 | |
| nam pol | 1.0 | 0.4 | 3.2 | 0.8 |
| nam res | | | 2.8 | 0.8 |
| nam tec | | | 4.5 | 0.8 |
| nam val | 1.4 | 4.1 | 9.0 | 4.5 |
| ind oas | 8.0 | | | |
| ind lan | 32.3 | 18.1 | | |
| ind wat | 17.6 | 31.7 | | |
| ind urb | 7.8 | 16.1 | | |
| ind rur | 5.2 | | | |
| ind mal | 39.1 | 37.3 | 0.1 | |
| ind mit | 26.4 | 15.0 | | |
| ind plu | 16.7 | 32.2 | | |
| ind com | 11.2 | 3.2 | | |
| ind con | 30.5 | 22.5 | | |
| ind ecl | 38.2 | 34.2 | 0.5 | |
| ind ecn | 35.1 | 32.7 | | |
| ind hea | 14.9 | 15.1 | 0.1 | |
| ind hum | 15.1 | 18.3 | 0.1 | |
| ind ins | 19.0 | 8.6 | | |
| ind dev | 18.1 | 11.3 | | |
| ind pol | 33.7 | 34.8 | 0.4 | |
| ind res | 10.7 | 2.7 | | |
| ind tec | 6.1 | 3.9 | | |
| ind val | 22.8 | 26.8 | 0.4 | |

*Table 5.3.2*   Continued

| | India | | UK | |
|---|---|---|---|---|
| | Express | Navbharat Times | Independent | Mirror |
| oas unk | | | 3.1 | |
| oas oec | 2.1 | | 2.0 | 0.4 |
| oas lan | 1.8 | | 4.4 | 3.4 |
| oas wat | 2.9 | | 1.0 | |
| oas dis | 2.4 | | 3.0 | 3.4 |
| oas mal | 10.2 | | 6.8 | 3.4 |
| oas mit | 2.2 | | 1.7 | |
| oas plu | 1.2 | 1.8 | 2.1 | |
| oas con | 2.2 | | 1.2 | |
| oas ecl | 8.6 | | 3.6 | 3.8 |
| oas ecn | 2.2 | | 2.0 | |
| oas hea | 6.7 | | 1.7 | 3.4 |
| oas ins | 0.6 | | 2.0 | |
| oas pol | 3.0 | 1.8 | 1.7 | |
| unk oec | | | 9.0 | |
| unk air | | | 5.5 | 1.8 |
| unk lan | | | 17.7 | 32.5 |
| unk wat | | | 9.7 | 7.5 |
| unk urb | | | 6.7 | 0.5 |
| unk rur | | | 3.7 | 3.9 |
| unk dis | | | 4.1 | 3.4 |
| unk mal | | | 17.8 | 19.4 |
| unk mit | 0.7 | | 12.9 | 2.8 |
| unk plu | | | 12.5 | 9.7 |
| unk com | | | 10.0 | 3.9 |
| unk con | | | 14.7 | 33.3 |
| unk ecl | | | 21.0 | 16.2 |
| unk ecn | 0.7 | | 26.4 | 41.4 |
| unk hea | | | 7.8 | 9.7 |
| unk hum | | | 4.5 | 11.2 |
| unk ins | 0.7 | | 18.4 | 21.7 |
| unk pol | 0.7 | | 23.5 | 8.8 |
| unk res | | | 7.8 | 6.6 |
| unk tec | | | 9.1 | 4.9 |
| unk val | 0.7 | | 20.4 | 33.2 |
| oec air | 1.3 | | 2.3 | |
| oec lan | | | 4.7 | 4.6 |
| oec wat | 1.0 | 3.5 | 3.1 | 5.1 |
| oec rur | 1.0 | | 2.2 | 0.1 |
| oec mal | 2.4 | 3.1 | 4.5 | 0.6 |
| oec mit | 1.1 | 3.1 | 2.0 | |
| oec plu | 2.2 | 3.5 | 5.2 | 4.5 |
| oec com | | | 3.6 | |
| oec con | 1.1 | 3.1 | 2.4 | |
| oec ecl | 1.1 | | 4.7 | 1.1 |
| oec ecn | 2.0 | 5.4 | 6.9 | 0.1 |
| oec hea | 0.1 | | 2.3 | |
| oec ins | 0.4 | 2.4 | 4.7 | 4.6 |

*Table 5.3.2* Continued

| | India | | UK | |
|---|---|---|---|---|
| | Express | Navbharat Times | Independent | Mirror |
| oec pol | 1.2 | 2.4 | 6.0 | 0.1 |
| oec tec | 1.1 | 1.2 | 2.0 | 0.1 |
| oec val | | 2.1 | 3.2 | 0.5 |
| sca mal | 1.3 | | 2.6 | |
| sca ecl | | | 3.1 | |
| sca ecn | 2.2 | | 1.9 | |
| sca hum | 1.3 | | 3.1 | |
| sca val | | | 3.2 | |
| air lan | | | 2.2 | 1.8 |
| air wat | 2.9 | | 1.9 | |
| air mal | 4.6 | 3.7 | 4.6 | 1.7 |
| air plu | 4.3 | 3.9 | 5.0 | 2.0 |
| air con | 1.1 | | 2.6 | 0.1 |
| air ecl | 1.7 | 1.0 | 4.7 | 1.7 |
| air ecn | 1.0 | | 2.9 | 1.8 |
| air val | 1.4 | 2.7 | 1.5 | 0.1 |
| lan wat | 9.6 | 9.6 | 4.3 | 6.9 |
| lan urb | 0.4 | 8.8 | 3.8 | |
| lan rur | 5.4 | 2.9 | 4.5 | 4.0 |
| lan dis | 4.6 | 2.7 | 3.6 | 5.0 |
| lan mal | 18.3 | 11.8 | 14.0 | 12.0 |
| lan mit | 7.9 | 7.4 | 3.1 | 1.7 |
| lan plu | 1.8 | | 4.0 | 6.3 |
| lan com | 7.6 | 0.6 | 1.4 | |
| lan con | 16.3 | 6.4 | 10.8 | 32.3 |
| lan ecl | 14.3 | 9.6 | 12.1 | 11.6 |
| lan ecn | 20.3 | 12.7 | 18.2 | 32.8 |
| lan hea | 6.6 | 2.9 | 2.6 | 5.2 |
| lan hum | 7.3 | 2.9 | 5.9 | 9.3 |
| lan ins | 7.9 | 4.4 | 7.1 | 8.5 |
| lan dev | 9.3 | 5.0 | 0.5 | 1.7 |
| lan pol | 11.5 | 7.0 | 12.1 | 0.1 |
| lan res | 2.8 | | 0.9 | |
| lan tec | | 1.2 | 2.2 | 0.1 |
| lan val | 3.7 | 7.6 | 6.4 | 11.6 |
| wat dis | 2.7 | 2.7 | 4.8 | 5.0 |
| wat mal | 9.6 | 16.3 | 4.5 | 1.2 |
| wat mit | 3.7 | 12.9 | 2.8 | 1.7 |
| wat plu | 3.6 | 14.8 | 5.7 | 7.3 |
| wat con | 7.6 | 5.7 | 2.5 | 2.4 |
| wat ecl | 10.7 | 14.5 | 4.1 | 3.1 |
| wat ecn | 13.0 | 20.6 | 5.9 | 2.4 |
| wat hea | 7.6 | 4.8 | 1.2 | 3.3 |
| wat hum | 8.8 | 6.0 | 0.8 | 0.5 |
| wat ins | 8.4 | 5.5 | 2.9 | 4.5 |
| wat dev | 2.8 | 8.9 | 0.3 | 1.7 |
| wat pol | 10.7 | 9.4 | 4.9 | 2.8 |
| wat tec | | 1.8 | 4.2 | |
| wat val | 2.4 | 3.6 | 3.3 | 0.8 |
| urb mal | 3.1 | 13.9 | 0.2 | |

*Table 5.3.2*   Continued

| | India | | UK | |
| --- | --- | --- | --- | --- |
| | Express | Navbharat Times | Independent | Mirror |
| urb mit | 5.2 | 7.2 | 1.2 | |
| urb plu | 7.3 | 10.2 | 0.4 | |
| urb con | 5.7 | 1.4 | 0.2 | |
| urb ecn | 3.7 | 12.7 | 2.6 | |
| urb hea | 0.8 | 8.4 | 2.3 | |
| urb ins | 0.4 | 4.4 | 2.5 | |
| urb dev | 5.2 | | | |
| urb pol | 1.3 | 9.1 | 2.6 | |
| urb val | 0.8 | 8.7 | 1.1 | |
| rur mal | 4.2 | 2.9 | 1.6 | 3.9 |
| rur con | | | 3.0 | 2.8 |
| rur ecl | | | 1.8 | 4.7 |
| rur ecn | 7.4 | 2.9 | 4.5 | 4.0 |
| rur hum | 4.5 | 2.9 | 0.0 | |
| rur ins | 2.0 | 2.9 | 0.1 | 3.9 |
| rur pol | | 2.9 | 1.8 | 0.1 |
| rur val | | | 1.7 | 3.6 |
| dis mal | 4.0 | 2.7 | 4.0 | 3.4 |
| dis com | | | 6.5 | |
| dis ecl | 2.7 | | 0.3 | 5.0 |
| dis ecn | 1.3 | | 2.0 | 1.9 |
| dis hea | 3.3 | | 1.6 | 3.4 |
| dis val | | | 7.1 | |
| mal mit | 8.5 | 18.4 | 5.4 | 1.0 |
| mal plu | 6.2 | 25.6 | 8.2 | 1.7 |
| mal com | 3.1 | | 4.1 | 4.9 |
| mal con | 7.0 | 16.3 | 3.7 | 8.7 |
| mal ecl | 29.5 | 16.2 | 18.1 | 11.7 |
| mal ecn | 15.9 | 20.4 | 15.1 | 9.8 |
| mal hea | 8.4 | 8.8 | 3.8 | 4.3 |
| mal hum | 9.7 | 8.0 | 5.9 | 2.6 |
| mal ins | 3.4 | 2.9 | 6.5 | 13.5 |
| mal dev | 5.3 | 5.7 | 1.4 | |
| mal pol | 15.1 | 15.2 | 14.3 | 0.3 |
| mal res | 6.3 | 0.5 | 4.7 | 3.6 |
| mal tec | 3.4 | 0.5 | 4.6 | 3.6 |
| mal val | 9.9 | 11.2 | 5.6 | 8.6 |
| mit plu | 9.7 | 13.9 | 4.2 | |
| mit com | 4.3 | | 2.6 | 1.0 |
| mit con | 17.9 | 5.7 | 8.6 | 1.7 |
| mit ecl | 6.3 | 0.4 | 4.1 | 2.7 |
| mit ecn | 8.9 | 8.4 | 5.5 | 0.4 |
| mit ins | 7.2 | 2.7 | 4.5 | |
| mit dev | 16.8 | 5.7 | 2.8 | 1.7 |
| mit pol | 15.3 | 9.0 | 5.9 | 1.2 |
| mit res | 6.8 | 2.0 | 6.3 | 2.8 |
| mit tec | 4.0 | 2.0 | 6.4 | 1.2 |
| mit val | 7.2 | 0.9 | 1.5 | 2.5 |

*Table 5.3.2* Continued

| | India | | UK | |
|---|---|---|---|---|
| | Express | Navbharat Times | Independent | Mirror |
| plu com | 2.1 | 0.9 | 2.1 | 0.6 |
| plu con | 6.4 | 3.1 | 2.2 | 0.1 |
| plu ecl | 2.0 | 4.6 | 7.5 | 1.7 |
| plu ecn | 7.3 | 11.4 | 7.9 | 6.9 |
| plu hea | 3.7 | 15.1 | 1.6 | 2.8 |
| plu hum | 2.4 | 14.1 | 1.2 | 1.7 |
| plu ins | 3.0 | | 6.2 | 10.2 |
| plu dev | 6.4 | 3.1 | 0.3 | |
| plu pol | 5.4 | 22.7 | 7.2 | 2.8 |
| plu res | 2.5 | | 2.1 | |
| plu tec | 2.5 | 1.8 | 3.3 | |
| plu val | 4.7 | 20.2 | 3.3 | 5.2 |
| com con | 4.1 | 0.6 | 2.3 | |
| com ecl | 5.3 | | 2.9 | 1.3 |
| com ecn | 4.1 | 0.6 | 3.0 | 3.6 |
| com hea | 0.3 | 0.3 | 2.1 | |
| com ins | 2.2 | | 3.1 | 4.2 |
| com pol | 5.0 | 2.3 | 9.3 | 0.3 |
| com res | 0.7 | | 0.6 | 3.6 |
| com tec | | | 2.0 | 3.6 |
| com val | 1.5 | 2.9 | 8.1 | 4.9 |
| con ecl | 12.4 | 15.4 | 6.5 | 8.2 |
| con ecn | 11.7 | 6.8 | 9.9 | 30.5 |
| con hea | 4.9 | | 0.4 | 1.8 |
| con hum | 3.2 | 0.3 | 3.3 | 7.6 |
| con ins | 9.1 | 1.1 | 4.9 | 8.9 |
| con dev | 16.8 | 8.1 | 2.8 | 1.7 |
| con pol | 14.8 | 5.6 | 6.0 | |
| con res | 3.5 | | 4.1 | |
| con tec | | | 2.7 | |
| con val | 6.5 | 0.9 | 8.5 | 9.1 |
| ecl ecn | 12.5 | 15.6 | 14.3 | 14.3 |
| ecl hea | 8.8 | 1.0 | 3.6 | 3.8 |
| ecl hum | 7.5 | 7.0 | 6.6 | 13.6 |
| ecl ins | 5.9 | 4.6 | 7.4 | 7.0 |
| ecl dev | 5.7 | 5.0 | 1.4 | 1.7 |
| ecl pol | 17.1 | 13.8 | 12.4 | 0.6 |
| ecl res | 2.8 | | 1.9 | |
| ecl tec | 1.6 | 2.5 | 1.7 | |
| ecl val | 13.0 | 10.2 | 13.3 | 15.7 |
| ecn hea | 7.4 | 2.9 | 4.4 | 2.3 |
| ecn hum | 14.1 | 12.0 | 6.5 | 18.2 |
| ecn ins | 17.0 | 12.3 | 8.6 | 12.8 |
| ecn dev | 4.3 | 8.6 | 0.5 | |
| ecn pol | 20.0 | 24.7 | 13.3 | 0.1 |
| ecn res | 6.4 | 2.3 | 7.7 | 4.2 |
| ecn tec | 2.6 | 2.3 | 6.6 | 3.8 |
| ecn val | 6.7 | 10.7 | 10.8 | 17.5 |

*Table 5.3.2*    Continued

| | India | | UK | |
| --- | --- | --- | --- | --- |
| | Express | Navbharat Times | Independent | Mirror |
| hea hum | 4.8 | 12.8 | 2.3 | 0.9 |
| hea ins | 5.5 | 2.9 | 0.9 | |
| hea pol | 5.0 | 9.2 | 2.8 | 7.0 |
| hea tec | | | 2.3 | |
| hea val | 2.9 | 9.0 | 3.7 | 4.2 |
| hum ins | 5.3 | 9.0 | 1.3 | |
| hum pol | 7.8 | 20.9 | 3.1 | 0.8 |
| hum res | 1.0 | | 2.4 | 0.8 |
| hum val | 5.4 | 15.1 | 5.4 | 8.6 |
| ins pol | 18.2 | 12.9 | 13.5 | |
| ins res | 2.4 | 2.3 | 1.2 | 3.6 |
| ins tec | 1.0 | 2.3 | 2.5 | 3.6 |
| ins val | 3.5 | 2.8 | 5.7 | 12.0 |
| dev pol | 7.5 | 5.0 | 0.6 | |
| pol res | 3.9 | 2.7 | 4.3 | 1.2 |
| pol tec | 2.8 | 2.7 | 6.0 | 1.3 |
| pol val | 12.8 | 21.3 | 13.0 | 6.1 |
| res tec | 3.0 | 2.7 | 6.7 | 4.9 |
| res val | 1.1 | | 4.3 | 5.1 |
| tec val | 1.4 | 0.5 | 1.9 | 4.9 |

*Note:* [1] Six sample weeks. For detailed explanation, see text.

*The Independent*, for example, only 5 per cent of its environmental stories make the front page. There was nothing obvious about what would make a story front-page news, although there was a certain flavour (human interest) to the *Mirror*'s approach which runs throughout our commentary. In the case of the *Express*, failings in Project Tiger, municipal water supply problems in Pune, and Indian law on performing animals all made the front page: in the case of the *Navbharat Times*, pollution in Bombay (more on this below) and a Narmada Dam update made the front page. *The Independent* included storm damage, rain and flooding in the UK, nuclear reprocessing at Sellafield, ozone depletion, satellite sensing technology, roads policy, and the environmental problems of the Chunnel (English Channel Tunnel) link. The *Mirror*'s stories were about beach pollution, anti-nuclear protesters climbing over the walls of Buckingham Palace, and the case of a drunken train driver in charge of a train carrying nuclear waste. The personal human-interest level, or the event-related nature, of story telling is explicit.

What are most of the whole set of stories about? Here for the moment we must just briefly go back to the methodology. We have chosen 32 words at $B^2$ under which to categorise all $B^1$ level words (which can belong to more than one $B^2$ category). The first 11 words are for geographical regions of the world. Since we are dealing with British and Indian media, two of these words are for these two countries – 'UK' and 'India'. In the tables they are represented by three letter codes, 'unk' and 'ind' respectively. The other ten regions are at larger scales. Here we list the 11 three-letter codes and the broad definition of the region:

unk:    United Kingdom
ind:    India

afr:    Africa
cis:    Former Soviet Union/Commonwealth of Independent States plus former Eastern Europe
int:    International/Global
nam:    North America (USA plus Canada)
nme:    Near and Middle East
poa:    Polar (either polar region)
oas:    Asia (excluding India, including Australia and New Zealand, and the Pacific, excluding Japan)
oec:    The advanced industrial countries minus UK/USA/Canada – this translates as western Europe plus Japan
sca:    Latin America plus Caribbean

There are 21 other B$^2$ words which deal with subject-matter categories, which will be considered after we have first looked at the geographical coverage.

## Geographical coverage of environmental items in the newspapers

It is apparent from Table 5.3.2 that the percentages for the geographical regions of the world add up to more than 100 because some stories can involve more than one region, for example US technical help to the CIS to clean up pollution from the Chernobyl disaster. A story about Britain and Europe is coded both as 'unk' and 'oec'.

There is very little coverage in the two countries of the polar regions, the Middle East, the CIS and eastern Europe. In both countries the coverage tends to be home-based, although *The Independent* has a fair amount on North America and on other advanced industrial nations. Neither country seems to include anything about the other – apart from a derisory level of coverage by *The Independent* on India and by the *Express* on the UK. The *Mirror* has 15 per cent on Africa but this turns out to be about wildlife safaris/holidays (see below) and not about issues such as desertification.

In the case of broadcast news (see below) it is possible to pick up some stories, usually leading international political/conflict issues such as the war in former Yugoslavia, which are covered in both countries. In the case of the environmental news in the newspapers, it is extremely hard to find any stories common to both countries. In our data set it was the small categories on Polar regions and Other Asia which revealed some matching. The *Express* (India) and *The Independent* (UK) both carried stories on ozone depletion, but two months apart and from different sources. The *Navbharat Times* (India) and *The Independent* (UK) both carried stories about the Soviet dumping of nuclear waste (submarine reactors) in the Arctic Ocean, at approximately the same time (two days apart). The *Mirror*'s (UK) one and only Polar story was as idiosyncratic as ever – about the health of huskies used in Antarctica.

Both the *Express* and *The Independent* carried stories about forest-fires in Australia, though once only in the *Express*, and endlessly and at length in *The Independent*. The *Mirror* carried some coverage of the fires, the *Navbharat Times* not at all.

As for the rest, the coverage in the two countries reflects different concerns. The flooding in Chichester is not covered in India, nor are the pollution and water supply problems in Bombay carried in the UK. The concerns are usually local, if not parochial.

**Subject coverage of environmental items in the newspapers**

The other B$^2$ words are for general subject matter and are as follows:

air: atmosphere, weather, climate etc.
lan: terrestrial environment – land
wat: aqueous environment – fresh water or maritime
urb: specifically urban
rur: specifically rural
dis: natural disaster – earthquake, typhoon, flood, etc.
mal: malfunction, e.g. industrial pollution, soil erosion
mit: mitigation, action to alleviate or prevent malfunction
plu: pollution
com: communications
con: conservation – of resources, of energy, etc.
ecl: ecology
ecn: economics/finance
hea: health – mostly of human beings but occasionally of animals
hum: human interest
ins: institutions
dev: development in the sense used in India
pol: politics
res: resources, e.g. oil, forests
tec: technology
val: values, e.g. morals, corruption, legal system, etc.

The part of Table 5.3.2 which follows the geographical codes shows the percentage of environmental coverage by these key words on their own. Several things are immediately apparent. Both Indian newspapers report a lot of things going wrong – malfunction – whereas for *The Independent* and *Mirror* less of the coverage is about things going wrong. But equally the Indian papers have more about putting things right – mitigation. This is the beginning of a hint of what may be a more active involvement with the environment, rather than just a contemplation of it. For three of the papers, the Environment is fairly political – but not for the *Mirror*. For the *Express* and for the *Navbharat Times* the environment is moderately connected with development – perhaps less than we had expected – whereas in *The Independent* and *Mirror* it is really not connected at all. This is not surprising given the geographical coverage of the latter two papers, and that, as pointed out in Chapter 1, the key word 'development' just does not apply to most stories in the North.

Note how Water, mostly in terms of shortages or pollution but also floods, is a bigger issue for both Indian papers than for the UK papers, and that Natural Disasters attract a bigger coverage in the UK papers than in the Indian papers. Note the *Express* has about the same level of coverage of urban and rural issues, but the *Navbharat Times* has much more on urban issues than rural.

The computer programs continue the analysis to give printouts for all combinations of key words – but the full table runs to hundreds of lines. Here we have had to restrict ourselves to showing all the pairs of words which have a significant loading for at least one newspaper. The first of the paired or coupled words comes after the end of the single words, the least three of which are res, tec and val. The first two lines of the paired words Africa+Ecology and Africa+Economics would not be there but for the *Mirror*. These reflect the same features – about eco-tourism (safaris) in Africa, and have the 'economic' label because of the tourist and leisure industry. The same paper has high loadings for UK+Land and UK+Conservation. These reflect two superficially different types of features: (1) many articles on gardening – again a leisure environmental activity, and (2) coverage of threatened wildlife species in the UK, and issues such as deer-hunting in particular and blood sports in general. Lower down the table the *Mirror* is again different – having virtually no combinations of Politics and Economics with Environment – whereas the other three do. The *Navbharat Times* is also noticeable for the connection between Pollution and Politics.

Table 5.3.3 gives a breakdown by type of environmental article – the idea of 'form' or 'traffic'. The top line shows the extent to which the *Mirror* and *The Independent* have more feature writing on the environment than the Indian papers, which mostly reflects the greater availability of Supplement or Magazine space. This again lends circumstantial evidence to the idea that the environment is complex, and not necessarily newsworthy in the ordinary sense – the stories move too slowly – they are processes, not events. There is better scope within feature articles to consider the problems. The next set of lines shows some interesting distinctions – the extent to which, for example, the feature writing in *The Independent* accounts for much of its environmental coverage of North America. In all cases, although the absolute numbers are rather small, letters to the editor stress the home country more than any other category, and for the Indian papers, letters are exclusively about India – which reinforces the points made by the editors and journalists (Chapter 4) that the audience was focused on India. In the next section on broadcast news, the point will be made that the environment in India is perhaps less politicised than in the UK. Here, in Table 5.3.3 it is possible to examine the extent to which a contrasting point made by the newspaper journalists actually stands scrutiny. Dina Vakil claimed that in India perhaps there was more political writing on the environment in the editorials, particularly in the English-language press. The line for politics shows that the highest figures do indeed come out for editorials (note that the *Mirror* in UK has none), although the *Navbharat Times* has an even higher figure than the *Express*.

With these hints in mind we have reviewed the full data set time and time again. From such reviews some clear impressions emerge which are consistent with the above comments. First, of the two Indian papers, the *Express* takes a more internationalist viewpoint, though still a restricted one, than the *Navbharat Times*. The *Express* also takes a more pan-Indian view, and is less urban oriented. The *Navbharat Times*, on the other hand, reflects more local concerns, about the pollution of both air and water in Bombay, of the health hazards that go with it, of traffic problems and hazards. We have not been able to test formally our hypothesis – which emerged from our project and did not precede it – that the vernacular press is more pro-development and the English press more pro-environment. But to the extent that the solutions to the problems that the *Navbharat Times* highlights are more technology, more safeguards, more enforcement, enhanced supplies (of water) – then clearly this material is pro-development. It is equally clear in Table 5.3.2 that the *Express* is stronger on issues of Conservation with Development, as perhaps we might have expected from the discussion about language in India and the development and environment debate.

Of the two UK papers, *The Independent* has greater coverage and more international coverage (but this is still much less than the coverage of UK-based issues), more in-depth coverage. In a sense it comes over as clearly committed to reporting the debates over these issues, even if trying not to espouse particular viewpoints.

Table 5.3.3 Breakdown of environmental items by form

| | Express | | | | | Navbharat Times | | | | | Independent | | | | | Mirror | | | | |
|---|---|---|---|---|---|---|---|---|---|---|---|---|---|---|---|---|---|---|---|---|
| | rep¹ | fea | ed | fol | let | rep¹ | fea | ed | fol | let | rep¹ | fea | ed | fol | let | rep¹ | fea | ed | fol | let |
| % of total data set | 11 | 4 | 1 | 2 | 1 | 3 | 2 | 1 | 0 | 0 | 30 | 18 | 1 | 4 | 2 | 5 | 8 | 1 | 4 | 0 |
| % of column total | | | | | | | | | | | | | | | | | | | | |
| afr | 5 | 3 | | | | | | | | | 1 | 7 | | 5 | 2 | 14 | 18 | | | 21 |
| cis | 3 | | | | | | | | | | 2 | | | | | 1 | 2 | | | |
| int | 16 | 3 | 19 | | | 10 | 15 | 43 | | 65 | 6 | 32 | 18 | 24 | 9 | 12 | 6 | 20 | 56 | 19 |
| nam | 6 | | | | | 8 | 7 | | | | 11 | | 20 | | 2 | 1 | | | 21 | |
| nme | | | | | | | | | | | 1 | | 11 | | 2 | | 2 | | | |
| poa | 0 | | | | | | | | | | 1 | 0 | | | | | | | | |
| ind | 80 | 89 | 80 | 82 | 100 | 79 | 78 | 57 | 100 | 100 | 12 | 8 | 9 | 5 | | 10 | | | 1 | |
| oas | 9 | 26 | | 7 | | | | 8 | | | 70 | 51 | 79 | 84 | 91 | 72 | 64 | | 56 | 79 |
| unk | 1 | | | | | | | | | | 22 | 6 | 13 | 13 | 8 | 3 | 9 | | 3 | |
| oec | 4 | 4 | | | | | | | | | | 6 | 12 | 3 | 7 | | | | | |
| sca | 5 | | | | | | | | | | 4 | 6 | | | | | | | | |
| air | 10 | 3 | 10 | 36 | 65 | 7 | | | 17 | | 8 | 5 | 15 | 18 | 11 | 25 | 3 | | | 1 |
| lan | 29 | 46 | 9 | 36 | | 22 | 19 | 11 | | | 31 | 31 | 4 | | 32 | 72 | 58 | | 56 | 65 |
| wat | 25 | 22 | 26 | 6 | 10 | 36 | 25 | 36 | 65 | 100 | 19 | 6 | 13 | | 11 | 12 | 18 | | 21 | 25 |
| urb | 10 | 4 | 13 | 3 | | 28 | 10 | 22 | 4 | 10 | 5 | 13 | | 30 | 9 | 3 | 1 | | | |
| rur | 10 | 13 | | | | 5 | | | | | 6 | 4 | | | 2 | 18 | 8 | | 11 | |
| dis | 7 | 6 | | 7 | | 5 | | | | | 10 | 24 | 9 | 3 | 11 | 18 | 3 | | 4 | |
| mal | 44 | 65 | 54 | 50 | 20 | 38 | 67 | 74 | 43 | | 34 | 30 | 41 | 24 | 16 | 43 | 16 | | 21 | 19 |

Table 5.3.3 Continued

| | Express | | | | | Navbharat Times | | | | | Independent | | | | | Mirror | | | | |
|---|---|---|---|---|---|---|---|---|---|---|---|---|---|---|---|---|---|---|---|---|
| | rep[1] | fea | ed | fol | let | rep | fea | ed | fol | let | rep | fea | ed | fol | let | rep | fea | ed | fol | let |
| % of total data set | 11 | 4 | 1 | 2 | 1 | 3 | 2 | 1 | 0 | 0 | 30 | 18 | 2 | 2 | 0 | 5 | 8 | | 4 | 0 |
| % of column total | | | | | | | | | | | | | | | | | | | | |
| mit | 27 | 19 | 36 | 34 | 35 | 16 | 40 | 41 | | | 15 | 17 | 20 | 52 | 14 | 4 | 7 | | 10 | |
| plu | 28 | 15 | 10 | 11 | | 40 | 54 | 71 | | | 25 | 11 | 24 | 3 | | 15 | 16 | | 21 | |
| com | 15 | | | 29 | 55 | 5 | | | | 100 | 16 | 27 | 35 | 17 | 17 | 12 | | | 2 | |
| con | 28 | 37 | 23 | 19 | 44 | 17 | 36 | 14 | | | 32 | 25 | | 41 | 39 | 11 | 55 | | 54 | |
| ecl | 40 | 71 | 32 | 40 | 50 | 41 | 37 | 21 | 79 | | 36 | 33 | 21 | 27 | 20 | 28 | 42 | 100 | 36 | |
| ecn | 36 | 46 | 49 | 28 | 45 | 45 | 18 | 65 | 65 | 100 | 13 | 30 | 20 | 19 | 42 | 38 | 63 | | 61 | |
| hea | 14 | 38 | | 16 | | 21 | 16 | 17 | 17 | | 8 | 7 | 14 | 34 | 9 | 23 | 8 | | | |
| hum | 24 | 9 | 26 | 9 | 50 | 24 | 16 | 26 | 13 | 100 | 27 | 17 | 21 | 2 | 11 | 5 | 31 | | 48 | |
| ins | 24 | 16 | 27 | 28 | 35 | 19 | | 18 | | | 1 | 12 | 28 | 18 | 18 | 37 | 18 | | 39 | |
| dev | 15 | 19 | 9 | 19 | 90 | 15 | 8 | 14 | 35 | | | 5 | | 18 | | | 3 | | 4 | |
| pol | 31 | 41 | 68 | 61 | | 55 | 31 | 75 | 53 | | 41 | 27 | 53 | 31 | 32 | 16 | 5 | | | 2 |
| res | 10 | 7 | 27 | 16 | | 5 | | | | | 11 | 11 | 8 | | 4 | 14 | 2 | | 5 | |
| tec | 10 | | 13 | 16 | | 11 | | | | | 15 | 21 | 11 | 25 | | 13 | | | | |
| val | 24 | 26 | 31 | 3 | | 38 | 25 | 49 | 18 | | 24 | 41 | 63 | 20 | 34 | 59 | 25 | | 33 | |

Note:[1] rep=news report: fea= feature article: ed=editorial: fol=follow-up: let=letters

For detailed explanation see text.

The *Mirror* has surprisingly extensive coverage on very specific things, but virtually all of these can be summed up by the conception that its readers could be personally involved – in gardening, in holidays abroad – or have some human empathy with animal subjects – huskies, dolphins and deer among them.

Not surprisingly, the advertisers seem to respond to this perception. In the *Express* during the six weeks there were two environmental adverts, one for investment in a Green company, and one for investment in a waste-recycling building materials company. In the *Navbharat Times* there are none. In *The Independent* there are three – two for NGOs (The Royal Society for the Protection of Birds (RSPB) and Earthwatch), and one for a Green washing-up liquid. In the *Mirror*, there is one, for the RSPB.

## 5.4 ANALYSIS OF NEWS ON TV AND RADIO

We attempted to cover all the main news broadcasts for the same six weeks, for television on the following channels: BBC1, BBC2 (Newsnight), ITV (News at Ten), Channel 4, and Doordarshan (the main evening news first in the Marathi bulletin then in the English bulletin); and for radio on BBC Radio 4, some Radio 1, and All-India Radio (main Evening News in English). The coverage is shown in Appendix III. For both countries the coverage of TV is near-perfect, and also for All-India Radio, but because of difficulty with a coder, coverage of Radio 4 and Radio 1 was less good.

All the news stories in each broadcast were noted, so the first stage of this analysis is not the same as the analysis of the newspapers. We are able to say what the other items which compete with the environment for news coverage are, what percentage of the whole broadcast news time the environment gets, and where it comes in terms of priority. Although the analytical procedures are the same as before, and although the geographical $B^2$ regions are the same as used for the newspapers, for this first stage of analysis the other $B^2$ words use a different set of categories, since we are covering any news about anything. The categories are the same as those used by Chapman (1991) in the *InterMedia* survey of news globally on a single day. They are:

conf:   conflict – meaning anything related to actual or potential armed conflict between states, i.e. including talks on disarmament as well as actual war
cri:    crime – for the purposes of our analysis and consistent with the definition just given, terrorism in Ireland and India (in Punjab and Kashmir for example) is deemed crime
cul:    culture
des:    destruction – any non-criminal loss of life or property, through cyclones, train crashes, etc.
ecn:    economics and finance
env:    environment
hea:    health
pol:    politics
spo:    sport
tec:    technology
val:    values, usually referring to educational systems, legal systems, moral debates about abortion, religion, etc.

Table 5.4.1 shows that all channels, from both countries and in both radio and TV, emphasise the home country to a large extent, largest of all in the case of

Doordarshan. In the case of the UK there is a moderately heavy exposure on North America and the CIS+East Europe. The overseas coverage by the Indian channels is far lower, and the first-ranking region is Other Asia – covering events in neighbouring states such as Pakistan, Bangladesh, Sri Lanka and China. There is no mention anywhere in any broadcast of Polar regions. The coverage of Latin (Central and South) America is nearly zero in both the UK and India – again this reflects findings both of the *InterMedia* survey and the findings of the International Television Flows project in 1979.

In terms of categories, the UK channels have a heavier coverage of Conflict than does India, reflecting amongst other things extensive coverage of Yugoslavia and the fact that the UK has a more engaged foreign policy, partly because of its history, partly as an attempt to keep its present Permanent Seat on the UN Security Council. Independent India has not yet attempted an extra-regional military role (although this may change) and only comparatively recently has she begun to contribute peace-keeping forces to overseas conflicts, such as in Angola and former Yugoslavia. In other ways the coverage of the channels in both countries tends to show similar proportions over the different categories. Politics ranks highest for all, with the exception of Radio 1 in the UK; there is a 5 to 10 per cent range for Sport, with the exception of BBC2 and Channel 4, where coverage is much less; and Economics is important to all except Radio 1. Then the Environment has a small coverage in the range 3–5 per cent (remarkably similar to the value calculated in *InterMedia* global survey of 1991), except for the much higher level of 12 per cent for Doordarshan – something we will investigate in more depth below. Lower down the table, among the pairs of words, India+Env has 12 per cent – so this means that virtually all the Environmental coverage on Doordarshan is about India. Again, we leave it to the reader to look at the lower lines of the table in more detail, but here are a few things which are perhaps worth noticing. The coverage of Channel 4 for Politics in other regions is about 12 per cent for North America and CIS+E Europe, for other regions 5 per cent, except for Latin America, which drops out of the table. In both India and the UK home coverage of the combination of Crime+Politics is quite high – mostly reflecting terrorism in Ireland and the 'disturbed' regions of India. Table 5.4.2 places the Environmental stories in a second column, showing what other categories are associated with them – and it is then possible to see the extent to which the pattern of categories differs. For example, under the category 'Crime', for nearly all Environmental stories the amount of Crime is much less than for general stories, except for Radio 5. (This is all relative: within the environmental coverage on a channel, which may be 3 per cent as on BBC1, or 12 per cent as on DD, then 7 per cent of that on BBC1 is related to Crime, etc.) On the other hand, for nearly all channels except Channel 4, Destruction is very high for Environmental stories – reflecting damage by storms and floods, for example.

Interestingly Conflict is also proportionately higher for Environmental stories on BBC1 than for all stories. The combination of Environment+Conflict does not occur for any other channel. This result, however, is partly reflecting the small number of Environmental stories among all stories in the first place – only 15 stories in all 42 bulletins of BBC1, and of those 15, just two stories have the Conflict+ Environment combination – but they happen to be fairly high up the bulletins and of long duration and together account for 20 per cent of all time for Environmental items on that channel. One story is about Saddam Hussein's draining of the marshes in Iraq to attack the Shi'ite Marsh Arabs, and the other is about Kampuchea, the Khmer Rouge and rainforest destruction. These are event-related environmental stories.

*Table 5.4.1*  Subject categories of all broadcast news[1]

| | BBC1 | BBC2 | ITV3 | CHA4 | DD | RAD4 | RAD1 | AIR |
|---|---|---|---|---|---|---|---|---|
| afr | 2 | 3 | 3 | 6 | 2 | 2 | 4 | 3 |
| cis | 16 | 19 | 16 | 17 | 3 | 10 | 9 | 2 |
| int | 6 | 14 | 6 | 7 | 4 | 6 | 6 | 3 |
| nam | 16 | 23 | 17 | 19 | 4 | 11 | 12 | 4 |
| nme | 9 | 10 | 4 | 6 | 2 | 4 | | 4 |
| ind | 1 | | 0 | | 82 | 1 | | 67 |
| oas | 2 | 6 | 3 | 4 | 10 | 3 | 1 | 12 |
| unk | 57 | 56 | 66 | 58 | 2 | 68 | 69 | 2 |
| oec | 9 | 7 | 8 | 8 | 3 | 8 | 3 | 3 |
| sca | 1 | 2 | 1 | 2 | 1 | | 2 | 1 |
| conf | 13 | 14 | 12 | 15 | 1 | 6 | 7 | 1 |
| cri | 23 | 18 | 23 | 19 | 15 | 30 | 23 | 12 |
| cul | 2 | 1 | 3 | 3 | 9 | 5 | 16 | 3 |
| des | 2 | 1 | 6 | 1 | 4 | 5 | 4 | 5 |
| ecn | 21 | 26 | 18 | 27 | 25 | 15 | 7 | 19 |
| env | 3 | 3 | 4 | 5 | 12 | 4 | 4 | |
| hea | 5 | 6 | 6 | 5 | 4 | 6 | 12 | 1 |
| pol | 52 | 65 | 33 | 58 | 50 | 45 | 19 | 52 |
| spo | 5 | 2 | 6 | 1 | 7 | 10 | 12 | 5 |
| tec | 1 | 2 | 3 | 2 | 6 | 2 | 1 | 3 |
| val | 18 | 36 | 30 | 25 | 19 | 27 | 37 | 6 |
| afr conf | 1 | 1 | 1 | 2 | 0 | 0 | | 0 |
| afr cri | 1 | 1 | 1 | 1 | 0 | 1 | 1 | 1 |
| afr pol | 1 | 2 | 2 | 5 | 1 | 1 | 3 | 1 |
| afr spo | | | 0 | | 0 | | | |
| cis int | 3 | 7 | 3 | 6 | 0 | 3 | 2 | |
| cis nam | 6 | 10 | 6 | 7 | 2 | 4 | 2 | 1 |
| cis unk | 3 | 4 | 5 | 3 | | 2 | 5 | |
| cis oec | 3 | 2 | 2 | 3 | 1 | 2 | | 1 |
| cis conf | 10 | 9 | 9 | 11 | | 6 | 4 | 0 |
| cis ecn | 1 | 5 | 3 | 1 | 1 | 0 | | 0 |
| cis pol | 12 | 11 | 8 | 12 | 3 | 6 | 5 | 2 |
| cis val | | | 5 | 3 | 2 | | 2 | |
| int conf | 3 | 6 | 3 | 6 | 0 | 3 | 0 | 0 |
| int cul | | | | | 0 | 0 | 1 | |
| int ecn | 2 | 7 | 1 | 1 | 1 | 0 | | 1 |
| int hea | 0 | 1 | | 1 | 1 | 1 | 2 | 0 |
| int pol | 5 | 9 | 3 | 5 | 2 | 2 | 2 | 1 |
| int spo | 0 | | 0 | | 1 | 2 | 1 | |
| int val | | 3 | 1 | | 1 | 0 | | 0 |
| nam unk | 3 | 5 | 5 | 3 | | 3 | 3 | |
| nam oec | 4 | 5 | 3 | 3 | 2 | 2 | | 1 |
| nam conf | 4 | 4 | 3 | 5 | 0 | 1 | 3 | 0 |
| nam cri | 5 | 4 | 7 | 6 | 0 | 4 | 7 | 1 |
| nam cul | 1 | 1 | 1 | | 0 | 1 | | |
| nam ecn | 2 | 7 | 3 | 4 | 1 | | | 0 |
| nam pol | 7 | 13 | 6 | 13 | 3 | 6 | 2 | 3 |
| nam val | 4 | 8 | 6 | 5 | 1 | 4 | 7 | |

*Table 5.4.1* Continued

| | BBC1 | BBC2 | ITV3 | CHA4 | DD | RAD4 | RAD1 | AIR |
|---|---|---|---|---|---|---|---|---|
| nme conf | 1 | 3 | 0 | 1 | 0 | | | 0 |
| nme cri | 2 | 2 | 0 | 3 | 0 | 0 | | |
| nme ecn | 1 | 0 | | 0 | 0 | | | 1 |
| nme pol | 8 | 7 | 4 | 6 | 2 | 3 | | 3 |
| ind cri | 0 | | | | 12 | 0 | | 9 |
| ind cul | | | | | 9 | | | 3 |
| ind des | | | | | 3 | 0 | | 5 |
| ind ecn | | | | | 24 | | | 17 |
| ind env | | | | | 12 | 0 | | 6 |
| ind hea | | | | | 4 | | | 1 |
| ind pol | 0 | | | 0 | 42 | | | 42 |
| ind spo | 0 | | | | 5 | 1 | | 3 |
| ind tec | | | | | 6 | | | 3 |
| ind val | 0 | | | | 18 | | | 6 |
| oas cri | | | | | 3 | 0 | | 2 |
| oas cul | | | | | 1 | 1 | 1 | 0 |
| oas pol | 1 | 5 | 1 | 4 | 7 | 1 | | 9 |
| oas spo | 0 | 0 | 0 | | 2 | 0 | | 2 |
| unk cri | 15 | 11 | 17 | 10 | 0 | 24 | 13 | 0 |
| unk cul | 1 | | 2 | 3 | 0 | 2 | 10 | |
| unk des | 2 | 1 | 4 | 1 | | 4 | 2 | |
| unk ecn | 15 | 17 | 15 | 23 | 0 | 9 | 5 | 0 |
| unk env | 2 | 2 | 4 | 3 | 0 | 4 | 4 | 0 |
| unk hea | 4 | 4 | 5 | 5 | | 6 | 8 | |
| unk pol | 28 | 37 | 19 | 37 | 0 | 31 | 13 | 1 |
| unk spo | 4 | 1 | 6 | 1 | 1 | 6 | 10 | 1 |
| unk tec | 0 | | 2 | 1 | 0 | 0 | 1 | 0 |
| unk val | 16 | 26 | 23 | 20 | 0 | 24 | 27 | |
| oec conf | 2 | 1 | 1 | 2 | | 2 | | 0 |
| oec ecn | 4 | 4 | 3 | 4 | 1 | 3 | 2 | 1 |
| oec pol | 6 | 6 | 3 | 6 | 2 | 7 | | 1 |
| oec spo | 0 | | 1 | | 0 | 0 | 1 | 0 |
| sca pol | 1 | 2 | | 2 | 0 | | | 0 |
| conf pol | 7 | 6 | 5 | 9 | 0 | 3 | 3 | 0 |
| cri pol | 10 | 10 | 4 | 12 | 12 | 13 | 3 | 9 |
| cri val | 4 | 6 | 11 | 7 | 2 | 9 | 8 | 0 |
| ecn pol | 12 | 17 | 7 | 16 | 11 | 6 | 1 | 10 |
| ecn val | 2 | 5 | 2 | 4 | 5 | 2 | 2 | 1 |
| env tec | 0 | | 0 | 1 | 2 | 0 | 1 | 0 |
| pol val | 8 | 20 | 8 | 14 | 6 | 11 | 8 | 3 |
| tec val | 0 | 2 | 0 | 0 | 1 | | | 0 |
| | | | | | | | | |
| afr conf pol | | 0 | 0 | 2 | | | | |
| afr cri pol | 1 | 1 | 1 | 1 | 0 | 0 | 1 | 1 |
| afr des hea | | | | | | | | 2 |
| afr pol val | | 0 | 1 | | 0 | | | 2 |
| cis int conf | 3 | 5 | 3 | 6 | | 3 | | 0 |
| cis int ecn | | 2 | 0 | 0 | 0 | | | |
| cis int pol | 3 | 4 | 2 | 4 | 0 | 1 | 2 | |
| cis nam conf | 3 | 3 | 3 | 4 | | 1 | 2 | 0 |

*Table 5.4.1*   Continued

| | BBC1 | BBC2 | ITV3 | CHA4 | DD | RAD4 | RAD1 | AIR |
|---|---|---|---|---|---|---|---|---|
| cis nam pol | 4 | 6 | 4 | 7 | 1 | 3 | 2 | 1 |
| cis unk conf | 2 | 1 | 3 | 2 | | 2 | 3 | |
| cis unk ecn | | 3 | 1 | 0 | | | | |
| cis unk val | 1 | | | 0 | 1 | | | |
| cis oec env | | | | 1 | | | | |
| cis conf cri | 1 | | | 1 | | | 2 | |
| cis conf hea | | 2 | 1 | | | 0 | | |
| cis conf pol | 7 | 4 | 4 | 7 | | 3 | 3 | |
| cis conf val | 1 | 2 | 1 | 2 | | 0 | 1 | |
| cis cri val | | | | | | | 2 | |
| cis ecn pol | 1 | 3 | 2 | 1 | 1 | 0 | | 0 |
| cis pol val | 0 | 3 | 1 | 1 | | | | |
| int nam ecn | 0 | 2 | 0 | 1 | 0 | | | 0 |
| int nme pol | 1 | | 0 | | 0 | 1 | | 1 |
| int oas pol | | 3 | 0 | 1 | 0 | | | |
| int unk ecn | 0 | 1 | 0 | 0 | | | | |
| nam nme cri | 0 | 1 | | 2 | | 0 | | |
| nam nme pol | 1 | | 1 | 3 | 1 | 1 | | 0 |
| nam ind pol | | | | | 0 | | | 1 |
| nam oas ecn | | 1 | 0 | 1 | 0 | | | 0 |
| nam oas pol | 0 | 2 | 1 | 1 | 1 | | | 1 |
| nam unk cri | 1 | 0 | 2 | 0 | | 3 | | |
| nam unk ecn | 0 | 2 | 1 | 1 | | | | |
| nam unk hea | 1 | 2 | 1 | 0 | | | | |
| nam unk pol | 2 | 2 | 0 | 2 | | 2 | 2 | |
| nam oec ecn | 1 | 2 | 2 | 2 | 1 | | | 1 |
| nam oec pol | 2 | 4 | 2 | 2 | 2 | 1 | | 1 |
| nam sca cri | | | | | | | 2 | |
| nam conf val | 1 | 2 | | 1 | | | 1 | |
| nam cri pol | 1 | 0 | | 3 | 0 | 2 | | 1 |
| nam cri val | 2 | 2 | 5 | 2 | 0 | 4 | 5 | |
| nam ecn pol | 1 | 6 | 2 | 3 | 1 | | | 1 |
| nam env pol | 0 | 0 | | | 0 | 2 | | |
| nam pol val | 0 | 2 | | 2 | 0 | 2 | | |
| nme ind pol | | | | | 0 | | | 1 |
| nme unk des | | | | | | 1 | | |
| nme conf pol | 0 | 3 | 0 | 1 | 0 | | | |
| nme cri pol | 2 | 1 | 0 | 3 | 0 | 0 | | |
| ind oas cri | | | | | 1 | | | 0 |
| ind oas ecn | | | | | 1 | | | 1 |
| ind oas pol | | | | | 4 | | | 4 |
| ind oas spo | | | | | 2 | | | 1 |
| ind unk spo | 0 | | | | 1 | | | 1 |
| ind cri pol | 0 | | | | 10 | | | 6 |
| ind cri val | 0 | | | | 2 | | | 0 |
| ind cul ecn | | | | | 1 | | | |
| ind cul pol | | | | | 2 | | | 1 |
| ind cul val | | | | | 2 | | | 1 |
| ind des env | | | | | 2 | 0 | | 3 |
| ind ecn env | | | | | 5 | | | 1 |

*Table 5.4.1* Continued

| | BBC1 | BBC2 | ITV3 | CHA4 | DD | RAD4 | RAD1 | AIR |
|---|---|---|---|---|---|---|---|---|
| ind ecn hea | | | | | 1 | | | 0 |
| ind ecn pol | | | | | 11 | | | 9 |
| ind ecn tec | | | | | 2 | | | 1 |
| ind ecn val | | | | | 5 | | | 1 |
| ind env pol | | | | | 3 | | | 1 |
| ind env tec | | | | | 2 | | | 0 |
| ind env val | | | | | 3 | | | 0 |
| ind hea val | | | | | 2 | | | 1 |
| ind pol val | 0 | | | | 6 | | | 3 |
| ind tec val | | | | | 1 | | | 0 |
| oas unk pol | 1 | 1 | 0 | 3 | 0 | 0 | | 0 |
| oas cri pol | | | | | 2 | | | 2 |
| oas ecn pol | | 3 | 0 | 1 | 1 | | | 1 |
| oas pol val | 0 | 1 | 0 | 1 | 0 | 0 | | 0 |
| unk oec ecn | 1 | 1 | 2 | 2 | 0 | | | |
| unk oec pol | 1 | 1 | 1 | 3 | | 1 | | 0 |
| unk oec spo | | 1 | | | | | 1 | |
| unk oec val | 1 | 1 | 1 | | | 0 | 0 | |
| unk conf val | 1 | | 2 | 1 | | | 2 | |
| unk cri pol | 6 | 7 | 3 | 7 | 0 | 11 | 2 | 0 |
| unk cri spo | | | | | | 2 | 2 | |
| unk cri val | 2 | 4 | 8 | 4 | | 8 | 1 | |
| unk cul ecn | 0 | | 1 | 1 | | | | |
| unk cul hea | | | | | | | 3 | |
| unk cul val | 0 | | 0 | 0 | 0 | | 2 | |
| unk des env | 1 | 0 | 1 | | | 0 | | |
| unk ecn hea | 1 | 1 | | 2 | | 0 | | |
| unk ecn pol | 8 | 11 | 5 | 13 | | 3 | 1 | |
| unk ecn val | 2 | 4 | 2 | 3 | 0 | 1 | 2 | |
| unk env hea | 0 | 1 | 1 | 1 | | | 1 | |
| unk env pol | 1 | 1 | 0 | 1 | | 3 | | 0 |
| unk env val | 0 | 1 | 2 | | | 3 | 2 | |
| unk hea pol | 2 | 1 | | 1 | | 3 | | |
| unk hea val | 1 | 1 | 2 | 1 | | 4 | | |
| unk pol val | 8 | 17 | 6 | 13 | | 11 | 7 | |
| oec cri pol | 1 | 1 | 0 | 1 | 0 | 1 | | 0 |
| oec ecn pol | 2 | 4 | 2 | 3 | 1 | 3 | | 1 |

*Note:* 1 Figures are percentages of channel totals, for six sample weeks.

The already close attention to India paid by DD and AIR in the general news is even more concentrated here: 98 per cent of Environmental news in both cases is about India herself. Polar regions get no mention anywhere, international environmental issues seem to have no airing, but in the UK the CIS+E Europe, North America, and OECD do get some coverage. As Chapter 3 noted, this might not be because these regions are intrinsically more interesting – in the case of the latter two at least, it may well be because the UK can relate to, can accept as of good enough quality and can afford news footage from or about these regions.

Table 5.4.3 shows the running order of stories where Environmental items occur. The numbers 1 to 30 at the left of the table indicate how many stories appeared

Table 5.4.2 Percentage distribution of broadcast subject matter for all news, and for environmental news[1]

| | BBC1 | env | BBC2 | env | ITV3 | env | C4 | env | DODA | env | RAD4 | env | RAD1 | env | AIR | env |
|---|---|---|---|---|---|---|---|---|---|---|---|---|---|---|---|---|
| Africa | 2 | | 3 | | 3 | 1 | 5 | 6 | 2 | | 2 | 4 | 4 | | 3 | 1 |
| CIS+E Europe | 16 | | 19 | 23 | 15 | | 18 | 32 | 3 | 2 | 10 | | 10 | | 2 | 5 |
| International | 7 | | 14 | | 6 | 1 | 8 | 6 | 4 | 6 | 6 | 6 | 6 | 2 | 3 | 2 |
| N America | 16 | 7 | 23 | 42 | 17 | 10 | 19 | 15 | 4 | 1 | 9 | 37 | 12 | | 4 | |
| Near+Mid East | 9 | 14 | 10 | 37 | 4 | | 6 | | 2 | | 4 | | 0 | | 4 | |
| Polar | 1 | | 0 | | 0 | | 0 | | | | | | | | | |
| India | 1 | 21 | 6 | 1 | 2 | 16 | 4 | 29 | 81 | 98 | 1 | 6 | 1 | | 68 | 98 |
| Other Asia | 2 | | 1 | | | | | | 10 | 10 | 3 | 13 | | | 12 | 4 |
| UK | 57 | 66 | 57 | 77 | 68 | 84 | 58 | 69 | 2 | 2 | 69 | 75 | 71 | 100 | 2 | 4 |
| OECD | 10 | 11 | 7 | | 7 | 18 | 8 | 32 | 3 | 5 | 9 | | 3 | | 3 | 2 |
| S and Cent Amer | 1 | 21 | 2 | | 1 | | 2 | 13 | 1 | | | | 2 | | 1 | |
| Conflict | 13 | 7 | 14 | 5 | 12 | 7 | 15 | 2 | 1 | 0 | 6 | | 7 | | 1 | 3 |
| Crime | 23 | | 18 | | 23 | | 19 | | 14 | 9 | 31 | 46 | 23 | | 13 | |
| Culture | 2 | | 1 | | 3 | 7 | 3 | 11 | 9 | | 5 | 7 | 16 | | 4 | |
| Destruction | 2 | 35 | 1 | 12 | 6 | 42 | 1 | 2 | 4 | 19 | 6 | 28 | 4 | 17 | 6 | 54 |
| Economic | 21 | 17 | 26 | 23 | 19 | 8 | 28 | 13 | 26 | 47 | 15 | | 7 | | 20 | 20 |
| Environment | 3 | 100 | 3 | 100 | 5 | 100 | 4 | 100 | 11 | 100 | 5 | 100 | 4 | 100 | 7 | 100 |
| Health | 5 | 12 | 6 | 37 | 6 | 11 | 5 | 14 | 4 | 15 | 7 | | 12 | 20 | 2 | 1 |
| Politics | 52 | 33 | 65 | 29 | 34 | 3 | 59 | 26 | 50 | 26 | 45 | 56 | 20 | 2 | 53 | 13 |
| Sport | 5 | | 2 | | 7 | | 1 | | 7 | | 10 | | 11 | | 6 | |
| Technology | 1 | 5 | 2 | | 3 | 6 | 2 | 33 | 6 | 16 | 2 | 6 | 1 | 19 | 3 | 8 |
| Values | 18 | 3 | 36 | 21 | 30 | 39 | 26 | 1 | 19 | 29 | 27 | 56 | 37 | 42 | 6 | 8 |

Note: 1 Six sample weeks.

Table 5.4.3  Running order of all stories and environmental stories

| Story Number | BBC1 | | BBC2 | | ITV3 | | CHA4 | | DD English | | DD Marathi | | RAD4 | | RAD1 | | AIR | |
|---|---|---|---|---|---|---|---|---|---|---|---|---|---|---|---|---|---|---|
| | all | env | all | env | all | env | all | env | all | env | all | env | all | env | all | env | all | env |
| 1 | 42 | | 36 | | 36 | | 33 | | 42 | 2 | 41 | 1 | 22 | 1 | 10 | | 42 | 3 |
| 2 | 42 | | 36 | | 36 | | 33 | | 42 | 5 | 41 | 4 | 22 | 4 | 10 | | 42 | 3 |
| 3 | 42 | | 36 | 2 | 36 | | 33 | | 42 | | 41 | 3 | 22 | 3 | 10 | | 42 | |
| 4 | 42 | 2 | 31 | 2 | 36 | 2 | 32 | 3 | 42 | 3 | 41 | 3 | 21 | 3 | 10 | | 42 | |
| 5 | 41 | | 26 | 2 | 34 | | 28 | | 42 | 3 | 41 | 3 | 20 | 1 | 10 | | 42 | |
| 6 | 41 | | 19 | | 34 | | 24 | | 42 | 3 | 41 | 5 | 18 | 5 | 10 | | 42 | |
| 7 | 39 | | 10 | | 33 | | 17 | | 42 | 5 | 41 | 7 | 18 | 2 | 10 | 2 | 42 | |
| 8 | 36 | | 8 | | 33 | | 17 | | 42 | 5 | 40 | 4 | 16 | 2 | 9 | 1 | 42 | 2 |
| 9 | 33 | | 6 | 2 | 27 | | 15 | 4 | 41 | 4 | 40 | 4 | 15 | 2 | 9 | 2 | 42 | 1 |
| 10 | 29 | 2 | 4 | | 26 | | 15 | | 41 | 4 | 40 | 5 | 15 | | 9 | | 42 | 2 |
| 11 | 25 | | 3 | 2 | 25 | | 13 | | 40 | 5 | 40 | 5 | 11 | | 8 | | 42 | |
| 12 | 20 | | 2 | | 17 | | 8 | | 39 | 2 | 40 | 4 | 10 | | 4 | | 42 | 5 |
| 13 | 15 | | 1 | | 12 | | 6 | | 35 | 2 | 35 | 2 | 9 | 2 | 3 | | 41 | 6 |
| 14 | 11 | | 1 | | 12 | | 5 | | 32 | | 34 | 1 | 7 | | 1 | | 39 | 3 |
| 15 | 11 | | | | 9 | | 5 | | 26 | 1 | 29 | 2 | 7 | 2 | 1 | | 37 | 1 |
| 16 | 5 | | | | 7 | | 4 | | 21 | | 27 | 1 | 6 | 2 | 1 | | 35 | 5 |
| 17 | 5 | | | | 7 | | 4 | | 13 | | 19 | 2 | 6 | | 1 | | 30 | 2 |
| 18 | 1 | | | | 4 | | 1 | | 10 | | 15 | 1 | 4 | | | | 28 | 5 |
| 19 | | | | | 1 | | 1 | | 7 | | 10 | 2 | 4 | | | | 25 | |
| 20 | | | | | 1 | | 1 | | 3 | | 7 | 1 | 3 | | | | 21 | |
| 21 | | | | | 1 | | 1 | | 1 | | 6 | | 2 | | | | 16 | |
| 22 | | | | | 1 | | 1 | | | | 4 | | 1 | | | | 13 | |
| 23 | | | | | 1 | | 1 | | | | 2 | | 1 | | | | 9 | |
| 24 | | | | | 1 | | | | | | 1 | | | | | | 7 | 1 |
| 25 | | | | | | | | | | | | | | | | | 5 | |
| 26 | | | | | | | | | | | | | | | | | 2 | |
| 27 | | | | | | | | | | | | | | | | | 2 | |
| 28 | | | | | | | | | | | | | | | | | 2 | |
| 29 | | | | | | | | | | | | | | | | | | |
| 30 | | | | | | | | | | | | | | | | | | |
| totals | 469 | 14 | 219 | 8 | 424 | 20 | 297 | 17 | 646 | 46 | 676 | 63 | 261 | 14 | 116 | 5 | 900 | 60 |
| % of stories on environment | 3.0 | | 3.7 | | 4.7 | | 5.7 | | 7.1 | | 9.3 | | 5.4 | | 4.3 | | 6.7 | |

in a news bulletin. For six weeks the broadcasts on all 7 days of the week were logged, giving a total of 42 individual news broadcasts. Take as an example the first column, for *BBC*1 News ('all'). All of the broadcasts have at least one item, so there are 42 stories which took first place in their bulletin. It can be seen that in fact 42 bulletins had at least 4 items, since all have a fourth story. From then on the items diminish fairly fast. Twenty bulletins had 12 stories, and finally one bulletin had 17 stories. BBC2 mostly does not have so many stories per bulletin, but Doordarshan both in English and in Marathi, has many more items on average. This reflects the fact that the bulletins rely very little on visuals and location reporting: many of the items are simply read out and are quite brief. Some of the reasoning behind this is the attempt every night to give quite a wide coverage of national news – almost something for every region – which is larger and more complex than Europe. Some of it has to do with inadequate resources and old technology.

The second column for each channel shows how many stories there were about the environment and where in the order they came. On BBC1 no story came higher than fourth, and the average seems to be around ninth or tenth. For BBC2 no story comes higher than second, and the average would be about fourth. One story on ITV leads, but the average position is about sixth. On Channel 4 one story makes the lead, in Doordarshan three do so. In general it is probably fair to say that the stories in India are a little more likely to occur in any place in the bulletin, and are not necessarily pushed lower down the order, whereas in the UK the tendency is for that to happen. What Environmental stories do make the lead? As there are so few of them, we can consider them individually. In the UK they concern a conservationist and development squabble over the derelict Battersea electricity generating station in London (ITV), Russian nuclear reactors dumped in the sea off Japan (Channel 4), and the anti-nuclear protesters climbing over the wall of Buckingham Palace (Radio 4). Doordarshan included drought relief in Orissa, an Indo-Chinese treaty that included environmental co-operation amongst many other issues, and a cyclone in Kerala which killed 36 people. The AIR broadcasts included drought relief in Orissa, landslides in Tamil Nadu, and a statement by the Prime Minister that resettlement of oustees (as they are known) from the Narmada Dam would get top priority. (See Chapter 4 on the Narmada project.)

Since this first part of the study includes all broadcast news stories, it is feasible to ask to what extent the Indian and UK channels offer the same stories. The answer is that there are just a few stories which may get into both countries' bulletins. These include issues about nuclear arms reduction negotiated between the USA and Russia, and the occasional environmental story, such as forest fires in Australia. In both cases the amount of coverage of the stories was greater in the UK than in India, and the Australian story ran for more nights. This reflects a difference in news values between Indian and UK broadcasting channels. In the former case very few stories apart from political ones will develop over several days, whereas in the UK non-political stories may also develop over several days.

**The environmental stories**

The next step is to look at the Environmental stories more closely. First, all such stories are abstracted from the total data set. Then, for these stories, we can break down the word 'environment' into the same categories as were used for the newspaper analysis. The geographical names are the same as those just discussed above, and are not mentioned again here for the summary solo data in the top lines of Table 5.4.4.

The solo words which follow below the geographical regions show some interesting contrasts. The pessimist stories carrying 'malfunction' seem mostly restricted to the UK channels, whereas the optimist stories carrying 'mitigation' are much more evident on the Indian channels. 'Development' is a category which shows only for India. Out of the *c*. 250 Environmental stories, 34 carry the word 'development' and are shown with both the raw descriptors and their backcloth aggregation in Table 5.4.5. In other words, in broadcast news in India, the preponderance of development and mitigation over malfunction is clear, and much more so than in the Indian press. The reflects the government's control and philosophy, and it is left to the élite English-language press to swim against this tide. The optimist scenario is stronger on Indian TV than on India Radio, perhaps reflecting the different histories of the two (see Chapter 1).

The next figures worth noting are those that show a discrepancy between the greater amount of 'Political' coverage on the environment in the UK than in India. There appears to be no *a priori* reason why this should be the case, and it is possible that there is a difference between the 'antennae' of the data coders in India and the UK, yet inspection of the stories which aggregate to Politics does suggest specific reasons. There are several strands to consider here. First there is the question of who 'owns' the environment or, in broader terms, who is responsible for the environment. Next, there is the question of who lobbies the media gatekeepers, the history of protest action in each country, and finally the question of the way that protest groups interface with the media.

In the UK many of the current environmental news stories implicitly ask the question (and explicitly in the case of the 1996 *Sea Empress* oil tanker disaster in South Wales) of who is responsible – not just for a mess but also for regulating it. This has to be seen in the context of the privatisation of public services and the imposition of EU standards and legislation, resulting in confusion, in the public mind at least, over how (for example) water quality and river quality are to be protected, to what standards by which agencies, and therefore in the context of a wider political debate which is covertly about 'who owns the environment'. Public broadcasting in Britain is more distanced from government than in India, and independent broadcasting offers strong competition. Protest groups know that media coverage is important, and are skilled in tactics which provide good visual news material. Greenpeace has often led the way by taking dramatic high-seas action in little speed-boats to spoil Japanese whaling expeditions – or protesters in the path of motorways hang themselves from the tree tops or from high tetrapods.

In India Doordarshan is much closer to government, and although it reports stories about, for example, agitators against the Narmada Dam, much of the news on such an issue is reporting official pronouncements. The fact that government ultimately has responsibility for development and for the environment is not usually the subject of debate; on the issue of responsibility there is therefore less political debate. The history of public protest is different from that in Britain. This century it has been linked most closely with Gandhi, who had to sway public opinion before mass broadcasting and before higher literacy levels permitted a strong vernacular press. Gandhi rejected the injury and destruction of violent protest, which in any event in India has often ended in communal violence (and that is mostly unconcerned with media coverage). For him protest events had to be able to spread rapidly in a simple mouth-to-ear bush telegraph. He chose the idea of the fast until death, a political weapon he used with skill. But it does not relate well to visual

*Table 5.4.4* Percentage distribution of environmental subject matter in broadcast news

|       | BBC1 | BBC2 | ITV3 | CHAN4 | DD | RAD4 | RAD1 | AIR |
|-------|------|------|------|-------|----|------|------|-----|
| afr   |      |      | 1    | 6     |    | 4    |      |     |
| cis   |      | 23   |      | 32    | 2  |      |      | 1   |
| int   |      |      | 1    | 7     | 6  | 6    | 2    | 5   |
| nam   | 7    | 42   | 10   | 15    | 1  | 37   |      | 2   |
| nme   | 14   | 37   |      |       |    |      |      |     |
| ind   |      |      |      |       | 99 | 6    |      | 98  |
| oas   | 21   | 1    | 16   | 29    | 10 | 13   |      | 4   |
| uk    | 66   | 77   | 84   | 69    | 2  | 75   | 100  | 4   |
| oec   | 11   |      | 18   | 32    | 5  |      |      | 2   |
| sca   |      |      |      | 13    |    |      |      |     |
| air   | 10   |      |      | 6     | 2  |      | 20   |     |
| lan   | 8    |      |      |       | 13 |      | 20   | 4   |
| wat   | 58   | 13   | 17   | 58    | 44 | 7    | 22   | 54  |
| urb   | 6    | 11   | 25   | 11    | 2  | 22   |      | 1   |
| rur   |      | 23   | 3    |       | 39 | 6    | 42   | 4   |
| dis   | 20   | 12   | 42   | 2     | 20 | 28   |      | 53  |
| mal   | 36   |      | 0    | 19    | 1  |      | 20   | 2   |
| mit   | 11   |      | 9    |       | 31 |      |      | 20  |
| plu   | 38   | 37   | 9    | 39    | 6  | 4    | 2    | 2   |
| com   | 11   | 21   |      | 1     | 3  | 15   | 36   | 11  |
| con   | 6    | 21   | 22   | 10    | 3  | 15   | 59   | 0   |
| ecl   |      |      | 21   | 9     | 4  | 21   |      | 1   |
| ecn   |      | 34   | 28   |       | 17 | 6    |      | 11  |
| hea   |      | 37   | 11   | 8     | 11 |      | 20   | 1   |
| ins   |      | 21   | 3    |       | 13 |      | 42   |     |
| dev   |      |      |      |       | 36 |      |      | 16  |
| pol   | 64   | 67   | 27   | 22    | 26 | 60   | 19   | 17  |
| res   | 8    |      |      |       | 12 | 10   |      | 3   |
| tec   | 5    |      |      | 33    | 10 | 10   | 19   | 7   |
| val   |      |      | 38   | 11    | 18 | 56   |      | 12  |
| cis oas |    |      |      | 19    | 1  |      |      |     |
| cis uk  |    |      |      | 19    |    |      |      |     |
| cis oec |    |      |      | 32    |    |      |      |     |
| cis wat |    |      |      | 32    |    |      |      |     |
| cis rur |    | 23   |      |       |    |      |      |     |
| cis plu |    |      |      | 32    | 1  |      |      | 1   |
| cis ecn |    | 23   |      |       |    |      |      |     |
| cis pol |    | 23   |      |       |    |      |      |     |
| int ind |    |      |      |       | 6  |      |      | 2   |
| int tec |    |      |      |       | 0  |      | 6    | 2   |
| nam nme |    | 37   |      |       |    |      |      |     |
| nam uk  | 7  | 42   | 6    | 15    |    | 37   |      |     |
| nam dis |    |      | 4    |       |    |      |      |     |
| nam plu | 7  | 37   |      |       |    |      |      |     |
| nam hea |    | 37   |      |       |    |      |      |     |
| nam pol | 7  | 42   | 6    | 2     | 1  | 37   |      |     |
| nam val |    |      | 6    |       |    | 37   |      | 2   |
| nme uk  |    | 37   |      |       |    |      |      |     |

*Table 5.4.4* Continued

| | BBC1 | BBC2 | ITV3 | CHAN4 | DD | RAD4 | RAD1 | AIR |
|---|---|---|---|---|---|---|---|---|
| nme plu | | 37 | | | | | | |
| nme hea | | 37 | | | | | | |
| nme pol | 14 | 37 | | | | | | |
| ind oas | | | | | 10 | | | 1 |
| ind oec | | | | | 5 | | | 2 |
| ind lan | | | | | 13 | | | 4 |
| ind wat | | | | | 44 | | | 54 |
| ind rur | | | | | 39 | 6 | | 6 |
| ind dis | | | | | 20 | 6 | | 53 |
| ind mal | | | | | 1 | | | 2 |
| ind mit | | | | | 30 | | | 20 |
| ind plu | | | | | 6 | | | 1 |
| ind com | | | | | 3 | | | 11 |
| ind ecl | | | | | 4 | 6 | | 1 |
| ind ecn | | | | | 17 | | | 11 |
| ind hea | | | | | 11 | | | 1 |
| ind ins | | | | | 13 | | | |
| ind dev | | | | | 36 | | | 16 |
| ind pol | | | | | 25 | | | 17 |
| ind res | | | | | 9 | | | 3 |
| ind tec | | | | | 10 | | | 7 |
| ind val | | | | | 17 | 6 | | 11 |
| oas uk | | | 5 | 28 | | | | |
| oas oec | 5 | | | 19 | 2 | | | |
| oas wat | | | 5 | 19 | 1 | | | |
| oas dis | 8 | 1 | 16 | 1 | | 13 | | 1 |
| oas plu | | | | 19 | | | | 1 |
| oas ecl | | | 11 | 9 | 1 | | | |
| oas pol | | | | 9 | | | | 1 |
| oas tec | 5 | | | 9 | 2 | | | |
| uk oec | 6 | | 18 | 19 | 0 | | | 2 |
| uk wat | 44 | 13 | 17 | 46 | 0 | 7 | 22 | |
| uk urb | 6 | 11 | 25 | 11 | | 22 | | |
| uk dis | 12 | 11 | 27 | 0 | | 8 | | |
| uk mal | 15 | | 0 | 13 | | | 20 | |
| uk plu | 38 | 37 | 9 | 26 | | 4 | 2 | |
| uk com | 11 | 21 | | 1 | | 15 | 36 | |
| uk con | 6 | 21 | 21 | 10 | 1 | 15 | 59 | |
| uk ecl | | | 10 | 9 | 1 | 15 | | |
| uk ecn | | 11 | 28 | | 1 | 4 | | 2 |
| uk hea | | 37 | 11 | 8 | | | 20 | |
| uk ins | | 21 | 3 | | | 4 | 42 | |
| uk pol | 50 | 44 | 27 | 17 | | 60 | 19 | 4 |
| uk tec | | | | 22 | | 4 | 19 | 2 |
| uk val | | | 38 | 11 | 0 | 45 | | |
| oec wat | | | | 32 | 1 | | | |
| oec urb | 6 | | 18 | | | | | |
| oec plu | | | | 32 | | | | |
| oec con | 6 | | 18 | | | | | |
| oec ecn | | | 18 | | | | | |

*Table 5.4.4*  Continued

|  | BBC1 | BBC2 | ITV3 | CHAN4 | DD | RAD4 | RAD1 | AIR |
|---|---|---|---|---|---|---|---|---|
| oec pol | 6 |  | 18 |  | 4 |  |  | 2 |
| oec val |  |  | 18 |  | 1 |  |  | 0 |
| lan wat |  |  |  |  | 5 |  | 20 |  |
| lan rur |  |  |  |  | 11 |  |  | 1 |
| lan mit |  |  |  |  | 12 |  |  | 3 |
| lan dev |  |  |  |  | 7 |  |  |  |
| wat rur |  |  |  |  | 25 |  |  | 6 |
| wat dis | 12 | 11 | 17 | 0 | 9 | 7 |  | 40 |
| wat mal | 29 |  |  | 13 |  |  | 20 |  |
| wat mit | 7 |  | 3 |  | 10 |  |  | 4 |
| wat plu | 22 |  |  | 38 | 1 |  | 2 |  |
| wat ecn |  | 11 | 2 |  | 4 |  |  | 6 |
| wat hea |  |  |  | 7 | 5 |  | 20 |  |
| wat dev |  |  |  |  | 28 |  |  | 12 |
| wat pol | 46 | 2 |  | 14 | 4 |  | 2 | 5 |
| wat tec |  |  |  | 13 | 4 |  |  | 1 |
| urb con | 6 |  | 18 | 1 |  | 15 |  |  |
| urb ecn |  | 11 | 18 |  |  |  |  |  |
| urb pol | 6 |  | 18 | 1 |  | 15 |  | 1 |
| urb val |  |  | 25 | 11 | 0 | 7 |  |  |
| rur dis |  |  |  |  | 4 | 6 |  | 5 |
| rur mit |  |  |  |  | 13 |  |  | 1 |
| rur ecn |  | 23 |  |  | 9 |  |  |  |
| rur ins |  |  | 3 |  | 10 |  | 42 |  |
| rur dev |  |  |  |  | 26 |  |  | 1 |
| rur pol |  | 23 | 3 |  | 8 |  |  | 1 |
| rur tec |  |  |  |  | 6 |  |  |  |
| rur val |  |  | 3 |  | 7 | 6 |  |  |
| dis mit |  |  | 3 |  | 8 |  |  | 10 |
| dis ecn |  | 11 | 2 |  | 5 | 1 |  | 4 |
| dis pol |  |  |  |  | 4 |  |  | 7 |
| dis val |  |  |  |  | 3 | 13 |  | 2 |
| mal pol | 29 |  |  | 6 |  |  |  |  |
| mit plu | 11 |  |  |  | 5 |  |  | 1 |
| mit hea |  |  | 6 |  | 5 |  |  | 1 |
| mit dev |  |  |  |  | 9 |  |  | 4 |
| mit pol | 7 |  |  |  | 4 |  |  | 5 |
| mit val |  |  |  |  | 6 |  |  | 7 |
| plu hea |  | 37 | 4 | 1 |  |  |  | 1 |
| plu pol | 34 | 37 |  | 6 | 1 | 4 | 2 |  |
| com con | 6 | 21 |  | 1 |  | 15 | 17 |  |
| com ins |  | 21 |  |  |  |  |  |  |
| com pol | 11 |  |  | 1 |  | 15 | 17 |  |
| con ecl |  |  | 4 | 9 | 3 | 15 |  | 0 |
| con ecn |  |  | 18 |  | 1 |  |  |  |
| con ins |  | 21 | 3 |  |  |  | 42 |  |
| con pol | 6 |  | 21 | 1 | 1 | 15 | 17 |  |
| con val |  |  | 21 | 1 |  |  |  |  |
| ecn dev |  |  |  |  | 7 |  |  | 6 |
| ecn pol |  | 23 | 18 |  | 7 | 4 |  | 9 |

*Table 5.4.4* Continued

| | BBC1 | BBC2 | ITV3 | CHAN4 | DD | RAD4 | RAD1 | AIR |
|---|---|---|---|---|---|---|---|---|
| ecn val | | 18 | | 3 | | | | |
| hea dev | | | | 5 | | | | |
| hea pol | 37 | | 7 | 0 | | | | |
| ins dev | | | | 10 | | | | |
| ins pol | | 3 | | 5 | 4 | | | |
| ins res | | | | 6 | | | | |
| dev pol | | | | 7 | | | | 4 |
| dev tec | | | | 7 | | | | 1 |
| pol res | | | | 6 | | | | |
| pol tec | | | | 6 | 4 | | | 2 |
| pol val | | 27 | 1 | 5 | 37 | | | 4 |
| res tec | | | 12 | 4 | | | | 1 |
| Samples of triples follow | | | | | | | | |
| ind wat rur | | | | 25 | | | | 6 |
| ind wat dis | | | | 9 | | | | 40 |
| ind wat mit | | | | 10 | | | | 4 |
| ind wat com | | | | 1 | | | | 9 |
| ind wat ecn | | | | 4 | | | | 6 |
| ind wat dev | | | | 28 | | | | 12 |

media coverage – since at best the strategy can provide a barely changing and usually motionless body being attended by supporters during the month or so it takes to approach death. Given the scant use of visuals in Indian TV news, and possibly the lower penetration of TV, a better strategy has not so far been necessary. Sunderlal Bahaguna's fast in 1995 was mentioned in Chapter 4. Medha Patkar offered a variation on this theme which had better visual possibilities, when she pronounced she would sit and await drowning by the rising waters of the Narmada Dam. (The police removed her.) Although some journalists (of the English language press in India, see Chapter 4) suggested that the NGOs were getting very skilled at influencing the news agenda, there is little evidence that they have yet had much impact on politicising TV and radio environmental news. The point is also made by participants in the Indian focus groups (Chapter 6), who were impressed by how English people organised themselves to make public protests (e.g. against a supermarket on a greenfield site). This also fits with the analysis of the questionnaires of public opinion (Chapter 7) in India, which suggests that 'disorganisation' is seen to be one of the contemporary problems of India. In sum, the pretexts, the means and the achievements of politicising 'the environment' in Indian broadcast news are so far not as great as in Britain.

Within the double categories, it can be seen that Channel 4's coverage of CIS-EE+Pollution at 32 per cent is the same as its coverage of the CIS-EE (solo figure 32 per cent above). This means that all the coverage of the CIS-EE is about pollution – painting a somewhat grim picture of eastern Europe and the ex-Soviet Union, and mirroring the focus groups' observations (Chapter 6) that 'the environment' was a problem that did not happen in Britain so much as in these territories and in the Third World.

In the figures for the UK TV channels, the UK+Pollution ranges somewhere around 30 per cent, which is less than half the total coverage of Environment in

*Table 5.4.5*  Broadcast environmental news stories in India which carry 'development'

---

TI    Ratnagiri District Bulletin:Literacy Movement:One Village 100%
      Literate:Health Schemes For All:New Scheme Launched:Drinking
      Water:Irrigation Projects:Small Savi
wat:hea:ind:dev
TI    Report On Sindhudurg District:Water Management
      Programmes:Forestation:Kolhpur Bunds:In Ghats Water Storing:101
      Temporary And 10 Permanent Bunds:Horticulture Suc
wat:mit:dev:lan:ind:rur
TI    Report On Thana District:New Irrigation Projects:Drinking Water Problem In
      Many Villages:Shahpur Combined Forest Convention:Red Cross Unit Of
      District Arranged
wat:hea:ins:dev:ind:rur
RI    Ind:Finance Minister:Fund Allocation:Sardar Sarovar Project
ind:wat:ecn:dev
RI    Ind:Water Resources:Link:Regional Grid
ind:tec:wat:dev
TI    India:Narmada Dam Project:Environment:Govt Committed To Completion Of
      Project:World Bank Loan Declined:Reliance On Own Resources:Resettlement
      Of People Affected
int:ind:ecn:wat:dev:pol:val:rur
TI    India:Earth Day:Environment Protection A Mass Movement:Sustainable
      Development:Environment Minister Speaks:Asian Perspective On
      Environment
ind:pol:val:dev
TI    India:Water:Interlinking Of Rivers:India A Surplus Water
      Nation:Redistribution Of Water Resources
ind:wat:tec:dev
TI    Narmada Dam Project:Coming Two And A Half Years 170 Million Dollars
      For Project:Gas Turbines From Japan Received:Central Water Minister
      Informs Parliament About
tec:wat:oas:ind:dev:pol:oec
TI    Sangali District Reportage:Literacy Level On Increase:RBI Oks Loan For
      Cooperatives:Milk Schemes:Bio Gas Programmes:Farmers Made Aware Of
      Technology:Help By Sta
ind:tec:dev:res:ins:pol:ecn:rur
TI    Sholapur District Reportage:Consumer Courts:839 Cases Disposed:Ambedkar
      Seminar:Conference On Water Conservation:Small Medium & Large
      Dams:Benefits
ind:wat:dev:rur
TI    Ahmednagar District:Development Work In Full Swing:Water For Farmers At
      Low Cost:New Schemes:Opening Of Stadium
ind:dev:wat:ecn:rur
RI    Ind:Punjab:River Water Project:Law & Order Situation:Role Of Police &
      Military
ind:wat:dev
RI    Ind:Sardar Sarovar Dam:Funds To Be Raised:Gujarat Maharashtra Madhya
      Pradesh To Form A Joint Venture Company
ind:pol:ecn:wat:dev
TI    Bhandara District:Report Continued:Cooperatives From The State To Provide
      Seeds:Rainfall Satisfactory:Rice & Wheat Ploughing Finished In District
ind:dev:wat:ins:rur
TI    Bid District Report:Horticulture Projects:State Govt Promoted:Water
      Management Programmes:133 Nala Bunding works Completed During
      Summer
ind:dev:wat:pol:rur
TI    Bid District Report:Continued:Small Bunds:District Literacy Movement:41%
      Matriculation Result:4 Children In Merit List
ind:wat:dev:val:rur

---

*Table 5.4.5* Continued

TI    Buldhana District Report:New Low Power Transmitter In District:Large Area
      Under Agriculture For Kharif Crops:Rain Satisfactory:Social Forestry
      Drive:Saplings Fr
ind:dev:wat:rur
TI    Chandrapur District Report:Development Plan:New District Plan:Farmers
      Workshop:Railway Overbridge Completed:Coal Mines Planted:70 Lakh
      Plants:For National Envir
ind:lan:mit:dev
TI    Dhulia District Bulletin:Narmada Dam Project:Tribals Given Land:For
      Tourism-Many Places In District Developed:Satisfactory Rain:Workshop For
      Farmers Arranged
ind:dev:wat:mit:rur
TI    Jalana District Report:253 Small Bunds Completed:Water Management
      Efficiently:Literacy Drive:Student Participation:100% Target Decided:Tree
      Plantation Drive
ind:wat:lan:dev:mit:val:rur
TI    India:Environment:Sardar Sarovar (Narmada) Project:State Governments To
      Raise Finances To Build Dam:Public Issue Of Debenture
ind:wat:dev:ecn
TI    India:Konkan Railway Project:Environment:Economy:Opposition Not
      Environmental But Sentimental Says Oza Group:Railway Minister Says
      Konkan Railway Will Benefit
ind:dev:ecn:com
TI    Chandrapur District Report:Coal India To Build Sports Complex In The
      Region:Rural Sportsmen To Get Scholarships:Small Bund:Work For
      Tribals:Water Management Project
ind:wat:dev
TI    Dhulia District Report:Fisheries Projects For Tribals:Satpura Paper Pulp
      Factory:Paper From Molasses:Chemical Recovery Plant:Less Pollution
ind:plu:mit:wat:dev
RI    Ind:Haryana:Sutaj River Project:2 Crore Allocated:CM Message
ind:wat:dev:ecn
RI    Ind:Planning Commission:Clears New Irrigation Projects
ind:pol:wat:dev:rur
TI    Dhulia District Report:Cultural Programmes:Upper Tapti Canal Project
      Complete:Help In Solving Water Problem For District:Agriculture:New
      Methods & New Seeds
ind:wat:dev:tec:rur
TI    Jalgaon District Report:Social Forestry Projects On Rise:Water Shed
      Management Effectively Conducted By Farmers With Help Of Local Govt
ind:wat:dev:ins:lan:mit:rur
RI    Ind:PM Speech:CM's Meeting:Top Priority To Be Given To People Affected
      By Sardar Sarovar Project
ind:val:mit:dev
RI    Ind:Cabinet Approves:River Action Plan On 28 Stretches Of 17 Rivers:Lake
      Conservation Plan Also To Be Considered
ind:wat:dev
TI    Nanded District Reportage:Social Forestry Drive:Literacy
      Mission:Innoculation Programmes In The District By Local Council:Education
      Minister's Visit To District
ind:hea:lan:mit:dev:ins:rur
TI    India:Meeting Of Chief Ministers On Sardar Sarovar Dam:Construction Of
      Dams:Environment Protection:Rehabilitation:Status Reports:PM On Need To
      Monitor Dam
ind:dev:wat:val
TI    India:Cabinet Of Ministers Approves Action Plan On Water Rivers Lakes
      Etc:Conservation Plan For Country:Environment:Natural Resources:17 Rivers
      Covered
ind:wat:dev:pol

the UK. It is, however, significantly higher than the two Indian+Pollution figures of 6 per cent and 1 per cent for Doordarshan and AIR respectively. Note also that India+Mitigation has very high figures (30 per cent and 20 per cent for the two Indian channels), whereas in the case of the UK, UK+Mitigation does not occur. (This to some extent reflects the idea that in India 'development' is often 'mitigating', whereas in the UK further development (= unbridled economic growth) may be suspected of causing environmental deterioration.)

Amongst the double values, 'India+Water' ranks highest. The stories that underlie these data are not just about water scarcity and water supply programmes, but also about floods during the monsoon. The same two elements – water supply and floods – also feature in the UK quite highly. The former, as already noted above, is connected more with the ownership and obligation of the water service companies than the level of its provision (although this would probably not be true for the bulletins in the summer of 1995), but the latter had a rare equivalence with India in 1995. The six sample weeks included one week in which the cumulative impact of a wet winter caused major flooding in Chichester town centre.

The triples lower down show a curious disjunction (also visible in simpler data lines above) between the radio channel AIR and the TV channel Doordarshan 1 in India. This is the greater prevalence of disaster reporting on radio contrasted with the greater prevalence of 'development' on TV – see the two figures for India+Water+Development and India+Water+Disaster in each case. We simply surmise that this may reflect the different histories of the two media, with TV starting later and with a definite 'development' angle as part of the nation's grand 'development project'.

Table 5.4.6 disaggregates the Indian broadcast news by language. The two TV bulletins monitored were the TV National Evening News in English and the (Bombay-based) TV regional evening news in Marathi. The latter is not the equivalent of regional news in England following the main national news: there, both programmes are in the same language and equally accessible to the audience. Here the Marathi programmes may have aspirations to be more 'national', but as we learnt in earlier chapters, they do not have the resources. The radio news coded is the main evening English broadcast of AIR.

The first lines of the table suggest that the English 'environmental' TV news, though still heavily focused on India, does indeed have a more internationalist slant than the other two. Indeed, the AIR main news seems 'surprisingly' un-international – unless one presupposes that the 'environment' is not an international issue. The TV English newscasts are heavily dominated by Politics, whilst the regional services and the AIR newscasts are far less so, but both of them mention 'water' a lot – in the case of AIR this is in relation to Disaster (floods). The regional Marathi TV newscasts are clearly heavily involved in Water, but also and simultaneously in Development and Mitigation – putting things right – and are clearly much more 'rural' than the other two. None of this suggests clear simple differences across a language divide: indeed there seem to be two divides operating simultaneously, one of which may be language, but the other is between the older and more parochial radio and the newer and more international and political TV.

## 5.5 TV AND RADIO GENERAL (NON-NEWS) PROGRAMMES

For general TV and radio programmes, the coders were asked to make an entry for every programme listed in programme schedules, detailing its title, language, start and end time, and time and place of the programme setting, together with

*Table 5.4.6* Broadcast environmental news in India: percentage distribution of time by subject matter

| Total stories | DD English[1] 46 | DD Marathi[2] 63 | AIR English[3] 60 |
|---|---|---|---|
| cis | 5 | | 1 |
| int | 8 | 5 | 5 |
| nam | 2 | | 2 |
| ind | 98 | 99 | 98 |
| oas | 25 | 2 | 4 |
| uk | 4 | | 4 |
| oec | 10 | 3 | 2 |
| air | | 2 | |
| lan | 3 | 19 | 4 |
| wat | 23 | 56 | 54 |
| urb | 1 | 3 | 1 |
| rur | 11 | 54 | 6 |
| dis | 28 | 16 | 53 |
| mal | 3 | 0 | 2 |
| mit | 17 | 39 | 20 |
| plu | 2 | 8 | 2 |
| com | 5 | 1 | 11 |
| con | 9 | | 0 |
| ecl | 12 | | 1 |
| ecn | 25 | 12 | 11 |
| hea | 5 | 14 | 1 |
| ins | 4 | 18 | |
| dev | 13 | 50 | 16 |
| pol | 50 | 13 | 17 |
| res | 8 | 10 | 3 |
| tec | 8 | 11 | 7 |
| val | 21 | 16 | 12 |

Notes:   [1] 9 p.m. News
         [2] 7 p.m. News
         [3] Main Evening News

the traffic (i.e. programme 'form') details (Fig. 5.5.1). This enabled us to build a profile of the kinds of general broadcasting output, to see the extent to which the environmental output was different. The coders were asked to view or listen to any programmes which from schedules or other information they suspected might contain relevant environmental material. If such a programme included one or more relevant items or was in its entirety relevant, then the coders completed a second sheet giving similar information and also details about the subject matter (Fig. 5.5.2). Quite a number of these are brief items within magazine programmes like Radio 4's Today. There are only a few full-length films that take up environmental issues – in either fictional or factual terms. The items for which content (backcloth) data are collected are referred to here as 'monitored' programmes.

It is up to us to aggregate the backcloth descriptions into a relevant environmental category. The techniques are the same as used in the analysis of broadcast news and the newspapers. Remember that the classification system used is not a partition (in which a lower level item would belong to one and only one higher level category). The same channels have been recorded and analysed as for broadcast news. The UK channels were logged in London and the Indian channels logged in Pune. There are a few extra B² words to indicate the time the programme is set on or about:

bcp:    distant past dated BC
hpa:    historical past, AD 0 to 1945
rpa:    recent past: 1945–90
now:    anything post-1990
fut:    future

The top line of Table 5.5.1 shows the total number of hours for all programmes which have been logged, and the total hours of the items deemed to have had an environmental interest and which have therefore had their backcloth monitored and included in the data set. The percentages suggest that there is proportionately more on the UK channels than on the Indian channels: this could be a result of the different attitudes of coders in India and the UK, but we doubt whether that sufficiently explains the large difference. Coders in the UK were trained by Fraser and in India by Kumar, using the same manuals. The next point to notice about the top lines of Table 5.5.1 is that the average length of programmes is longer in the UK, although the difference between UK and Indian TV is not marked. The average length for AIR is much shorter. Also, for all channels the average length of a monitored item is significantly less than the average programme length: this suggests that in both countries these items are either complete programmes but shorter, or that they are single items within magazine-type programmes like Radio 4's Today or Science Now. More will be said about the length of programmes below.

The other lines of Table 5.5.1 depict traffic – the types of programme form, distributed by percentages. DD in India has a higher percentage of studio locations – exactly as suspected from casual viewing. The monitored environmental output is heavily oriented to Documentary and Discussion. Proportionately, the UK channels rely less on documentaries, even include a substantial drama element, and seem to avoid studio locations. Another clear distinction between the UK channels and the Indian channels, in both radio and TV, is the greater extent to which environmental output is associated with Series in the UK and the extent to which programmes in general are in Series of some sort. If this is the case, then it suggests that the process of defining a channel format and then exploiting parts of that format for particular messages may be more developed and more significant in the UK. If these vehicles for expression do not exist in India to the same extent, then perhaps what matters there is the overriding 'tone' of the channels, which in the case of DD is still set by government and the national goal in their own version of a Reithian aura.

Table 5.5.2 shows the backcloth: for time and place this is available for all programmes, but other details of backcloth are available only for the monitored programmes. As before, the distribution is shown as a percentage of the total hours for each column. The geographical focus for much of the material in gross terms reflects the kinds of concentrations seen before: the UK shows most about itself, and then some about North America and other OECD nations (these may well be imported programmes) but there are some interesting small divergencies: for most channels the percentage for Africa and Other Asia is higher amongst the environmental programmes than the general ones – and higher in every case than the percentage level in news broadcasts. This clearly suggests that both regions might not be conventionally newsworthy – in terms of politics or economics or conflict –

Table 5.5.1 Lengths and types of all and monitored programmes

| | BBC1 all | BBC1 mon | BBC2 all | BBC2 mon | ITV all | ITV mon | C4 all | C4 mon | DODA all | DODA mon | RAD1 all | RAD1 mon | RAD4 all | RAD4 mon | AIR all | AIR mon |
|---|---|---|---|---|---|---|---|---|---|---|---|---|---|---|---|---|
| Total in hrs and mins | 746.55 | 37.14 | 734.11 | 65.22 | 964.40 | 40.1 | 818.19 | 49.42 | 477.29 | 5.7 | 153.30 | 0.0 | 645.34 | 22.43 | 453.9 | 4.14 |
| Monitored as % | 5.0 | | 8.9 | | 4.1 | | 6.1 | | 1.1 | | | | 3.5 | | 0.9 | |
| No. of items | 1232 | 208 | 1289 | 189 | 1306 | 233 | 1168 | 139 | 1145 | 28 | 71 | | 1228 | 187 | 2110 | 38 |
| Av. mins | 36 | 11 | 34 | 21 | 44 | 10 | 42 | 21 | 25 | 11 | | | 32 | 7 | 13 | 7 |
| Fiction | 27 | 11 | 22 | 8 | 28 | 22 | 27 | 14 | 16 | 0 | 1 | 0 | 18 | 2 | 0 | 0 |
| Magazine | 7 | 2 | 1 | 1 | 18 | 12 | 7 | 0 | 5 | 0 | | 0 | 7 | 2 | 0 | 0 |
| Report | 35 | 40 | 29 | 37 | 37 | 41 | 19 | 26 | 23 | 24 | 25 | 0 | 46 | 70 | 33 | 10 |
| Discussion | 35 | 64 | 51 | 83 | 44 | 49 | 35 | 62 | 39 | 99 | 19 | 0 | 62 | 75 | 19 | 55 |
| Documentary | 6 | 33 | 10 | 38 | 4 | 13 | 9 | 38 | 13 | 86 | | 0 | 4 | 16 | 0 | 0 |
| Drama | 15 | 3 | 13 | 3 | 24 | 26 | 18 | 3 | 4 | 16 | | 0 | 15 | 5 | 2 | 0 |
| Education | 18 | 52 | 28 | 74 | 20 | 43 | 16 | 74 | 32 | 76 | | 0 | 20 | 47 | 36 | 96 |
| Cinema | 10 | 0 | 9 | 0 | 6 | 0 | 13 | 8 | 13 | 0 | | 0 | 0 | 0 | 0 | 0 |
| Tv movie | 2 | 0 | 6 | 3 | 2 | 0 | 4 | 0 | 0 | 0 | | 0 | 0 | 0 | 0 | 0 |
| Games/competition | 6 | 6 | 9 | 1 | 11 | 3 | 7 | 0 | 1 | 0 | | 0 | 1 | 0 | 0 | 0 |
| Music | 11 | 30 | 8 | 30 | 18 | 10 | 12 | 29 | 43 | 12 | 98 | 0 | 11 | 4 | 49 | 0 |
| Child audience | 13 | 21 | 6 | 10 | 12 | 18 | 14 | 10 | 5 | 0 | | 0 | 0 | 0 | 3 | 0 |
| Special audience | 9 | 19 | 13 | 16 | 15 | 17 | 10 | 11 | 13 | 26 | | 0 | 27 | 39 | 10 | 0 |
| Viewer participation | 8 | 8 | 1 | 6 | 16 | 11 | 9 | 14 | 1 | 0 | 59 | 0 | 10 | 10 | 2 | 0 |
| Animation | 7 | 8 | 3 | 4 | 14 | 13 | 13 | 7 | 2 | 0 | | 0 | 0 | 0 | 0 | 0 |
| Serial | 10 | 5 | 2 | 2 | 13 | 18 | 6 | 0 | 3 | 0 | | 0 | 9 | 7 | 1 | 0 |
| Series | 23 | 55 | 35 | 60 | 24 | 28 | 39 | 53 | 9 | 0 | 5 | 0 | 28 | 49 | 4 | 0 |
| Studio | 3 | 6 | 1 | 0 | 5 | 5 | 4 | 1 | 25 | 12 | | 0 | 3 | 9 | 0 | 0 |

**THE MASS MEDIA AND GLOBAL ENVIRONMENTAL LEARNING
OUTLINE PAGE**

T or R  | T |

Starting Time | 930705 |

Coder's Name | SUE PEARLMAN |

Channel | BBC 1 |

| | hour | min. | sec. | | | hour | min. | sec. | | | min. | sec. |
|---|---|---|---|---|---|---|---|---|---|---|---|---|
| Starting Time | 07 | 00 | 38 | End Time | | 09 | 04 | 53 | Adverts | | – | – |

New ✓ or Repeat ☐

Title | BREAKFAST NEWS |

Subtitle | |

Country/ies of origin | GB |

Company/ies of origin | BBC |

Original spoken language(s) | ENGLISH |   Dubbed into | |

Subtitled into | |   Voice over in | |

Traffic

| NEW | : | REP | : | WEA | : | ANC | : | ROR |
|---|---|---|---|---|---|---|---|---|
| EXP | : | MUS | : | GRA | : | | : | |
| | : | | : | | : | | : | |

Place | GB & WORLD |   Time | NOW |

*Figure 5.5.1*   Sample data sheet for general programmes

but are nevertheless environmentally interesting. In the case of both regions the material is shown to be associated with the idea of 'wildlife' and the 'natural world' – which for the outside West seem to be distinguishable from society at large (as in most Attenborough films). For India nearly everything is about India – in the case of the monitored environmental items it is 100 per cent – and virtually everything has been recorded as 'now' in time. The marginal historical spread of the UK programmes is not repeated.

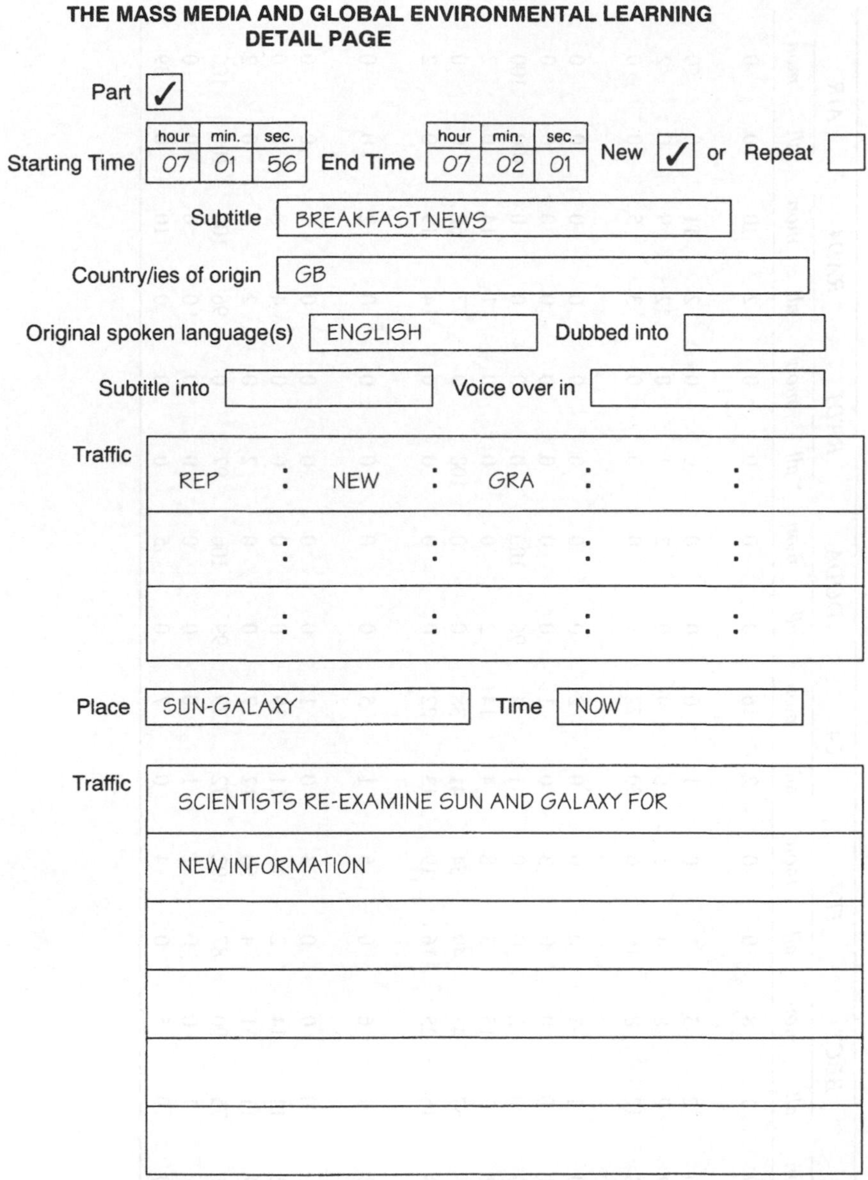

**THE MASS MEDIA AND GLOBAL ENVIRONMENTAL LEARNING**
**DETAIL PAGE**

Part [ ✓ ]

| | hour | min. | sec. | | | hour | min. | sec. | | | | |
|---|---|---|---|---|---|---|---|---|---|---|---|---|
| Starting Time | 07 | 01 | 56 | End Time | | 07 | 02 | 01 | New [ ✓ ] | or | Repeat [ ] | |

Subtitle  [ BREAKFAST NEWS ]

Country/ies of origin  [ GB ]

Original spoken language(s) [ ENGLISH ]   Dubbed into [ ]

Subtitle into [ ]   Voice over in [ ]

Traffic

| REP | : | NEW | : | GRA | : | | : |
|---|---|---|---|---|---|---|---|
| | : | | : | | : | | : |
| | : | | : | | : | | : |

Place [ SUN-GALAXY ]   Time [ NOW ]

Traffic

SCIENTISTS RE-EXAMINE SUN AND GALAXY FOR

NEW INFORMATION

*Figure 5.5.2*   Sample data sheet for environmental material

Comparison between the UK and India of some of the other values for the back-cloth data in Table 5.5.2 is a little difficult because of the small number (4 or 5 hours) of monitored output in the case of Indian TV and radio. Comparison between DD and AIR can be made cautiously: AIR seems to have both more emphasis on malfunction and mitigation and more on development. Table 5.5.3 shows the significant impact of language on the Indian broadcasts. The two lead national languages – Hindi and English – which have the greatest resources (see Chapters 1 and 4)

Table 5.5.2 Backcloth of all and monitored programmes

| | BBC1 | | BBC2 | | ITV | | C4 | | DODA | | RAD1 | | RAD4 | | AIR | |
|---|---|---|---|---|---|---|---|---|---|---|---|---|---|---|---|---|
| | all | mon | all | mon | all | mon | all | mon | all | mon | all | mon | all | mon | all | mon |
| Africa | 1 | 9 | 1 | 8 | 0 | 0 | 2 | 16 | 0 | 0 | 0 | 0 | 2 | 10 | 0 | 0 |
| CIS+E | | | | | | | | | | | | | | | | |
| Europe | 1 | 0 | 2 | 3 | 1 | 0 | 1 | 0 | 0 | 0 | 0 | 0 | 2 | 11 | 0 | 0 |
| International | 11 | 19 | 2 | 8 | 4 | 1 | 6 | 9 | 0 | 2 | 1 | 0 | 12 | 9 | 0 | 2 |
| N America | 13 | 12 | 12 | 8 | 16 | 6 | 30 | 22 | 1 | 0 | 0 | 0 | 3 | 5 | 0 | 0 |
| Near+Mid East | 0 | 0 | 1 | 3 | 0 | 0 | 0 | 2 | 0 | 0 | 0 | 0 | 0 | 0 | 0 | 0 |
| Polar | 0 | 1 | 0 | 0 | 0 | 3 | 0 | 1 | 0 | 0 | 0 | 0 | 0 | 0 | 0 | 0 |
| India | 0 | 0 | 0 | 2 | 0 | 0 | 1 | 4 | 96 | 100 | 0 | 0 | 0 | 0 | 100 | 100 |
| Other Asia | 3 | 12 | 3 | 15 | 3 | 8 | 4 | 14 | 2 | 0 | 0 | 0 | 1 | 14 | 0 | 2 |
| UK | 64 | 56 | 57 | 47 | 49 | 54 | 41 | 38 | 0 | 0 | 100 | 0 | 77 | 62 | 0 | 0 |
| OECD | 13 | 18 | 16 | 25 | 16 | 19 | 13 | 22 | 0 | 0 | 0 | 0 | 14 | 19 | 0 | 2 |
| S and Cent Amer | 0 | 2 | 1 | 6 | 0 | 5 | 1 | 5 | 0 | 0 | 0 | 0 | 0 | 2 | 0 | 0 |
| Prehistoric past | 0 | 1 | 0 | 0 | 0 | 3 | 0 | 1 | 0 | 0 | 0 | 0 | 0 | 1 | 0 | 0 |
| Historic past | 5 | 12 | 11 | 14 | 2 | 11 | 11 | 19 | 0 | 0 | 0 | 0 | 4 | 8 | 0 | 0 |
| Recent past | 8 | 9 | 11 | 11 | 4 | 50 | 12 | 17 | 0 | 0 | 2 | 0 | 2 | 0 | 0 | 2 |
| Now | 87 | 93 | 75 | 90 | 87 | 92 | 72 | 79 | 99 | 100 | 97 | 0 | 96 | 100 | 100 | 100 |
| Future | 1 | 1 | 1 | 0 | 0 | 2 | 1 | 0 | 0 | 0 | 0 | 0 | 0 | 6 | 0 | 0 |
| Air | 0 | 10 | 0 | 5 | 0 | 4 | 0 | 7 | 0 | 5 | 0 | 0 | 0 | 10 | 0 | 39 |

Table 5.5.2 Continued

| | BBC1 | | BBC2 | | ITV | | C4 | | DODA | | RAD1 | | RAD4 | | AIR | |
|---|---|---|---|---|---|---|---|---|---|---|---|---|---|---|---|---|
| | all | mon | all | mon | all | mon | all | mon | all | mon | all | mon | all | mon | all | mon |
| Land | 0 | 25 | 0 | 43 | 0 | 18 | 0 | 38 | 0 | 81 | 0 | 0 | 0 | 49 | 0 | 67 |
| Water | 0 | 25 | 0 | 19 | 0 | 3 | 0 | 19 | 0 | 45 | 0 | 0 | 0 | 17 | 0 | 8 |
| Urban | 0 | 4 | 0 | 6 | 0 | 3 | 0 | 6 | 0 | 32 | 0 | 0 | 0 | 6 | 0 | 0 |
| Rural | 0 | 14 | 0 | 13 | 0 | 8 | 0 | 9 | 0 | 40 | 0 | 0 | 0 | 24 | 0 | 0 |
| Disaster | 0 | 6 | 0 | 6 | 0 | 1 | 0 | 12 | 0 | 32 | 0 | 0 | 0 | 3 | 0 | 2 |
| Malfunction | 0 | 28 | 0 | 21 | 0 | 21 | 0 | 11 | 0 | 36 | 0 | 0 | 0 | 28 | 0 | 59 |
| Mitigation | 0 | 11 | 0 | 6 | 0 | 2 | 0 | 7 | 0 | 37 | 0 | 0 | 0 | 14 | 0 | 51 |
| Pollution | 0 | 15 | 0 | 9 | 0 | 3 | 0 | 12 | 0 | 28 | 0 | 0 | 0 | 15 | 0 | 22 |
| Communication | 0 | 8 | 0 | 7 | 0 | 1 | 0 | 8 | 0 | 11 | 0 | 0 | 0 | 12 | 0 | 0 |
| Conservation | 0 | 15 | 0 | 25 | 0 | 14 | 0 | 22 | 0 | 44 | 0 | 0 | 0 | 26 | 0 | 69 |
| Ecology | 0 | 42 | 0 | 35 | 0 | 21 | 0 | 40 | 0 | 45 | 0 | 0 | 0 | 29 | 0 | 70 |
| Economics | 0 | 27 | 0 | 45 | 0 | 26 | 0 | 31 | 0 | 64 | 0 | 0 | 0 | 55 | 0 | 22 |
| Health | 0 | 13 | 0 | 10 | 0 | 8 | 0 | 10 | 0 | 15 | 0 | 0 | 0 | 7 | 0 | 2 |
| Human | 0 | 19 | 0 | 21 | 0 | 14 | 0 | 23 | 0 | 36 | 0 | 0 | 0 | 20 | 0 | 2 |
| Institutions | 0 | 5 | 0 | 9 | 0 | 2 | 0 | 2 | 0 | 12 | 0 | 0 | 0 | 3 | 0 | 0 |
| Development | 0 | 6 | 0 | 1 | 0 | 0 | 0 | 2 | 0 | 24 | 0 | 0 | 0 | 7 | 0 | 49 |
| Politics | 0 | 8 | 0 | 18 | 0 | 10 | 0 | 10 | 0 | 24 | 0 | 0 | 0 | 13 | 0 | 29 |
| Resources | 0 | 5 | 0 | 9 | 0 | 8 | 0 | 11 | 0 | 10 | 0 | 0 | 0 | 6 | 0 | 2 |
| Technology | 0 | 23 | 0 | 10 | 0 | 7 | 0 | 9 | 0 | 0 | 0 | 0 | 0 | 10 | 0 | 0 |
| Values | 0 | 26 | 0 | 24 | 0 | 9 | 0 | 26 | 0 | 35 | 0 | 0 | 0 | 21 | 0 | 47 |

Table 5.5.3 Languages and broadcasting, sample weeks, sample locations[1]

| | BBC1 | | BBC2 | | ITV | | C4 | | DODA | | RAD1 | | RAD4 | | AIR | |
|---|---|---|---|---|---|---|---|---|---|---|---|---|---|---|---|---|
| | all | mon | all | mon | all | mon | all | mon | all | mon | all | mon | all | mon | all | mon |
| Total in hrs and mins | 746 55 | 37 14 | 734 11 | 65 22 | 964 40 | 40 31 | 818 19 | 49 42 | 387 37 | 4 11 | 153 30 | 0 | 645 34 | 22 43 | 453 9 | 4 14 |
| English | 100 | 100 | 100 | 100 | 100 | 100 | 100 | 100 | 38 | 24 | 100 | 0 | 100 | 100 | 16 | 0 |
| French | 0 | 5 | 1 | 1 | 1 | 0 | 1 | 1 | 0 | 0 | 0 | 0 | 0 | 0 | 0 | 0 |
| Hindi | 0 | 0 | 0 | 0 | 0 | 0 | 1 | 0 | 50 | 5 | 0 | 0 | 0 | 0 | 15 | 2 |
| Kanada | 0 | 0 | 0 | 0 | 0 | 0 | 0 | 0 | 0 | 0 | 0 | 0 | 0 | 0 | 1 | 0 |
| Marathi | 0 | 0 | 0 | 0 | 0 | 0 | 0 | 0 | 22 | 71 | 0 | 0 | 0 | 0 | 54 | 98 |
| Oriya | 0 | 0 | 0 | 0 | 0 | 0 | 0 | 0 | 1 | 0 | 0 | 0 | 0 | 0 | 0 | 0 |
| Other | 0 | 6 | 4 | 1 | 0 | 0 | 2 | 0 | 1 | 0 | 0 | 0 | 0 | 0 | 0 | 0 |
| Punjabi | 0 | 0 | 0 | 0 | 0 | 0 | 0 | 0 | 0 | 0 | 0 | 0 | 0 | 0 | 1 | 0 |
| Sanskrit | 0 | 0 | 0 | 0 | 0 | 0 | 0 | 0 | 0 | 0 | 0 | 0 | 0 | 0 | 2 | 0 |
| Spanish | 0 | 0 | 0 | 0 | 0 | 0 | 0 | 0 | 0 | 0 | 0 | 0 | 0 | 0 | 0 | 0 |
| Tamil | 0 | 0 | 0 | 0 | 0 | 0 | 0 | 0 | 1 | 0 | 0 | 0 | 0 | 0 | 0 | 0 |
| Zanskari | 0 | 0 | 0 | 1 | 0 | 0 | 0 | 0 | 0 | 0 | 0 | 0 | 0 | 0 | 0 | 0 |

Note: [1] Figures for individual languages are percentages of total broadcast time. Any language used at least once in a programme will have the whole programme time attributed to it. This leads to double counting and the sum of percentages may therefore exceed 100.

Table 5.5.4 Language and programme format in India: percentages

|  | DD: Mar | | DD: Hin | | DD: Eng | | AIR: Mar | | AIR Hin | | AIR: Eng | |
|---|---|---|---|---|---|---|---|---|---|---|---|---|
|  | all | mon | all | mon | all | mon | all | mon | all | mon | all | mon |
| Total in hrs and mins | 109 6 | 3 5 | 236 35 | 1 2 | 171 13 | 0 59 | 242 56 | 4 9 | 66 33 | 0 5 | 71 26 | 0 0 |
| Serious | 8 | 0 | 10 | 0 | 2 | 0 | 5 | 0 | 5 | 0 | 0 | 0 |
| Light | 10 | 0 | 11 | 0 | 1 | 0 | 9 | 0 | 2 | 0 | 0 | 0 |
| Fact | 6 | 1 | 3 | 5 | 6 | 49 | 0 | 0 | 0 | 0 | 0 | 0 |
| Fiction | 21 | 0 | 21 | 0 | 3 | 0 | 0 | 0 | 0 | 0 | 0 | 0 |
| Magazine | 4 | 0 | 4 | 0 | 7 | 0 | 0 | 0 | 0 | 0 | 0 | 0 |
| Report | 5 | 33 | 22 | 0 | 40 | 0 | 20 | 10 | 60 | 0 | 90 | 0 |
| Discussion | 31 | 98 | 37 | 87 | 66 | 100 | 24 | 56 | 21 | 0 | 27 | 0 |
| Documentary | 7 | 81 | 11 | 87 | 21 | 100 | 0 | 0 | 0 | 0 | 0 | 0 |
| Drama | 9 | 18 | 3 | 11 | 0 | 0 | 4 | 0 | 0 | 0 | 0 | 0 |
| Education | 39 | 84 | 31 | 79 | 44 | 51 | 56 | 96 | 28 | 100 | 19 | 0 |
| Cinema | 13 | 0 | 17 | 0 | 2 | 0 | 0 | 0 | 0 | 0 | 0 | 0 |
| TV movie | 0 | 0 | 1 | 0 | 0 | 0 | 0 | 0 | 0 | 0 | 0 | 0 |
| Games/competition | 0 | 0 | 0 | 0 | 1 | 0 | 0 | 0 | 0 | 0 | 0 | 0 |
| Music | 49 | 0 | 47 | 8 | 18 | 49 | 55 | 0 | 18 | 0 | 0 | 0 |
| Child audience | 15 | 0 | 2 | 0 | 4 | 0 | 5 | 0 | 0 | 0 | 0 | 0 |
| Special audience | 18 | 33 | 11 | 5 | 21 | 0 | 17 | 0 | 8 | 0 | 1 | 0 |
| Viewer participation | 0 | 0 | 1 | 0 | 1 | 0 | 4 | 0 | 3 | 0 | 0 | 0 |
| Animation | 1 | 3 | 1 | 13 | 5 | 0 | 0 | 0 | 0 | 0 | 0 | 0 |
| Serial | 3 | 0 | 4 | 0 | 1 | 0 | 1 | 0 | 0 | 0 | 0 | 0 |
| Series | 8 | 0 | 10 | 0 | 2 | 0 | 5 | 0 | 5 | 0 | 0 | 0 |
| Studio | 25 | 16 | 16 | 0 | 31 | 0 | 0 | 0 | 0 | 0 | 0 | 0 |

Table 5.5.5  Indian broadcasting by language and backcloth: percentages

| | DD: Mar | | DD: Hin | | DD: Eng | | AIR: Mar | | AIR Hin | | AIR: Eng | |
|---|---|---|---|---|---|---|---|---|---|---|---|---|
| | all | mon | all | mon | all | mon | all | mon | all | mon | all | mon |
| Total in hrs and mins | 109 6 | 3 5 | 236 35 | 1 2 | 171 13 | 0 59 | 242 56 | 4 9 | 66 33 | 0 5 | 71 26 | 0 0 |
| Africa | 0 | 0 | 0 | 0 | 0 | 0 | 0 | 0 | 0 | 0 | 0 | 0 |
| CIS+E Europe | 0 | 0 | 0 | 0 | 0 | 0 | 0 | 0 | 0 | 0 | 0 | 0 |
| International | 0 | 0 | 0 | 8 | 0 | 0 | 0 | 0 | 0 | 100 | 0 | 0 |
| N America | 0 | 0 | 0 | 0 | 3 | 0 | 0 | 0 | 0 | 0 | 0 | 0 |
| Near+Mid East | 0 | 0 | 0 | 0 | 0 | 0 | 0 | 0 | 0 | 0 | 0 | 0 |
| Polar | 0 | 0 | 0 | 0 | 0 | 0 | 0 | 0 | 0 | 0 | 0 | 0 |
| India | 99 | 100 | 98 | 100 | 93 | 100 | 100 | 100 | 100 | 100 | 99 | 0 |
| Other Asia | 0 | 0 | 1 | 0 | 1 | 0 | 0 | 0 | 0 | 100 | 0 | 0 |
| UK | 0 | 0 | 0 | 0 | 1 | 0 | 0 | 0 | 0 | 0 | 0 | 0 |
| OECD | 0 | 0 | 0 | 0 | 0 | 0 | 0 | 2 | 0 | 0 | 0 | 0 |
| S and Cent Amer | 0 | 0 | 0 | 0 | 0 | 0 | 0 | 0 | 0 | 0 | 0 | 0 |
| Prehistoric past | 0 | 0 | 0 | 0 | 0 | 0 | 0 | 0 | 0 | 0 | 0 | 0 |
| Historic Past | 0 | 0 | 0 | 0 | 0 | 0 | 0 | 0 | 0 | 0 | 0 | 0 |
| Recent Past | 0 | 0 | 0 | 0 | 0 | 0 | 0 | 2 | 0 | 0 | 0 | 0 |
| Now | 99 | 100 | 100 | 100 | 99 | 100 | 100 | 100 | 100 | 100 | 99 | 0 |
| Future | 0 | 0 | 0 | 0 | 0 | 0 | 0 | 0 | 0 | 0 | 0 | 0 |
| Air | 3 | 3 | 18 | 18 | 0 | 0 | 40 | 40 | 0 | 0 | 0 | 0 |
| Land | 92 | 92 | 76 | 76 | 51 | 51 | 66 | 66 | 100 | 100 | 0 | 0 |

Table 5.5.5 Continued

| | DD: Mar | | DD: Hin | | DD: Eng | | AIR: Mar | | AIR Hin | | AIR: Eng | |
|---|---|---|---|---|---|---|---|---|---|---|---|---|
| | all | mon | all | mon | all | mon | all | mon | all | mon | all | mon |
| Water | | 40 | | 53 | | 51 | | 8 | | 0 | | 0 |
| Urban | | 33 | | 13 | | 51 | | 0 | | 0 | | 0 |
| Rural | | 49 | | 0 | | 51 | | 0 | | 0 | | 0 |
| Global | | 3 | | 0 | | 0 | | 2 | | 100 | | 0 |
| Disaster | | 37 | | 45 | | 0 | | 0 | | 100 | | 0 |
| Malfunction | | 23 | | 11 | | 100 | | 58 | | 0 | | 0 |
| Mitigation | | 43 | | 52 | | 0 | | 52 | | 0 | | 0 |
| Pollution | | 6 | | 27 | | 100 | | 22 | | 0 | | 0 |
| Communication | | 3 | | 47 | | 0 | | 0 | | 0 | | 0 |
| Conservation | | 58 | | 45 | | 0 | | 70 | | 0 | | 0 |
| Ecology | | 35 | | 26 | | 100 | | 72 | | 0 | | 0 |
| Economics | | 68 | | 63 | | 51 | | 22 | | 0 | | 0 |
| Health | | 22 | | 11 | | 0 | | 2 | | 0 | | 0 |
| Human | | 38 | | 13 | | 51 | | 2 | | 0 | | 0 |
| Institutions | | 4 | | 45 | | 0 | | 0 | | 0 | | 0 |
| Development | | 24 | | 45 | | 0 | | 50 | | 0 | | 0 |
| Politics | | 19 | | 65 | | 0 | | 28 | | 100 | | 0 |
| Resources | | 16 | | 0 | | 0 | | 2 | | 0 | | 0 |
| Technology | | 0 | | 0 | | 0 | | 0 | | 0 | | 0 |
| Values | | 44 | | 44 | | 0 | | 46 | | 100 | | 0 |

*Table 5.5.6*  Backcloth of broadcast environmental programmes

|  | BBC1 | BBC2 | ITV3 | CHAN4 | DD | RAD4 | AIR |
|---|---|---|---|---|---|---|---|
| afr | 9 | 8 | 0 | 13 |  | 10 |  |
| cis |  | 3 |  |  |  | 12 |  |
| int | 19 | 8 | 1 | 10 | 2 | 8 | 2 |
| nam | 12 | 8 | 7 | 23 |  | 5 |  |
| nme | 0 | 3 |  | 2 |  |  |  |
| poa | 1 |  | 3 | 1 |  | 1 |  |
| ind |  | 2 |  | 4 | 100 |  | 100 |
| oas | 12 | 15 | 8 | 15 | 0 | 13 | 2 |
| uk | 56 | 47 | 56 | 40 |  | 63 |  |
| oec | 18 | 25 | 20 | 23 |  | 19 | 2 |
|  |  |  |  |  |  |  |  |
| sca | 2 | 6 | 1 | 5 |  | 2 |  |
| bcp | 1 |  | 4 | 1 |  | 1 |  |
| hpa | 11 | 14 | 11 | 20 |  | 7 |  |
| rpa | 9 | 11 | 51 | 14 |  |  | 2 |
| now | 93 | 90 | 91 | 83 | 100 | 100 | 100 |
| future | 1 | 0 | 2 |  |  | 6 |  |
|  |  |  |  |  |  |  |  |
| air | 10 | 5 | 5 | 7 | 5 | 10 | 39 |
| lan | 25 | 43 | 19 | 40 | 81 | 48 | 67 |
| wat | 25 | 19 | 3 | 20 | 45 | 17 | 8 |
| urb | 4 | 6 | 3 | 6 | 32 | 6 |  |
| rur | 14 | 13 | 9 | 10 | 40 | 23 |  |
| dis | 6 | 6 | 1 | 12 | 32 | 3 | 2 |
| mal | 28 | 21 | 19 | 12 | 36 | 27 | 59 |
| mit | 11 | 6 | 2 | 8 | 37 | 14 | 51 |
| plu | 15 | 9 | 3 | 13 | 28 | 15 | 22 |
| com | 8 | 7 | 1 | 8 | 11 | 13 |  |
| con | 15 | 25 | 14 | 23 | 44 | 26 | 69 |
| ecl | 42 | 35 | 18 | 42 | 45 | 29 | 70 |
| ecn | 27 | 45 | 24 | 32 | 64 | 53 | 22 |
| hea | 13 | 10 | 8 | 9 | 15 | 7 | 2 |
| hum | 18 | 21 | 11 | 22 | 36 | 19 | 2 |
| ins | 4 | 8 | 2 | 2 | 12 | 3 |  |
| dev | 6 | 1 | 0 | 2 | 24 | 7 | 49 |
| pol | 9 | 18 | 10 | 11 | 24 | 13 | 29 |
| res | 5 | 9 | 8 | 11 | 10 | 6 | 2 |
| tec | 24 | 10 | 7 | 10 |  | 11 |  |
| val | 26 | 22 | 9 | 28 | 35 | 21 | 47 |
|  |  |  |  |  |  |  |  |
| afr int | 2 | 2 |  | 0 |  | 1 |  |
| afr nme | 0 | 3 |  | 1 |  |  |  |
| afr oas | 0 | 1 |  | 2 |  | 4 |  |
| afr uk | 1 |  |  |  |  | 1 |  |
| afr oec |  | 1 |  | 2 |  | 4 |  |
| afr sca | 0 |  |  | 4 |  | 1 |  |
| afr hpa | 1 |  | 0 | 2 |  | 2 |  |
| afr rpa | 1 |  | 0 | 3 |  |  |  |
| afr now | 9 | 8 | 0 | 9 |  | 10 |  |
| afr air | 2 | 1 | 0 | 0 |  | 1 |  |

*Table 5.5.6* Continued

| | BBC1 | BBC2 | ITV3 | CHAN4 | DD | RAD4 | AIR |
|---|---|---|---|---|---|---|---|
| afr lan | 3 | 4 | | 7 | | 5 | |
| afr wat | 3 | 2 | | 4 | | 0 | |
| afr urb | 0 | 1 | | 1 | | | |
| afr rur | 1 | 1 | | 3 | | 2 | |
| afr dis | 1 | 2 | | 2 | | | |
| afr mal | 2 | 3 | | 5 | | 4 | |
| afr plu | 0 | | | 1 | | 0 | |
| afr con | 0 | 2 | | 3 | | 3 | |
| afr ecl | 4 | 3 | | 6 | | 6 | |
| afr ecn | 2 | 4 | | 5 | | 6 | |
| afr hea | 2 | 2 | | 2 | | | |
| afr hum | 1 | 2 | | 4 | | 4 | |
| afr ins | 1 | 2 | | | | 0 | |
| afr pol | 1 | | | 1 | | 1 | |
| afr tec | 4 | | | 0 | | 2 | |
| afr val | 2 | | | 3 | | 1 | |
| cis nam | | 1 | | | | 0 | |
| cis oec | | 2 | | | | 1 | |
| cis hpa | | 1 | | | | | |
| cis now | | 2 | | | | 12 | |
| cis future | | | | | | 5 | |
| cis air | | | | | | 3 | |
| cis rur | | | | | | 2 | |
| cis mal | | 1 | | | | 2 | |
| cis com | | 1 | | | | | |
| cis con | | 1 | | | | 1 | |
| cis ecl | | 1 | | | | 5 | |
| cis ecn | | 3 | | | | 1 | |
| cis hum | | | | | | 1 | |
| cis pol | | 2 | | | | 0 | |
| cis val | | | | | | 4 | |
| int nam | 2 | | | 1 | | 1 | |
| int ind | | | | | 2 | | 2 |
| int oas | 1 | 1 | | | | 2 | 2 |
| int uk | 5 | 3 | 1 | 3 | | 1 | |
| int oec | 6 | 2 | | 2 | | 0 | |
| int rpa | 2 | | | 3 | | | |
| int now | 17 | 8 | | 8 | 2 | 8 | 2 |
| int future | | | 1 | | | | |
| int air | 4 | 0 | 1 | 0 | | 1 | |
| int lan | 2 | 4 | | 1 | 2 | 2 | 2 |
| int wat | 3 | 3 | 1 | 3 | 2 | 0 | |
| int dis | 0 | 1 | | 2 | | | 2 |
| int mal | 4 | 2 | 1 | 0 | | 4 | 2 |
| int mit | | 0 | | 1 | | 1 | |
| int plu | 3 | 0 | 1 | 3 | | 1 | |
| int com | | 2 | | | | 1 | |
| int ecl | 6 | 5 | 0 | 4 | 2 | 2 | |
| int ecn | 1 | 1 | 1 | 3 | | 3 | |
| int hea | 2 | 3 | | 0 | | 1 | |

*Table 5.5.6*  Continued

|          | BBC1 | BBC2 | ITV3 | CHAN4 | DD | RAD4 | AIR |
|----------|------|------|------|-------|----|------|-----|
| int hum  | 2    | 1    |      | 2     |    | 4    |     |
| int ins  | 0    | 2    |      |       |    | 0    |     |
| int pol  | 1    | 1    |      | 1     |    | 3    | 2   |
| int res  | 1    | 0    |      | 2     |    |      |     |
| int tec  | 13   | 1    |      | 3     |    | 2    |     |
| int val  | 3    | 2    | 1    | 2     | 2  | 6    | 2   |
| nam oas  | 0    | 2    |      | 3     |    | 2    |     |
| nam uk   | 6    | 1    | 0    | 2     |    | 1    |     |
| nam oec  | 0    | 3    |      | 5     |    | 1    |     |
| nam hpa  | 0    | 4    | 3    | 2     |    |      |     |
| nam rpa  | 0    |      | 4    | 4     |    |      |     |
| nam now  | 11   | 4    | 7    | 23    |    | 5    |     |
| nam future | 1  |      |      |       |    |      |     |
| nam air  | 1    |      |      | 2     |    | 2    |     |
| nam lan  | 0    | 4    | 0    | 8     |    | 3    |     |
| nam wat  | 4    | 1    |      | 7     |    | 2    |     |
| nam urb  | 1    |      |      | 0     |    |      |     |
| nam rur  |      | 3    |      | 1     |    | 0    |     |
| nam dis  | 0    | 1    |      |       |    |      |     |
| nam mal  | 7    | 4    | 0    | 2     |    | 4    |     |
| nam mit  | 3    | 1    | 0    | 1     |    | 1    |     |
| nam plu  | 7    |      |      | 2     |    | 2    |     |
| nam con  | 3    | 0    |      | 5     |    | 1    |     |
| nam ecl  | 8    | 1    | 0    | 16    |    | 1    |     |
| nam ecn  | 3    | 6    |      | 4     |    | 1    |     |
| nam hea  | 1    |      |      | 2     |    | 1    |     |
| nam hum  | 3    | 3    | 0    | 3     |    | 3    |     |
| nam dev  | 2    |      |      |       |    |      |     |
| nam pol  | 1    | 4    | 0    |       |    | 1    |     |
| nam res  | 1    | 1    | 0    | 1     |    | 0    |     |
| nam tec  | 0    | 3    | 0    | 1     |    | 1    |     |
| nam val  | 5    | 2    | 0    | 4     |    | 2    |     |
| nme oec  |      | 0    |      | 1     |    |      |     |
| nme bcp  |      |      |      | 1     |    |      |     |
| nme hpa  |      |      |      | 1     |    |      |     |
| nme rpa  | 0    |      |      | 1     |    |      |     |
| nme now  | 0    | 3    |      | 1     |    |      |     |
| nme lan  |      | 3    |      | 2     |    |      |     |
| nme con  |      |      |      | 1     |    |      |     |
| nme ecl  |      | 0    |      | 1     |    |      |     |
| nme ecn  |      | 3    |      | 2     |    |      |     |
| poa now  | 1    |      | 3    | 1     |    | 1    |     |
| poa rur  |      |      | 2    |       |    |      |     |
| poa mal  |      |      | 3    | 0     |    | 0    |     |
| poa ecl  | 1    |      | 2    | 1     |    | 0    |     |
| poa val  | 0    |      | 2    | 0     |    |      |     |
| ind oas  |      | 2    |      | 4     |    |      | 2   |
| ind uk   |      |      |      | 4     |    |      |     |
| ind oec  |      | 1    |      |       |    |      | 2   |
| ind hpa  |      |      |      | 4     |    |      |     |

*Table 5.5.6* Continued

| | BBC1 | BBC2 | ITV3 | CHAN4 | DD | RAD4 | AIR |
|---|---|---|---|---|---|---|---|
| ind rpa | | | | | | | 2 |
| ind now | | 2 | | | 100 | | 100 |
| ind air | | | | | 5 | | 39 |
| ind lan | | 2 | | | 81 | | 67 |
| ind wat | | | | 4 | 45 | | 8 |
| ind urb | | | | | 32 | | |
| ind rur | | 1 | | | 40 | | |
| ind dis | | | | 4 | 32 | | 2 |
| ind mal | | 1 | | | 36 | | 59 |
| ind mit | | | | | 37 | | 51 |
| ind plu | | | | | 28 | | 22 |
| ind com | | | | | 11 | | |
| ind con | | | | | 44 | | 69 |
| ind ecl | | 1 | | | 45 | | 70 |
| ind ecn | | 1 | | | 64 | | 22 |
| ind hea | | | | 4 | 15 | | 2 |
| ind hum | | 1 | | | 36 | | 2 |
| ind ins | | | | | 12 | | |
| ind dev | | | | | 24 | | 49 |
| ind pol | | | | | 24 | | 29 |
| ind res | | | | | 10 | | 2 |
| ind val | | 2 | | | 35 | | 47 |
| oas uk | 3 | 1 | 4 | 5 | | 0 | |
| oas oec | 2 | 5 | 2 | 6 | | 5 | |
| oas sca | 0 | 1 | | 2 | | 1 | |
| oas hpa | | | 1 | 7 | | | |
| oas rpa | 2 | 4 | 5 | 3 | | | |
| oas now | 12 | 14 | 4 | 9 | 0 | 13 | 2 |
| oas air | 1 | | | | | 3 | |
| oas lan | 5 | 8 | 4 | 4 | 0 | 9 | 2 |
| oas wat | 5 | 3 | | 4 | 0 | 2 | |
| oas urb | | 1 | | | | | |
| oas rur | 4 | 3 | 1 | 2 | | 3 | |
| oas dis | | 2 | | 4 | | 0 | 2 |
| oas mal | 6 | 4 | 3 | 3 | 0 | 6 | 2 |
| oas mit | 1 | 2 | 0 | 2 | | 0 | |
| oas plu | 0 | | 1 | | | 2 | |
| oas com | 2 | | | | | 2 | |
| oas con | 1 | 1 | 0 | 5 | | 5 | |
| oas ecl | 6 | 3 | 4 | 6 | 0 | 5 | |
| oas ecn | 6 | 6 | 2 | 5 | | 8 | |
| oas hea | 4 | 2 | | 4 | | 1 | |
| oas hum | 7 | 4 | 2 | 3 | | 4 | |
| oas ins | 0 | 1 | | | | 0 | |
| oas dev | 1 | | 0 | 2 | | | |
| oas pol | 2 | 1 | 2 | 2 | | 1 | 2 |
| oas res | | 2 | | 2 | | 0 | |
| oas tec | 2 | 3 | | | | 1 | |
| oas val | 3 | 3 | 2 | 3 | | 2 | 2 |
| uk oec | 6 | 6 | 2 | 3 | | 11 | |

*Table 5.5.6*   Continued

| | BBC1 | BBC2 | ITV3 | CHAN4 | DD | RAD4 | AIR |
|---|---|---|---|---|---|---|---|
| uk sca | 1 | 2 | | 1 | | | |
| uk bcp | 1 | | | | | 1 | |
| uk hpa | 7 | 7 | 1 | 12 | | 4 | |
| uk rpa | 5 | 3 | 28 | 0 | | | |
| uk now | 53 | 43 | 54 | 30 | | 63 | |
| uk future | | | 1 | | | 1 | |
| uk air | 6 | 2 | 5 | 1 | | 3 | |
| uk lan | 17 | 19 | 13 | 16 | | 35 | |
| uk wat | 12 | 7 | 2 | 9 | | 14 | |
| uk urb | 2 | 3 | 2 | 5 | | 6 | |
| uk rur | 9 | 5 | 4 | 4 | | 17 | |
| uk dis | 3 | 0 | 1 | 10 | | 3 | |
| uk mal | 20 | 6 | 13 | 2 | | 15 | |
| uk mit | 9 | 4 | 2 | 1 | | 12 | |
| uk plu | 13 | 5 | 3 | 9 | | 11 | |
| uk com | 6 | 6 | 1 | 3 | | 9 | |
| uk con | 11 | 16 | 12 | 10 | | 19 | |
| uk ecl | 27 | 17 | 9 | 15 | | 17 | |
| uk ecn | 21 | 21 | 20 | 13 | | 40 | |
| uk hea | 8 | 6 | 8 | 7 | | 4 | |
| uk hum | 10 | 8 | 10 | 7 | | 7 | |
| uk ins | 3 | 5 | 1 | 2 | | 3 | |
| uk dev | 3 | 0 | 0 | | | 7 | |
| uk pol | 4 | 9 | 9 | 7 | | 10 | |
| uk res | 5 | 7 | 8 | 4 | | 5 | |
| uk tec | 6 | 4 | 7 | 5 | | 8 | |
| uk val | 18 | 14 | 5 | 12 | | 7 | |
| oec sca | | | 0 | 2 | | 1 | |
| oec bcp | | | 4 | 1 | | | |
| oec hpa | 5 | 2 | 6 | 2 | | 2 | |
| oec rpa | 2 | 4 | 6 | 2 | | | |
| oec now | 15 | 25 | 17 | 19 | | 19 | 2 |
| oec air | 1 | 1 | | 5 | | 1 | |
| oec lan | 5 | 8 | 5 | 13 | | 10 | |
| oec wat | 2 | 4 | 1 | 2 | | 0 | |
| oec urb | 0 | 1 | 1 | 1 | | 2 | |
| oec rur | 4 | 1 | 4 | 4 | | 5 | |
| oec dis | | 2 | 0 | | | | |
| oec mal | 3 | 4 | 3 | 9 | | 7 | |
| oec mit | 2 | 3 | 0 | 4 | | 1 | 2 |
| oec plu | 1 | 4 | | 3 | | 2 | |
| oec com | 0 | 3 | | 6 | | 4 | |
| oec con | 4 | 5 | 1 | 7 | | 7 | 2 |
| oec ecl | 6 | 7 | 7 | 11 | | 7 | |
| oec ecn | 5 | 12 | 4 | 13 | | 12 | 2 |
| oec hea | 2 | 3 | | 1 | | 1 | |
| oec hum | 0 | 4 | 0 | 6 | | 3 | |
| oec ins | | 4 | 1 | | | 1 | |
| oec dev | 1 | | | | | | |
| oec pol | 1 | 5 | 2 | 1 | | 5 | |

*Table 5.5.6* Continued

|         | BBC1 | BBC2 | ITV3 | CHAN4 | DD | RAD4 | AIR |
|---------|------|------|------|-------|----|------|-----|
| oec res | 1    | 4    | 0    | 3     |    | 0    | 2   |
| oec tec | 6    | 5    | 0    | 1     |    | 0    |     |
| oec val | 1    | 4    | 1    | 6     |    | 1    |     |
| sca hpa |      |      |      | 3     |    |      |     |
| sca now | 2    | 6    | 1    | 5     |    | 2    |     |
| sca lan | 1    | 4    | 1    | 2     |    | 1    |     |
| sca wat | 1    | 2    |      |       |    | 0    |     |
| sca rur |      | 0    |      | 2     |    | 0    |     |
| sca mal | 0    | 3    |      | 2     |    | 2    |     |
| sca plu | 0    | 1    |      |       |    |      |     |
| sca con | 0    | 2    | 0    | 2     |    |      |     |
| sca ecl | 1    | 4    | 1    | 2     |    | 1    |     |
| sca ecn | 1    | 3    | 0    | 2     |    | 2    |     |
| sca hum | 0    | 0    | 0    | 1     |    | 2    |     |
| sca pol | 1    | 2    |      |       |    | 1    |     |
| sca val | 1    | 2    |      | 1     |    | 1    |     |
| bcp hpa |      |      | 4    | 1     |    |      |     |
| bcp rpa |      |      |      | 1     |    |      |     |
| bcp now | 1    |      |      |       |    | 1    |     |
| bcp lan |      |      |      | 1     |    |      |     |
| bcp mal | 1    |      |      |       |    |      |     |
| bcp con |      |      |      | 1     |    |      |     |
| bcp ecl | 1    |      |      | 1     |    | 1    |     |
| bcp ecn |      |      |      | 1     |    |      |     |
| hpa rpa | 2    | 1    | 3    | 3     |    |      |     |
| hpa now | 8    | 6    | 7    | 8     |    | 7    |     |
| hpa lan | 2    | 5    | 1    | 7     |    | 1    |     |
| hpa wat |      | 1    |      | 4     |    |      |     |
| hpa urb |      | 0    |      | 2     |    |      |     |
| hpa rur | 2    | 3    |      | 3     |    |      |     |
| hpa dis | 1    | 1    |      | 4     |    |      |     |
| hpa mal | 1    | 3    | 1    | 0     |    | 1    |     |
| hpa com |      | 0    |      |       |    | 1    |     |
| hpa con |      | 2    |      | 2     |    |      |     |
| hpa ecl | 2    |      | 1    | 2     |    |      |     |
| hpa ecn | 2    | 8    | 1    | 8     |    | 1    |     |
| hpa hea | 2    | 2    |      | 4     |    |      |     |
| hpa hum | 1    | 4    | 1    | 3     |    | 1    |     |
| hpa pol | 0    | 4    | 1    | 2     |    |      |     |
| hpa res | 0    | 1    |      | 4     |    | 1    |     |
| hpa tec |      | 0    |      | 2     |    |      |     |
| hpa val | 0    | 1    |      | 4     |    |      |     |
| rpa now | 5    | 10   | 48   | 7     |    |      | 2   |
| rpa air | 1    |      |      |       |    |      |     |
| rpa lan | 5    | 2    | 1    | 2     |    |      |     |
| rpa wat | 0    |      |      | 2     |    |      |     |
| rpa urb |      | 1    |      |       |    |      |     |
| rpa rur | 5    | 1    |      |       |    |      |     |
| rpa mal | 3    |      | 1    |       |    |      |     |
| rpa mit | 1    |      |      | 0     |    |      |     |

*Table 5.5.6* Continued

| | BBC1 | BBC2 | ITV3 | CHAN4 | DD | RAD4 | AIR |
|---|---|---|---|---|---|---|---|
| rpa plu | 0 | | 1 | 0 | | | |
| rpa con | 2 | 2 | | 1 | | | |
| rpa ecl | 5 | | 1 | 2 | | | |
| rpa ecn | 5 | 2 | 1 | 4 | | | |
| rpa hea | 3 | 1 | | 0 | | | |
| rpa hum | 0 | 2 | 2 | 3 | | | |
| rpa dev | 1 | | | | | | |
| rpa pol | 3 | 1 | 1 | 3 | | | |
| rpa res | 0 | | | 2 | | | |
| rpa tec | 2 | | | 0 | | | |
| rpa val | 2 | 1 | 1 | 3 | | | |
| now future | | 0 | 1 | | | 6 | |
| now air | 7 | 4 | 3 | 7 | 5 | 10 | 39 |
| now lan | 23 | 38 | 19 | 32 | 81 | 48 | 67 |
| now wat | 24 | 18 | 2 | 14 | 45 | 16 | 8 |
| now urb | 4 | 6 | 3 | 6 | 32 | 6 | |
| now rur | 12 | 9 | 9 | 7 | 40 | 23 | |
| now dis | 4 | 6 | 1 | 8 | 32 | 3 | 2 |
| now mal | 27 | 18 | 17 | 11 | 36 | 27 | 59 |
| now mit | 10 | 6 | 2 | 7 | 37 | 14 | 51 |
| now plu | 14 | 9 | 1 | 12 | 28 | 15 | 22 |
| now com | 8 | 7 | 1 | 7 | 11 | 13 | |
| now con | 14 | 23 | 14 | 21 | 44 | 26 | 69 |
| now ecl | 38 | 34 | 18 | 40 | 45 | 29 | 70 |
| now ecn | 25 | 37 | 23 | 24 | 64 | 53 | 22 |
| now hea | 11 | 8 | 8 | 4 | 15 | 7 | 2 |
| now hum | 17 | 17 | 10 | 18 | 36 | 19 | 2 |
| now ins | 4 | 7 | 2 | 2 | 12 | 3 | |
| now dev | 5 | 1 | 0 | 2 | 24 | 7 | 49 |
| now pol | 7 | 14 | 10 | 6 | 24 | 13 | 29 |
| now res | 5 | 8 | 8 | 7 | 10 | 6 | 2 |
| now tec | 23 | 10 | 7 | 7 | | 11 | |
| now val | 25 | 21 | 7 | 23 | 35 | 21 | 47 |
| future air | | | 1 | | | | |
| future wat | | | 1 | | | | |
| future mal | | | 1 | | | | |
| future mit | 1 | | | | | | |
| future plu | 1 | | 1 | | | | |
| future con | 1 | 0 | | | | | |
| future ecn | | 0 | 1 | | | | |
| future dev | 1 | | | | | | |
| future pol | 1 | | | | | | |
| future val | 1 | | 1 | | | | |
| air lan | 3 | 2 | | 2 | | 3 | 20 |
| air wat | 2 | 1 | 1 | 2 | | 4 | |
| air urb | 0 | | | | 1 | 0 | |
| air dis | 1 | | | | | | |
| air mal | 6 | 1 | 5 | 4 | 2 | 4 | 26 |
| air mit | 2 | 1 | | 3 | 0 | 0 | 6 |
| air plu | 2 | 2 | 1 | 0 | 4 | 2 | 6 |

*Table 5.5.6*   Continued

| | BBC1 | BBC2 | ITV3 | CHAN4 | DD | RAD4 | AIR |
|---|---|---|---|---|---|---|---|
| air com | 0 | | | 3 | 4 | 0 | |
| air con | 2 | 2 | 3 | 2 | | 1 | 24 |
| air ecl | 8 | 3 | 3 | 4 | | 7 | 33 |
| air ecn | 3 | 2 | 1 | 2 | | 1 | 6 |
| air hea | 2 | 0 | | | 2 | 1 | |
| air hum | 4 | 1 | 0 | 5 | 3 | 4 | |
| air dev | 0 | 1 | | 0 | | | 6 |
| air pol | 0 | 0 | | | 2 | | |
| air res | 0 | | | 3 | | | |
| air tec | 1 | | | 0 | | 1 | |
| air val | 2 | 1 | 1 | 3 | 5 | 5 | |
| lan wat | 8 | 11 | 0 | 7 | 43 | 8 | 6 |
| lan urb | 0 | 6 | 3 | 6 | 30 | 2 | |
| lan rur | 14 | 12 | 6 | 9 | 40 | 21 | |
| lan dis | 2 | 4 | 0 | 2 | 32 | 1 | 2 |
| lan mal | 9 | 13 | 5 | 7 | 25 | 14 | 53 |
| lan mit | 2 | 3 | 1 | 1 | 32 | 8 | 35 |
| lan plu | 2 | 1 | 0 | 4 | 12 | 6 | 6 |
| lan com | 0 | 2 | 0 | 4 | 5 | 3 | |
| lan con | 6 | 16 | 8 | 17 | 43 | 20 | 39 |
| lan ecl | 14 | 16 | 8 | 22 | 36 | 10 | 57 |
| lan ecn | 18 | 33 | 16 | 24 | 62 | 42 | 12 |
| lan hea | 6 | 5 | 2 | 1 | 11 | 4 | 2 |
| lan hum | 7 | 13 | 3 | 6 | 33 | 11 | 2 |
| lan ins | 2 | 1 | 1 | 2 | 12 | 1 | |
| lan dev | 0 | 0 | 0 | 0 | 22 | 5 | 35 |
| lan pol | 4 | 8 | 3 | 2 | 22 | 6 | 19 |
| lan res | 1 | 3 | 0 | 4 | 10 | 4 | |
| lan tec | 5 | 3 | 0 | 4 | | 3 | |
| lan val | 9 | 7 | 3 | 9 | 29 | 5 | 37 |
| wat urb | 1 | 4 | | | 18 | | |
| wat rur | 3 | 3 | | 1 | 18 | 5 | |
| wat dis | 5 | 5 | 0 | 6 | 32 | 3 | |
| wat mal | 9 | 7 | 1 | 1 | 10 | 7 | 6 |
| wat mit | 3 | 2 | 0 | 1 | 23 | 1 | |
| wat plu | 5 | 1 | 1 | 2 | 11 | 3 | 8 |
| wat com | 3 | 1 | | | | | |
| wat con | 4 | 7 | 0 | 1 | 23 | 2 | |
| wat ecl | 16 | 10 | 2 | 14 | 21 | 6 | |
| wat ecn | 8 | 9 | 1 | 6 | 27 | 7 | 6 |
| wat hea | 3 | 4 | 0 | 6 | | | |
| wat hum | 6 | 7 | | 3 | 20 | 2 | |
| wat ins | 0 | 4 | 0 | 0 | 12 | | |
| wat dev | 1 | | | | 23 | 1 | |
| wat pol | 0 | 5 | 0 | 2 | 9 | 0 | 2 |
| wat res | 1 | 1 | 0 | 2 | | | |
| wat tec | 4 | 1 | 0 | 2 | | 3 | |
| wat val | 10 | 3 | 1 | 4 | 3 | 3 | 2 |
| urb rur | 0 | 4 | 1 | 2 | 30 | | |
| urb dis | 1 | 1 | | | 9 | | |

*Table 5.5.6* Continued

| | BBC1 | BBC2 | ITV3 | CHAN4 | DD | RAD4 | AIR |
|---|---|---|---|---|---|---|---|
| urb mal | 1 | 3 | | 1 | 21 | 2 | |
| urb mit | 2 | 0 | 0 | 1 | 1 | 2 | |
| urb plu | 2 | 1 | 0 | 2 | 12 | 1 | |
| urb com | 0 | 1 | | 1 | 3 | 1 | |
| urb con | 1 | 4 | 2 | 2 | | 5 | |
| urb ecl | 2 | 1 | 0 | 4 | 10 | 3 | |
| urb ecn | 1 | 6 | 3 | 3 | 31 | 3 | |
| urb hea | 0 | 1 | 2 | | 11 | 2 | |
| urb hum | 0 | 6 | | 4 | 30 | 0 | |
| urb ins | 0 | 1 | 1 | 2 | | | |
| urb dev | 0 | 0 | 0 | | | 2 | |
| urb pol | 0 | 3 | 1 | | 13 | 0 | |
| urb res | 1 | | | | | 2 | |
| urb val | 1 | 2 | 1 | 2 | 14 | 0 | |
| rur dis | 1 | 1 | | | 9 | 0 | |
| rur mal | 4 | 6 | 4 | 4 | 21 | 4 | |
| rur mit | 1 | 1 | | 1 | 10 | 1 | |
| rur plu | 0 | 0 | 0 | 1 | 10 | | |
| rur con | 4 | 3 | | 2 | 10 | 4 | |
| rur ecl | 5 | 4 | 6 | 4 | 20 | 6 | |
| rur ecn | 14 | 12 | 6 | 10 | 40 | 21 | |
| rur hea | 3 | 0 | 0 | 1 | 11 | 1 | |
| rur hum | 3 | 8 | 1 | 3 | 30 | 3 | |
| rur ins | 2 | 0 | 1 | | | 1 | |
| rur pol | 3 | 4 | 1 | | 11 | 4 | |
| rur res | 0 | 1 | | 2 | 10 | 0 | |
| rur tec | 2 | 0 | 0 | 1 | | 2 | |
| rur val | 4 | 3 | 5 | 0 | 11 | 4 | |
| dis mal | 4 | 5 | 0 | 0 | | 1 | 2 |
| dis mit | 0 | | | | 22 | 1 | |
| dis plu | | 1 | | 6 | 2 | | |
| dis con | 1 | 2 | 0 | | 22 | 1 | |
| dis ecl | 1 | 1 | 0 | 2 | 10 | | |
| dis ecn | 2 | 3 | | | 18 | 1 | |
| dis hea | | 2 | 0 | 4 | | | |
| dis hum | 1 | 3 | | 2 | 10 | | |
| dis ins | | 2 | 0 | | 12 | | |
| dis dev | | | | | 22 | 1 | |
| dis pol | | 1 | | 4 | 9 | | 2 |
| dis tec | 1 | 1 | 0 | 2 | | 1 | |
| dis val | 0 | 1 | 0 | 2 | 2 | | 2 |
| mal mit | 2 | 0 | 1 | 4 | 1 | 2 | 27 |
| mal plu | 8 | 2 | 2 | 1 | 20 | 7 | 12 |
| mal com | 3 | 2 | 0 | 4 | 2 | 7 | |
| mal con | 3 | 6 | 3 | 3 | 1 | 5 | 31 |
| mal ecl | 13 | 7 | 10 | 7 | 22 | 11 | 43 |
| mal ecn | 11 | 14 | 10 | 7 | 24 | 13 | 12 |
| mal hea | 6 | 3 | 6 | 1 | 11 | 3 | 2 |
| mal hum | 7 | 9 | 7 | 8 | 24 | 11 | 2 |
| mal ins | 2 | 1 | 0 | | | 0 | |

*Table* 5.5.6 Continued

| | BBC1 | BBC2 | ITV3 | CHAN4 | DD | RAD4 | AIR |
|---|---|---|---|---|---|---|---|
| mal dev | 2 | | 0 | 0 | 1 | 0 | 27 |
| mal pol | 4 | 7 | 8 | 1 | 11 | 6 | 19 |
| mal res | 1 | 1 | 6 | 3 | | 3 | |
| mal tec | 5 | 2 | 6 | 0 | | 4 | |
| mal val | 8 | 4 | 4 | 5 | 14 | 5 | 23 |
| mit plu | 5 | 1 | | 1 | 3 | 6 | 14 |
| mit com | 1 | 0 | | 3 | 3 | 2 | |
| mit con | 9 | 3 | 1 | 3 | 34 | 10 | 51 |
| mit ecl | 6 | 2 | 0 | 1 | 20 | 2 | 35 |
| mit ecn | 4 | 3 | 0 | 2 | 21 | 11 | 10 |
| mit hea | 0 | 0 | | | 2 | 2 | |
| mit hum | 1 | 1 | 0 | 6 | | 2 | |
| mit ins | 0 | 1 | 0 | | 12 | 0 | |
| mit dev | 6 | 1 | 0 | 2 | 24 | 7 | 49 |
| mit pol | 2 | 0 | 1 | | 9 | 2 | 25 |
| mit res | 3 | 4 | 2 | 5 | 10 | 3 | 2 |
| mit tec | 2 | 2 | 0 | | | 4 | |
| mit val | 5 | 0 | 0 | 5 | 2 | 3 | 43 |
| plu com | 3 | 4 | 0 | 3 | 7 | 5 | |
| plu con | 4 | 0 | | 1 | 0 | 3 | 14 |
| plu ecl | 6 | 4 | 1 | 3 | 20 | 3 | |
| plu ecn | 5 | 4 | 1 | 3 | 11 | 9 | 14 |
| plu hea | 4 | 2 | 0 | 2 | 4 | 2 | |
| plu hum | 4 | 3 | 1 | 3 | 13 | 4 | |
| plu ins | 1 | 2 | | | | | |
| plu dev | 1 | 0 | | | | 1 | 14 |
| plu pol | 3 | 1 | 1 | 4 | 2 | 3 | 10 |
| plu res | 2 | 1 | | 1 | | 0 | |
| plu tec | 1 | 1 | 0 | 3 | | 2 | |
| plu val | 8 | 2 | 3 | 6 | 8 | 6 | 10 |
| com con | 1 | 0 | 0 | 1 | | 1 | |
| com ecl | 3 | 3 | 0 | 3 | 4 | 2 | |
| com ecn | | 2 | | 3 | 4 | 8 | |
| com hea | 4 | 1 | | | 4 | | |
| com hum | 2 | 1 | | 3 | 3 | 2 | |
| com ins | 0 | 2 | 0 | | | 0 | |
| com pol | 1 | 1 | 1 | 1 | 4 | 4 | |
| com res | | | | 3 | | 2 | |
| com tec | 2 | | | | | 1 | |
| com val | 3 | 2 | 0 | 5 | 7 | 2 | |
| con ecl | 9 | 9 | 6 | 14 | 21 | 7 | 53 |
| con ecn | 6 | 16 | 8 | 12 | 30 | 19 | 14 |
| con hea | | 2 | 2 | 1 | | 2 | |
| con hum | 3 | 8 | 1 | 6 | | 4 | |
| con ins | 2 | 3 | 0 | 2 | 12 | 2 | |
| con dev | 6 | 1 | 0 | 2 | 24 | 7 | 49 |
| con pol | 2 | 5 | 1 | 0 | 9 | 2 | 25 |
| con res | 2 | 2 | 0 | 1 | 10 | 2 | 2 |
| con tec | 1 | 1 | 0 | | | 1 | |
| con val | 6 | 5 | 1 | 6 | 11 | 1 | 43 |

*Table 5.5.6*  Continued

| | BBC1 | BBC2 | ITV3 | CHAN4 | DD | RAD4 | AIR |
|---|---|---|---|---|---|---|---|
| ecl ecn | 10 | 11 | 6 | 14 | 22 | 10 | 6 |
| ecl hea | 8 | 4 | 0 | 3 | | 3 | 2 |
| ecl hum | 7 | 3 | 2 | 6 | 11 | 6 | |
| ecl ins | 1 | 5 | | 2 | | 2 | |
| ecl dev | 3 | 1 | 0 | 0 | 10 | 0 | 35 |
| ecl pol | 1 | 5 | 2 | 3 | 2 | 3 | 17 |
| ecl res | 2 | 1 | 0 | 3 | 10 | 1 | |
| ecl tec | 3 | 2 | 0 | 2 | | 5 | |
| ecl val | 15 | 8 | 5 | 9 | 5 | 10 | 35 |
| ecn hea | 4 | 5 | 8 | 2 | 11 | 3 | |
| ecn hum | 10 | 17 | 9 | 8 | 31 | 11 | |
| ecn ins | 3 | 3 | 2 | | 9 | 1 | |
| ecn dev | 0 | 0 | | | 10 | 6 | 8 |
| ecn pol | 5 | 14 | 8 | 5 | 21 | 8 | 8 |
| ecn res | 3 | 7 | 7 | 6 | 10 | 5 | 2 |
| ecn tec | 3 | 3 | 6 | 2 | | 4 | |
| ecn val | 7 | 7 | 5 | 7 | 24 | 6 | 8 |
| hea hum | 6 | 5 | 6 | 2 | 13 | 3 | |
| hea ins | 1 | 3 | | | | 1 | |
| hea dev | | | | | | 2 | |
| hea pol | 1 | 4 | 6 | | 12 | 0 | |
| hea res | 0 | | 6 | | | 2 | |
| hea tec | 2 | | 6 | 0 | | 1 | |
| hea val | 3 | 3 | | 3 | 14 | 2 | |
| hum ins | 1 | 1 | | 2 | | 1 | |
| hum dev | | 0 | | 2 | | 2 | |
| hum pol | 0 | 9 | 7 | 3 | 11 | 1 | |
| hum res | 1 | 1 | 6 | 5 | | 3 | |
| hum tec | 1 | 1 | 6 | 3 | | 2 | |
| hum val | 4 | 7 | 2 | 17 | 16 | 7 | |
| ins dev | 0 | | | | 12 | | |
| ins pol | 2 | 2 | 2 | 0 | 9 | 2 | |
| ins val | 1 | 2 | 1 | 2 | | 2 | |
| dev pol | 1 | | 0 | | 9 | 0 | 25 |
| dev res | 0 | | | | | 2 | |
| dev tec | 1 | | | | | 1 | |
| dev val | 3 | 0 | | 2 | 1 | 0 | 43 |
| pol res | 1 | 2 | 6 | 2 | | 1 | |
| pol tec | 3 | 0 | 6 | 0 | | 1 | |
| pol val | 4 | 6 | 2 | 4 | 15 | 4 | 29 |
| res tec | 1 | 2 | 6 | 1 | | 1 | |
| res val | 2 | 1 | 0 | 5 | | 1 | |
| tec val | 4 | 5 | 0 | 6 | | 5 | |

Table 5.5.7 Sample of environmental programmes, UK, ranked by duration

| T/R | Duration in mins | Title | Language | Start time | Backcloth/subject matter |
|---|---|---|---|---|---|
| T | 119 | Fighting for Gemma | Eng | 2000 | unk:now:sellafield:government:BNFL:prem death human:nuclear reprocessing:protest dra:doc:mus |
| T | 87 | Two Hands Hanging | Eng | 2320 | sca:now:habitat loss:Amazon:poverty:industry ser:non:edu:rep:dis:doc |
| T | 85 | Where's Wally? | Eng | 1545 | oecbcp  chi:ani:edu:fic |
| T | 70 | London Zoo | Eng | 1020 | unk:now: zoo:endangered species:GB:birds edu:rep:sei:non:spe:dis |
| R | 62 | Burleigh House | Eng | 1718 | unk:now:UK:farming:fishing rep:dis:spe:ser |
| T | 60 | Fields Of Dreams | Eng | 2300 | sca:afrnow doc:dis:non:rep:edu:sei |
| T | 60 | New Nightmares: Part 2: Nature S | Eng | 2101 | unk:int:now:intergenerational equity:genetic engineering: pollution:disaster edu:non:doc:ani:pat |
| T | 59 | The Last Kingdom Of The Elephant | Eng | 1700 | afr:now:rare fauna:drought:fresh water habitat doc:dis:edu:ser:sei:newnew |
| T | 58 | Tree Police | Eng | 1900 | unk:now:forestry:city:volunteer action:conservation:values:NGO:UK doc:non:sei:edu:ser |
| T | 57 | Wolves With Timothy Dalton | Eng | 2100 | poa:now:Arctic:endangered species:wilderness ser:edu:newnew:doc:sei:spe |
| T | 57 | The Guerrilla Base Of The King | Eng | 1901 | unk:hpa:land:scientific research:lifestyles:land use doc:spe:edu:sei:non:dis |
| T | 56 | Wild River No More | Eng | 2002 | nam:now:US:aquatic life:wildlife ser:dis:edu:non:sei:mus |
| T | 55 | Call Of The Wild | Eng | 1415 | nam:now dra:ser:sei:edu:fic |

Table 5.5.7  Continued

| T/R | Duration in mins | Title | Language | Start time | Backcloth/subject matter |
|---|---|---|---|---|---|
| T | 55 | Hell In The Pacific | Eng | 2100 | oas:hpa:rpamining:Australia:human rights rep:dis:non |
| T | 54 | Secrets Of The Billabong | Eng | 931 | oec:nam:now:Australia:wild life:habitat edu:mus:non:doc |
| T | 53 | Empires Of Oil | Eng | 1951 | hpa:industry:CIS:US:politics doc:ser:dis:non:edu |
| T | 53 | Wild Dogs: A Tale Of Two Sisters | Eng | 1958 | afr:now:wildlife:Africa doc:mus:dis:non:edu |
| T | 50 | Horizon | Eng | 1711 | afr:nmenow Egypt:building edu:spe:dis:sei:non:doc:mus |
| T | 50 | The New Alchemists | Eng | 2000 | unk:oas:nam:now:energy conservation:building:Japan:scientific research doc:rep:dis:spe:non:edu |
| T | 50 | Living With Technology: Energy | Eng | 730 | unk:now:England:Sweden:energy:Netherlands:finance:politics :industry rep:edu:sei:newnew |
| T | 50 | Living With Drought | Eng | 1410 | afr:now:Niger:drought:deforestation:urban migration:NGO:soil erosion:farming:education:children:env management:env awareness rep:doc:non:sei:edu:spe:dis |
| T | 32 | Miracle At Tendaho | Eng | 1555 | afr:rpa:irrigation:Ethiopia:protest:indigenes:water doc:dis:non:sei:edu |
| T | 30 | The Carbon Cycle | Eng | 1200 | unk:afr:now:rainforest:greenhouse effect:food:science doc:rep:edu |
| T | 30 | Countryfile | Eng | 1230 | unk:now:UK:env awareness:land:forests:water rep:edu |
| R | 29 | Litmus Test | Eng | 1225 | unk:now:marine-life:endangered-species:flora newnew:ser:lig |

Table 5.5.7 Continued

| T/R | Duration in mins | Title | Language | Start time | Backcloth/subject matter |
|---|---|---|---|---|---|
| R | 29 | Gardeners' Question Time | Eng | 1400 | unk:now:gardening:UK stu:sei:lig:ser:newnew:edu:pat |
| T | 29 | Once Upon A Time: Space | Eng | 955 | nam:now:space:science education chi:ani:non:fic:ser:lig |
| T | 29 | Living In A Material World | Eng | 1936 | glo:now:evolution:industry:building dis:doc:ser:non |
| T | 29 | My Wilderness Reprieved | Eng | 1130 | unk:now:UK:wildlife:nature:moral values doc:sel:dis |
| T | 29 | Alligator Hole | Eng | 2030 | nam:now:US:aquatic life:endangered species doc:ser:non:dis |
| R | 29 | Discursive Excursions | Eng | 2300 | afrnow dis:edu:non |
| R | 29 | The Parts | Eng | 1901 | unk:now:industry:env awareness:technology:waste disposal: UK:costs:cleaning rep:edu:dis:spe |
| T | 28 | Backstreet Bandits | Eng | 2000 | nam:now Canada:animals:city doc:newnew:sei:ser:edu:dis:mus |
| T | 28 | Emmerdale | Eng | 1859 | oecnow fic:sel:dra |
| T | 28 | Wildlife Showcase | Eng | 2030 | afr:now:endangered species:wildlife:Africa dis:doc:ser:non |
| T | 28 | Flowers To Seed | Eng | 1557 | unk:now:chemical pollution:city:plant life:transport non:ser:edu:dis:pat |
| T | 28 | Arctic Nursery | Eng | 1200 | nam:now:arctic:animals:evolution ser:non:dis:edu |
| T | 28 | Spain On A Plate | Eng | 1130 | oec:now:Spain:UK:gardening:organic farming ser:non:edu |

Table 5.5.7 Continued

| T/R | Duration in mins | Title | Language | Start time | Backcloth/subject matter |
|---|---|---|---|---|---|
| T | 27 | Oasis | Eng | 1640 | oec:now:farming:city:local government:building:values:env awareness dra:sei:edu:fic:sel:chi:newnew |
| T | 26 | Captain Planet | Eng | 800 | unk:fut:Ozone:water pollution:air pollution:greenhouse effect:env awareness:consumerism ani:ser:chi |
| T | 20 | World Studies: The Politics Of | Eng | 1018 | nam:afrnowfarming:poverty:food:economics: dis:sei:edu:non:pat |
| T | 20 | Science In Focus: Exemplars | Eng | 1059 | unk:now:industry newnew:edu:sei:non:mus:dis |
| T | 19 | The Geography Programme | Eng | 1209 | unk:hpa:education:Wales:mining:industry rep:doc:newnew:edu:dis:sei:sel |
| T | 19 | History File | Eng | 1205 | unk:hpa:industry:lifestyles:health ser:sei:dra:dis:spe:mus:ani |
| T | 19 | The Geography Programme | Eng | 1205 | unk:hpa:Wales:industry:land:mining ser:sei:newnew:edu:dis:mus |
| T | 19 | Science Challenge | Eng | 1055 | unk:now:children:energynewnew:dra:lig:sei:ser:edu:ani:non |
| T | 19 | Landmarks: Coping With Climate | Eng | 1340 | unk:intnowdrought:rainforest:desertification:international:evolution :sei:ser:mus:edu:rep:dis |
| T | 10 | Stop Look Listen: Hedgehogs | Eng | 1117 | unk:now:transport:fauna:GB:vehicles:conservation:habitat dis:edu:non:ser:sei:newnew |
| R | 10 | Michael Parkinson | Eng | 1246 | unk:now:blood sports:National Trust:protest dis:newnew:sei:pat |
| T | 10 | Food For Thought | Eng | 1950 | oec:now:health:organic farming:UK:disease dis:edu:sei:non |
| T | 10 | Maurice Mosquito's & A Nasty C | Eng | 2147 | oec:afr:int:rpa:Mediterranean:disease:water:insects:pesticides:food pollution non:ser:lig:dis:edu |

Table 5.5.7 Continued

| T/R | Duration in mins | Title | Language | Start time | Backcloth/subject matter |
|-----|------------------|-------|----------|-----------|--------------------------|
| R | 10 | Today | Eng | 632 | unk:now:UK:floods:env scheme:technology rep:sei |
| R | 10 | Letter From America | Eng | 919 | nam:now:US:tourism:parks:forest fire :ser:sei:dis |
| T | 10 | Countryfile News | Eng | 1250 | oec:now:farming:organic farming:UK:recycling:government:env legislation:finance rep:non |
| R | 9 | Costing The Earth | Eng | 1936 | sca:now:deforestation:indigenes:displaced persons:mining:South America edu:ser:sei:dis:doc:rep |
| R | 9 | Today | Eng | 810 | unk:now:Channel Tunnel:finance:land degradation:politics :rep:dis |
| T | 8 | This Morning | Eng | 1152 | unk:now:gardening:UK edu:mus |
| T | 8 | Your World | Eng | 1119 | now:rainforest:animals:plant life:water mus:newnew:dis:pat:ser:non |
| R | 8 | Science Now | Eng | 2004 | unk:now:animals:marine life:science education edu:dis:rep:sei:newnew:ser:spe |
| R | 8 | You & Yours | Eng | 1203 | unk:now:animal rights:food:farming:UK:NGO rep:dis:edu:sei |
| R | 8 | Any Questions | Eng | 1319 | unk:now:Channel Tunnel:transport:land degradation stu:spe::sei:ser:edu:pat |
| T | 8 | Blue Peter | Eng | 1706 | unk:now:UK:floods:damage chi:sel:non:mag:rep:mus:edu:dis:gam |
| T | 8 | Blue Peter | Eng | 1722 | afr:now:organic farming:Africa:water:drought:wildlife:animals domestic chi:sel:rep:mus:edu:dis:gam |

Table 5.5.7 Continued

| T/R | Duration in mins | Title | Language | Start time | Backcloth/subject matter |
|---|---|---|---|---|---|
| T | 7 | New Yogi Bear Show | Eng | 1625 | nam:now USA:animals ani:chi:non:fic:mus:ser:lig |
| T | 7 | Countryfile | Eng | 1247 | unk:now: animal rights:farming rep:dis:sei:edu |
| R | 7 | Natural History Programme | Eng | 1131 | unk:afr:now West-Africa:zoos:forests:env-management rep:dis:ser:sei |
| R | 7 | Tea Junction | Eng | 1735 | unk:now:tourism:parks:London:vehicle pollution:birds dis:sei:ser:lig:spe |
| T | 7 | Garden Club | Eng | 2120 | oec:now:gardening:UK :ser:sei:edu:dis |
| R | 4 | Today | Eng | 723 | unk:now:fresh water habitat:UK:industry:pollution rep |
| T | 4 | Blue Peter Compilation | Eng | 1726 | sca:now:wildlife:aquatic life:Antarctica:env awareness:conservation doc:rep:dis:chi |
| T | 4 | Newsroom South East | Eng | 1839 | unk:now:noise pollution:deforestation:UK:habitat loss:transport:protest rep:dis |
| R | 4 | Going Places | Eng | 1855 | unk:now:UK:transport:env scheme:vehicle pollution:city rep:dis:sei:mag |
| T | 3 | Party Political Comment | Eng | 1954 | unk:now:privatisation:forests:politics:conservation dis:edu:sei |
| R | 2 | Pick Of The Week | Eng | 1133 | unk:now:gardening:UK:health dis |

Table 5.5.8 Sample of environmental programmes, India, ranked by duration

| T/R | Duration in mins | Title | Language | Start time | Backcloth/subject matter |
|---|---|---|---|---|---|
| R | 44 | Forestry | Mar | 0716 | ind:now:forests:deforestation:values:endangered species: afforestation:env scheme:government:India |
| T | 35 | Documentary On Environment | Mar | 1800 | ind:now:urban migration:famine:poverty:government:land degradation:India:moral values |
| T | 30 | Programme On Social Forestry | Mar | 1830 | ind:now:sustainable forestry:India:moral values |
| T | 30 | Amachi Mati Amachi Manase | Mar | 1830 | ind:now:organic farming:Renewable energy:vermiculture |
| T | 30 | Vikas Gawacha | Mar | 1800 | ind:now:Pune:drought:water:env scheme:forests |
| T | 30 | Doc On Environment | Eng | 2245 | ind:now:deforestation:water pollution:urban migration:industrial pollution:forests |
| T | 29 | Nature & Man | Eng | 1900 | ind:now:damage:Nature:evolution:pollution |
| T | 28 | Tihari Dam Project | Hin | 0737 | ind:now:Dam:India:env scheme:drought:NGO:protest |
| T | 26 | Campaign Against Grazing & Cut | Mar | 1903 | ind:now:India:urban migration:drought:farming:water |
| R | 15 | Navneet: News | Mar | 1815 | ind:now:water shortage:soil erosion:industry:chemical pollution |
| R | 15 | Substitute For CFC | Mar | 0745 | ind:now:CFC:ozone depletion:env scheme |
| R | 10 | Preservation Of Birds | Mar | 1815 | ind:now:deforestation:birds:sustainable forestry:conservation |
| T | 8 | | Mar | 1950 | ind:now:Maharashtra:drought:env scheme:water:NGO |
| R | 5 | Environmental Programme | Mar | 0700 | ind:now:industry:deforestation:habitat loss:global warming |
| R | 5 | Himalaya | Hin | 0700 | ind:now:Himalayas:international politics:spiritual values:natural disaster:soil erosion |
| R | 5 | Environment | Mar | 0700 | ind:now:population:deforestation:Maharashtra |
| R | 5 | Science News | Mar | 0700 | ind:now:Japan:renewable energy:research investment |
| R | 5 | Bhimashankar | Mar | 0700 | ind:now:health:forests:soil erosion |
| R | 5 | Bhimashankar | Mar | 0700 | ind:now: |

Table 5.5.8  Continued

| T/R | Duration in mins | Title | Language | Start time | Backcloth/subject matter |
|---|---|---|---|---|---|
| T | 5 | Documentary | Hin | 1925 | ind:now:env awareness:media:government:India:plant life |
| T | 5 | Dependence | Hin | 1625 | ind:now:education:env awareness:plant life:water:forests:international |
| T | 5 | Environmental Documentary | Mar | 1720 | ind:now:children:education:env awareness:India:global warming |
| R | 5 | Understanding Jungles | Mar | 0700 | ind:now:afforestation:animal rights:conservation:India |
| R | 5 | Understanding Jungles | Mar | 0700 | ind:now:afforestation:animal rights:conservation:India |
| R | 5 | Understanding Jungles | Mar | 0700 | ind:now:afforestation:animal rights:conservation:India |
| T | 5 | Watershed Development | Mar | 1925 | ind:now:water:env scheme:Maharashtra |
| R | 5 | Reading The Book Ranvata (Road | Mar | 0700 | ind:now:animal rights:deforestation:env scheme |
| R | 5 | Reading The Book Ranvata (Road | Mar | 0700 | ind:now:animal rights:deforestation:env scheme |
| R | 5 | Ecomark | Mar | 0700 | ind:now:India:water pollution:chemical pollution:lifestyles:government |
| R | 5 | Ecomark Or Environment Friend | Mar | 0700 | ind:now:consumerism:India:government:env scheme:pollution control |
| R | 5 | Ecomark Or Environment Friend | Mar | 0700 | ind:now:consumerism:env awareness:env scheme:government:pollution control:India |
| R | 5 | Ecomark Or Environment Friend | Mar | 0700 | ind:now:India:pollution control:government:consumerism:env schem:env legislation |
| T | 5 | Water Pollution | Mar | 1925 | ind:now:water pollution:drought:volunteer action |
| R | 5 | Saras Krounch – The Book | Mar | 0700 | ind:now:birds:Conservation:nature |
| R | 5 | Saras Krounch – The Book | Mar | 0700 | ind:now:birds:nature:Conservation |

Table 5.5.8 Continued

| T/R | Duration in mins | Title | Language | Start time | Backcloth/subject matter |
|---|---|---|---|---|---|
| R | 5 | Saras Krounch – The Book | Mar | 0700 | ind:now:birds:nature :Conservation |
| R | 5 | Reading The Book Saras Krunch | Mar | 0700 | ind:now:birds:forests:endangered species |
| R | 5 | Reading The Book Saras Krunch | Mar | 0700 | ind:now:birds:forests:endangered species |
| R | 5 | Reading The Book Saras Krunch | Mar | 0700 | ind:now:birds:forests:endangered species |
| T | 4 | Documentary On Environment | Mar | 1411 | ind:now:deforestation:transport:building:population growth:habitat loss |
| T | 4 | Air Pollution | Hin | 2025 | ind:now:air pollution:vehicle pollution:India:env legislation:city |
| T | 4 | Air Pollution | Hin | 1925 | ind:now:air pollution:vehicle pollution:children:health:education:env awareness |
| T | 3 | Environmental Documentary | Hin | 2142 | ind:now:deforestation:transport:building |
| T | 3 | Air Pollution | Hin | 1927 | ind:now:air pollution:health:India:vehicle pollution:env legislation |
| T | 2 | Vanadevi: Goddess Of Forests | Mar | 1945 | ind:now:India:sustainable forestry:env awareness |
| T | 2 | Documentary On Air Pollution | Hin | 1925 | ind:now:vehicle pollution:India:trees:env awareness |
| T | 2 | Social Forestry | Mar | 1927 | ind:now:deforestation:afforestation:India:education:symbiosis:sustainable forestry |

dominate TV and push the regional language, Marathi, into third place. But for radio the reverse is the case: there Marathi is dominant. But for both TV and radio Marathi dominates the monitored environmental output – suggesting very closely that this is something taken up as a local concern. English does not feature at all for the monitored radio output.

Tables 5.5.4 and 5.5.5 show the same traffic and backcloth breakdowns as before, but for the DD and AIR broadcasts also broken down by language. The figures seem to be getting too small to talk of characteristic frequencies: we do not think it is possible from these data to say whether or not English language or vernacular broadcasts diverge significantly in their treatment of environmental and developmental issues.

Table 5.5.6 shows the percentage frequencies for different channels for combinations of regions and backcloth words, for monitored environmental programmes. Here it is possible to see the extent to which 'Africa' is associated on the British channels with Land, with Malfunction and with Ecology – but not with Politics. The programmes on North America also emphasise Ecology, and have very little to do with Politics. Even the programmes on the UK seem to have a low loading on Politics. We wonder whether the comments made in Chapter 3, where some of the UK editors seem to indicate that the environment is not a very political issue, also apply to the non-news programmes, to perhaps an even greater degree.

Table 5.5.7 has been prepared by listing some of the details for the monitored programmes for UK channels, rank ordering the list by duration of programme, and then taking a 10 per cent random sample of the whole. It gives a 'flavour' of the whole data set. Note that there are a number of fairly lengthy items at the top of the list, the second two of which are fictional movies. But less than 20 per cent of items are more than 10 minutes long. The list also indicates the variety of programme types – including animation with Captain Planet and others – and a number of obvious programme 'slots' – including Gardeners' Question Time, Science Now, Country-file – but also some non-obvious slots like Letter from America. Table 5.5.8 has been prepared in the same way from the Indian data but has a 50 per cent sample of a much smaller set. There are few long programmes – but the two TV programmes in English within this set both occur near the top. Most of the items are below 5 minutes: many of these are little public information films. Overall, 45 per cent mention trees, forestry or deforestation, a theme which picked up in Chapter 4 and which continues in the remaining chapters too.

## CONCLUSIONS

Since the mass media are perceived to be important in helping to shape worldviews, there is some legitimacy in asking how much coverage 'the environment' gets as an indication of the importance media managers give it in either leading or following public concerns. But it is also a rather vague question since 'the environment' can appear as either or both backcloth and traffic, as content and/or programme category (format). As a category it is weakly developed in the UK in the newspapers, where there are a few dedicated correspondents, but hardly at all in equivalent terms in the Indian newspapers, where correspondents dedicated to a range of developmental issues may promote some items. On TV in general programming it is quite strongly developed in the UK in the 'natural history' format, and that too is used a little in India. In the UK it is quite well adapted to the greater diversity of programming formats, and fits neatly as items with magazine programmes like Science Now. There is more than a suspicion that in India the narrower width and variety of programming formats, let us say the absence of many of the sophisticated formats associated with a more mature broadcasting system, probably gives less opportunity for a complex and more subtle inclusion of the terms of the debate. Recurrent short public information films target simple themes, whether afforestation or atmospheric pollution by vehicles.

As subject matter it can intrude anywhere – and, as we have said many times it actually intrudes implicitly everywhere. Any answer to the question 'how much?' is therefore bound to be messy and qualified. With respect to broadcast news we can answer with some more robust figures, which we have given here. In the UK the figure hovers around 3 per cent – it clearly trails as a category far behind categories such as Politics. In India it is higher on TV news – we have come out with a figure about 12 per cent – but still far behind Politics and other categories, but the problem is that it is so tightly intertwined with 'development' that one cannot say that it is being isolated as a category which has independent meaning. For both countries it is abundantly clear that coverage in any sense is rarely if ever concerned with global matters, although the British media are more prepared to cover non-UK foreign stories, and that in India the environment is recognised in eponymous terms, as 'India's environment'. There are differences in India between the English media and the vernacular, both printed and broadcast, which suggest that they are more likely to be pan-Indian and less provincial than the vernacular media – a hint of the 'post-nationalist' audience described in Chapter 4.

The reasons for the differences are to be found partly in the different histories and stages of development of the media in the two countries, but also as much in the different appreciation by media managers of the expectations of the different audiences. The differing attitudes of the managers we have dealt with in the preceding chapter. The different reactions of the audiences we deal with in the next.

## REFERENCES

Atkin, R. (1974) *Mathematical Structure in Human Affairs*, London: Heinemann.
Chapman, G.P. (1991) 'TV: the world next door?' *InterMedia* 20(1): 30–3 (reprinted in *Geographical Magazine*, October 1992).
—— (1996) 'Capturing and analysing holistic lifestyle data', in Saraswati, B. (ed.) *Computerising Cultures*, New Delhi: Indira Gandhi National Centre for the Arts.
Gould, P.R., Johnson, J.H. and Chapman, G.P. (1984) *The Structure of Television*, London: Pion.

# 6 Decoding the images
## Interpretations of environmental documentaries by British and Indian audiences

## 6.1 THE USE OF FOCUS GROUPS

The use of focus groups has its origins in media research. In 1941, Lazarsfeld and Merton asked audiences at Columbia University, USA, to react to a recorded radio programme by discussing the negative and positive feelings evoked by what they had just listened to (Merton, Fiske and Curtis, 1946). In recent years, communication researchers have employed the method to gauge public opinion on social and political issues (see Lindlof (1987), Jensen and Jankowski (1991), Schlesinger (1992) for a more detailed account of the use of focus groups in media research). The goal in using focus groups is 'to get closer to participants' understandings of the researcher's topic of interest. ... Focus groups are useful when it comes to investigating what participants think, but they excel at uncovering why participants think as they do' (Morgan, 1988, cited in Livingstone and Hunt, 1994). In India, the focus group is widely used by social workers to animate community discussions on local problems.

The focus group interview of today takes many forms and is used not only in social science research but also in recruitment drives and more widely in marketing and publicity campaigns. Despite modifications to the techniques used by Merton in the post-war period, there remain many similarities between these and contemporary 'in-depth group interviews' – of which focus groups are one variety. Goldman (1962) (quoted in Stewart and Shamdasani (1990: 10)) defines this method of research technique by simple definition of the words 'group', 'depth' and 'interview':

> A group is 'a number of interacting individuals having a community of interest'; depth involves 'seeking information that is more profound than is usually accessible at the level of inter-personal relationships'; and interview implies the presence of a moderator who 'uses the group as a device for eliciting information'.

The assumption is that views expressed in a group situation are more representative of the public than those stated in a personal interview or through a questionnaire. In any case, it is assumed that people are part of social groups, ('interpretative communities', according to Lindlof (White, 1994)) and that their understanding of an issue or question is in terms of their communities and their cultures.

Today's focus group normally consists of between six and twelve people with a moderator, or animator, whose job it is to allow participants to create their own dialogue, whilst keeping the discussion focused on a particular topic. The topic is defined and explained to the group before the discussion begins and the amount of direction given by the moderator varies according to the requirements of the broader research agenda. The aim of focus groups is to facilitate a better understanding of the ambiguities and multiple layers of meaning often hidden behind the limited responses to questionnaire surveys; it is important to allow participants the opportunity to qualify and contradict themselves, and each other, as a discussion progresses.

Whereas a focus group held to discuss a new cleaning product may be quite tightly directed by the moderator to elicit responses to specific, easily defined issues – such as price, packaging, efficacy, etc. – a focus group discussing such a nebulous subject as 'Nature' might easily be expected to play a much more important role in defining topics as the discussion progresses. In any case, the very presence of a moderator and the knowledge that he or she is there to gauge their reactions to a particular stimulus, is usually enough for group participants to bring themselves back into line when they feel they are digressing from the subject. This element of freedom from control of the discussion is itself an important feature of focus groups when the research is trying to ascertain what participants themselves see as relevant to a discussion on a given subject. When participants discussing pollution in India, for example, stop themselves from talking at length about the role of religion in the country and dismiss it as irrelevant to the subject, there is an implicit denial here of any connection between the two issues.

Stewart and Shamdasani (1990) identify one of the advantages which is at the same time a limitation provided by the interaction between participants and moderator in focus groups. Whilst the presence of a moderator allows him or her to probe responses in a way denied by the more closed nature of questionnaire surveys, he may also knowingly or unknowingly elicit specific responses through cues in the way in which questions are put to participants (1990: 17). In fact the question of verbal but non-transcribable cues occurring in focus groups is one which merits careful attention. Ideally, the additional information provided by this level of communication – nuances of intonation, laughter, groans of despair, etc. – from the participants should be analysed as carefully as the more readily transcribable text of the discussion. Its importance should not be overlooked either in terms of the way in which it fashions the responses of participants or as it relates to the language used by the moderator in asking questions and probing responses.

Added to this is the additional problem of the moderator being viewed by participants as a 'resident expert' in the subject being researched. Hesitancy can arise amongst participants for fear of somehow giving the 'wrong answers' if they think there is someone present who knows all the right ones! This fear can be extended to the interaction between participants themselves, where some participants may view themselves or be viewed by others as more or less capable of contributing to the discussion.

The above limitations – and others – of focus groups must be borne in mind not only when choosing the technique as a research method but throughout its use, from the recruitment of participants and moderator, the conducting of the group to the final analysis of the discussion. As focus groups are usually, although not always, used in conjunction with other methods of research – such as larger scale questionnaire surveys – this itself will influence all stages of the way in which groups are organised and analysed. As discussed below, the intended outcome of the use of focus groups in a broader research agenda and their relationship to the other parts of the research will to a large extent determine the way in which the groups are conducted and analysed. This and the role of the moderator are perhaps the two most crucial aspects to the use of focus groups in any research project.

The primary objective of conducting the focus groups in India and Britain was to understand the perceptions and awareness levels of small groups on both global and local issues as they relate to the environment. The public surveys discussed in Chapter 7 probed awareness levels on issues like acid rain, the ozone hole and global warming. The focus groups probed further on other overlapping issues like the lack of drinking water, the threat to green spaces, industrial and traffic pollution, the environment movement, with the help of documentaries made in India and Britain.

The second objective was cross-cultural. How do environment documentaries made for one cultural group transfer to another cultural group? What sense do small British groups make of documentaries made in India primarily for screening to Indian viewers? What do Indians in Tamil Nadu and Maharashtra make of BBC or Channel 4 documentaries on the environment? How, again, do viewers of these two varied cultural settings react to the issues discussed in the documentaries on the environment and environment-related issues? What form or structure did the discussion take? What is the nature of people's 'talk' about environment documentaries from home and abroad?

The focus groups were held over a 12-month period from May 1993 to May 1994, overlapping with the period of data coding (February 1993 – January 1994) and straddling the week of the questionnaire survey, in order to attempt to make linkages between these phases of the research project.

## 6.2 FOCUS GROUP STIMULI – TWO INDIAN AND TWO UK ENVIRONMENTAL DOCUMENTARIES

The stimuli used for the focus groups were four television documentaries dealing with an environmental topic. Of the four films used, two were made in India and two in the UK. The criterion used for selecting the films was that one from each country should deal with a global environmental problem and the other with an environmental issue more specific to the country of origin.

### Indian documentaries

Both the Indian films used are, like the majority of terrestrial documentaries broadcast in the country, produced by the Films Division of the Government of India. Entitled *Insult to Civilisation* and *A Matter of Life and Death*, they deal respectively with the problem of pollution in cities and with India's lack of clean drinking-water.

Both films are less than 20 minutes long – 11 and 16 minutes respectively – and are typical of home-produced Indian television documentaries. There is one straight commentary throughout in English. As most Government of India (GoI) Films Division productions are originally made in English, neither film required to be dubbed for UK purposes. Despite the apparent documentary-style format, there are no genuine interviews with members of the public or 'experts' in either of the films, the only 'interviews' being with actors playing the roles of members of the public – rather like a washing-powder or other similar advert on British television.

The very simple narrative devices used are reinforced by equally simple visual imagery with, for example, no or very few graphics. *A Matter of Life and Death* includes a sequence of stills showing newspaper cuttings and official documents to illustrate Government of India action on water treatment. There is also just over a minute of film of a diagram showing a water treatment system.

Although *Insult to Civilisation* purports to be about the problems of over-consumption and urban pollution in cities throughout the world, all the footage is Indian and some of the clips of polluted waterways are common to both films.

*Synopsis of Indian documentaries*

*Insult to Civilisation*

The film starts with shots of an Indian city – modern high-rise buildings, commuters on trains, slums and traffic jams – whilst the commentary makes generalised statements about the 'great cities of the world ... Babylon, Athens, Rome, Vijaynagar' and asserts that cities have 'always been the cradles of great civilisations'.

We are then told about the negative effects of consumerism on our quality of life and that the 'energy crisis is the result of the mindless use of Nature'. The second half of the film concentrates on the problems of pollution which result from our wasteful urban lifestyles – air pollution from vehicles, water pollution and pollution from industrial development. Finally the possible solution of nuclear power – as introduced by two of the actors-cum-members of the Indian public – is dismissed as dangerous because of the risks of radiation. The final moments of the film consist of a series of horrific close-ups of aborted foetuses in jars of formaldehyde with grating horror film-type music and a commentary which asks us to think of the future of our unborn children.

*Matter of Life and Death*

A discussion of the problem of water pollution in India, this film can be divided into two main parts: the first part explains why there is a lack of potable water and the second looks at what steps the Government of India is taking to rectify the problem. First we are told that industry is largely responsible for polluting water supplies and that people do not take adequate action – despite what they have been told – to avoid drinking contaminated water. This is followed by an explanation of GoI measures taken to improve water supplies, with reference to legislative changes and technical innovation. Simple scientific terms are used in the commentary to accompany footage of scientific analysts carrying out laboratory tests and graphics of water treatment showing the use of 'filters, pumps and chlorination'. This culminates, however, rather confusingly in a diagram of leaking pipes! The film's conclusion is that the most important step that can be taken is to safeguard a water supply at source.

**UK documentaries**

One of the two UK films is an edition of the BBC2 consumer series *Advice Shop*, which was shown in January 1993. It deals with the problem of local opposition to out-of-town superstores being built in two different regions in the UK. The second film is a documentary film in the Viewpoint '93 series which was made for Central Independent Television in association with the Television Trust for the Environment. The Advice Shop programme is 25 minutes long and can be divided into four main sections: two reports on the activities of local residents' groups, a recorded studio discussion with two planning experts and finally a recorded interview with the opposition spokesperson on the environment. During the first part of the programme there are interviews with representatives of the local groups, footage of their campaign protests and 'fly-on-the-wall' coverage of their activities. There is also a section in which actors play the role of businessmen planning a new building development. Some existing newsreel footage from local television is used in one of the reports on the residents' groups.

*Growing Up* is a 50-minute long documentary which includes nine case studies filmed in different countries of the world by different producers. Each case study includes interviews which have been dubbed into English, where this is not the original language. A commentary in English by a well-known British actor, which links the various sections and opens and closes the documentary, is interspersed with speeches from politicians and other participants at the Earth Summit in Rio.

There is a recurrent graphic device – a spinning top and the turning pages of a photograph album – used to introduce each new case study and strong, rather haunting, theme music used throughout. As well as using original music, the film also uses music written and performed by well-known western and African musicians.

## Synopsis of UK documentaries

### Advice Shop

The documentary starts with 50 seconds of footage of rather hysterical scenes of protesters against an out-of-town superstore development in Bristol, in the west of England, screaming and crying as four 100-year-old lime trees are being chain-sawed to the ground. This is followed by the regular presenter in the studio giving a brief explanation of what has just been shown. The rest of the first part of the programme follows the ultimately unsuccessful Bristol campaign. Two planning experts then explain to the studio presenter how to organise such a campaign successfully. We then follow the different stages of a more successful campaign in another part of England, in which local people organise themselves against the apparently sinister forces of big business – portrayed in the film by actors smoking cigars, making top-secret decisions in dark rooms . . . In a short interview, the opposition spokesperson on the environment explains changes that are needed in the planning system to help protect the environment against unnecessary large-scale developments.

### Growing Up

The first ten-minute section of the documentary opens with film of the fertilisation of a human ovum, the early stages of development of the embryo and the birth of a human baby. This is followed by film of a young couple in the UK who are expecting their first child – one of the case studies which is also used to conclude the film. Accompanying this is commentary which introduces the documentary's theme of children throughout the world.

This is followed by some spectacular aerial shots of Rio de Janeiro with the voices, followed by film of, Boutros Boutros-Ghali, UN Secretary-General, Fidel Castro and George Bush speaking at the conference. From this we return to the main theme and it is explained that the film will be looking at the lives of nine children all born within a year of the Earth Summit held in Rio, June 1992.

The children are from Brazil, India, China, Latvia, Papua New Guinea, Kenya, USA, South Africa and UK. Using the case studies from the South, we are shown how poverty leads to environmental degradation and to further poverty. The issues covered include problems of rapid urban migration, child labour, lack of potable water, over-population, industrial pollution, disease, desertification, deforestation and armed conflict. Pollution in Latvia is used to illustrate the environmental problems facing eastern Europe – untreated waste being pumped into the sea, contaminated water supplies, industrial waste from chemical plants and nuclear radiation. The two families in the UK and USA are both suffering the consequences of their countries' practice of unsustainable industrial development. A miner and a logger, the fathers of the families, are at risk of unemployment with the closure of their respective industries.

The film of each family is introduced and concluded by a summary commentary on the main issues explored in the case study, with reference to how the issues were commented on by world leaders at Rio. The documentary concludes with the comment that 'All our children face a world racing to disaster' and that for something to be done to halt this impending disaster, those who participated in the Rio Summit must turn their words into action – and sign cheques as well as documents.

### 6.3 THE COMPOSITION OF UK FOCUS GROUPS

Seven focus groups were held in the UK – four in the urban centres of London and Glasgow and three in rural communities in Gloucestershire and Shropshire. The urban groups, held in central London and central Glasgow, were composed of people living and/or working in the cities. Rural groups were conducted in the village of Newent, 10 miles from the shire-town of Gloucester and in the Shropshire village of Church Stretton, 12 miles south of Shrewsbury.

Although it had originally been hoped to include a significant number of farm workers in the rural UK groups to facilitate comparison with the large proportion of participants from the agricultural sector expected in Indian rural groups, this objective was not pursued as the inclusion of farmers in the Indian sample was significantly lower than expected. However, a significant proportion of rural group participants in the UK belonged to households where the main wage earner has a job in a sector strongly linked to farming and land use – surveyor, blacksmith, gardener – even when not directly involved in farming the land.

The average focus group consisted of 8 members, the largest having 12 and the smallest 6. No participant took part in more than one focus group as it was found that a 100 per cent consistent participation rate would have been impossible to achieve. Participants were not briefed beforehand on the contents of the film, only being told that they were going to be asked to watch and comment on a TV programme about 'a topical subject'. They were also aware when agreeing to participate that the discussion was to be audio-taped.

### Urban groups

- A gender mix was maintained in both rural and urban groups, although the urban groups had significantly more women than men (64 per cent, 36 per cent). The average age of the urban groups was 45 years.
- Urban groups were recruited with a bias towards participants from the professional and managerial social classes with 67 per cent of participants coming from these classes, compared with a national UK average of 27 per cent.
- In terms of educational attainment, urban groups had a significantly higher than average education level, with 67 per cent of participants holding a university degree, compared with a national average of 27 per cent of the population who have studied to A level and above.

**Rural groups**

- Rural groups had only slightly more women than men (57 per cent, 43 per cent) and an average age of 36.
- Amongst rural groups the sample had a higher percentage of households in which the main wage earner is employed in the semi-skilled manual or non-skilled sector.
- In rural groups, 67 per cent of rural participants had not studied beyond GCSE or O level.

Taken together, these variables suggest a higher predominance of A, B and C1 classes in the urban groups with a bias towards social classes C2 and D in the rural groups. The unemployment rate was 3.5 per cent compared with the UK average of 9.6 per cent (in 1994). There were, however, a number of unwaged participants – women looking after the house and/or family (12 per cent) and students (3.5 per cent).

## 6.4 THE RESPONSES IN THE UK

Techniques used for analysing focus groups range in complexity from post-group discussion amongst interested parties without using any transcript of the focus group discussion to tailor-made computer-assisted analysis of transcripts complemented by analysis of videotapes of the discussion. The technique used here falls somewhere between the two extremes. Each focus group discussion was audio-taped and transcribed and the transcriptions then analysed. As the use of focus groups here was intended to complement the other phases of the research, there were certain predetermined themes which were introduced by the moderator into the discussion and then explored more closely in the post-discussion analysis. These themes were largely determined by the content of the films watched by the groups – the stimulus to the discussion – and also by preliminary results emerging from the data analysis of UK print and broadcast media and by the topics identified for the questionnaire survey. What follows, then, is a largely descriptive analysis of the focus groups held in the UK between May 1993 and May 1994.

**First reactions**

Each group discussion began with participants being given the opportunity to volunteer their immediate reactions to the film they had just been watching. Having been told that the research team was interested in their reactions both to the film's content and to its form, there were clear differences in initial reactions to the UK films and to the Indian films. Whereas after viewing the UK films participants commented immediately on what they had understood to be the message:

> We've got no rights whatsoever.

> We're fighting the big boys all the time.

> The same old problem. Do we want these supermarkets?

> I'm very interested in this because I was one of the people who helped to start up BAMTAG [a local environmental campaign].

> (Advice Shop)

A collection of world problems.

The abuse of the planet.

Most of the parents were thinking about the future and hoping that the future for their children was going to be much rosier.

I can't remember when it [the Earth Summit] was.

I'm not saying it was a waste of time but . . . it has had no direct results so far.

*(Growing Up)*

Viewers of the Indian films expressed a reaction first of all to the way in which the message had been presented.

Could have been more to the point.

A bit sort of airy fairy.

It was like a school programme.

It doesn't say what we can do to help.

*(Matter of Life and Death)*

A shock . . . bit of a shock treatment.

Cheap trick really.

Very broad sweeping gestures.

Completely negative.

Patronising and offensive.

*(Insult to Civilisation)*

There were strong negative comments on the production of the Indian films at the beginning of the discussions, with people commenting on the dated appearance of the films. The accent of the presenters was also commented on: 'The film is scratchy, the sound is appalling, the accents are hard to understand' (L/IC)[1] and in two different groups people made the observation that they had been reminded of a contemporary television advert for a well-known phone company:

He sounds like the guy who does the Mercury 'phone ads.

(N/MLD)

Reminded me of the Mercury ads.

(L/IC)

Several participants described the films as 'patronising' and the imagery at the end of *Insult to Civilisation* as 'offensive'. Although the Indian films followed a much simpler format than a comparable UK documentary – with no or very simple graphics, a single almost uninterrupted commentary and no interviews with members of the public – participants found it difficult to understand what message the film was trying to convey to its audience:

I don't think we're very sure about what he was saying.

(CS/IC)

I found it a bit jumpy – I got lost on one or two occasions.

(G/MLD)

Bits put together and they didn't always seem to connect immediately.

(G/MLD)

Well, I don't know what the film was for.

(CS/IC)

What message were they trying to get across?

(N/MLD)

It was very unclear what the film was aiming to do.

(L/IC)

Those who viewed the UK films made no observations of this kind. Although the production of both UK films was much more complex and included much greater diversity – particularly in the case of *Growing Up* – in the issues covered, examples used to illustrate them and geographical locations, nobody expressed any difficulty in understanding the arguments being made.

Another frequent reaction to the Indian films was that they did not offer any solution to the problems they presented:

If that was being presented in this country you'd have possibly a similar sort of beginning ... but the end would have been turned round to sort of show the solutions in a more positive way.

(G/MLD)

It doesn't actually say what we can do to help.

(N/MLD)

It didn't say much about what we as normal people could do perhaps about preventing it.

(N/MLD)

It didn't really give alternatives.

(L/IC)

The proposals were all negatives – don't have kids, don't want blenders, don't drive cars.

(L/IC)

They're not offering any solution.

(G/MLD)

It's all very negative instead of a positive thing.

(L/IC)

Although *Growing Up* in fact presented the viewer with a much wider range of problems than either of the two Indian films – and just as few solutions – participants watching it viewed it more as 'entertainment' than as 'a public information film', which is how one viewer described *Insult to Civilisation* (S/IC):

I found it very thought-provoking obviously. I won't say I learnt a great deal I didn't know before but it brought it vividly in front of me, brought back things which I'd already known about.

(G/GU)

I thought the film generally was quite high on hope and emotion and a little low on facts and figures particularly on a national scale, a continental scale or global scale.

(G/GU)

Advice Shop, with the relatively easily recognisable format of a consumer advice programme – resident studio presenter, experts asked to comment, solutions offered – was accepted without question by UK viewers as typical of its genre. The two Indian films, however, were sufficiently far removed from what UK viewers expect to see when they sit down to watch a programme about the environment that participants were unable to react to the contents of the film without reacting first to the way in which it was produced and presented.

## Identifying environmental problems: local over global

While they were not asked directly during the discussion what they identified as the main environmental problems locally, nationally or globally, participants' comments on these issues show a marked tendency towards identifying specific problems ahead of others. The catchphrase of one of the UK's largest environmental NGOs, Friends of the Earth, urges people to 'Think globally, act locally'. From the discussions of UK focus groups watching films from India and the UK on local and global issues, a more appropriate slogan to describe the current state of mind of UK television viewers might be 'Watching globally, thinking locally'.

Within this general state, there are myriads of factors which influence people's interpretation of television programmes on an environmental issue. In the focus groups held in the UK the amount of discussion devoted to local and national environmental problems – as opposed to global issues – varied to a large extent according to two main factors: the subject of the film being watched and the background – rural or urban – of the group participants.

The film which provoked most comments on local and national issues was *Advice Shop*. Participants from both the urban and the rural group identified clearly with the issues being explored in the film and in both groups there were participants who had had personal experience of similar situations.

In the London group, several participants had themselves been involved in campaigns – successful and unsuccessful – against road building and commercial developments close to where they lived in southwest London:

> The South Circular was clearly a totally unsuitable highway so the government conceived the idea of building really a motorway right around South London and we decided that this was a bad idea and we decided to go to war with the government.
>
> (L/AS)

> Since the motorway project across Barnes was put forward and then deleted, we've had the proposal to redevelop the Harrods depository site that's now a derelict warehouse. ... It was stopped by a coalition between the local council and the local people. It was stopped very effectively.
>
> (L/AS)

> But in Hammersmith you see we didn't manage that. We had the complete opposite ... and when I look back I don't know why we did fail really. I think it was because we didn't know enough about what options were available to us.
>
> (L/AS)

They also had clear ideas of why they had taken action against plans to change their local environment:

> The reduction of property values is of pretty significant importance around here!
>
> (L/AS)

Five thousand people driving down your road every day that didn't do it before.

(L/AS)

That horrible building that's gone up. I mean, what are the repercussions now. How are these lorries going to load and unload?

(L/AS)

and they sympathised with the emotional reaction of the campaigners in the film:

The biggest impact there for me was having the trees cut down. After a while you start switching off to some of the people but those trees coming down ... seeing a whole tree being carried off in the air.

(L/AS)

There's a big difference between building on existing sites and ripping up green sites. ... There are hundreds of thousands of acres of derelict land in England that could be put back into life.

(L/AS)

In the Gloucestershire group, participants had already experienced the kind of debate which surrounds plans to build a large supermarket on a greenfield site – one such scheme to build a large supermarket in their village had been abandoned several years before after local opposition. Unlike the urban group, they were able to see more of the advantages of such developments:

I must admit I am a little bit one-sided with the out-of-town supermarkets. You can't really have big supermarkets in the town. There's nowhere to put them, is there? Problems with the lorries and delivery and everything like that. So they've got to build them outside. No matter where they put them they are going to be in front of somebody's house and some trees are going to have to be cut down.

(N/AS)

Well, it's only a field there now, isn't it? The farmer is going to retire. So think, you know, of what it'll bring to our people in Credgley.

(N/AS)

That is one instance that brought a lot of trade, brought an awful lot of work and that. It was only a field like.

(N/AS)

Years ago Woolworth's wanted to build here on this very spot. And Newent Council came up in arms. 'We're not having no big stores in here' blah, blah, blah. But it would have been the making of Newent.

It would have brought work to Newent. It would have brought so many people into Newent.

(N/AS)

Their feelings towards the demonstrators as shown in the film were also less than sympathetic:

They didn't come across very well I think in the beginning when they were all shouting. Made me think 'build it'. Just to annoy them.

(N/AS)

I didn't like the fact that they actually used the trees as the argument. The argument was that they just didn't want it. They hid behind the fact that the trees were there. And I think they used the trees.

(N/AS)

The Council can be a bit hasty putting preservation orders on trees.

(N/AS)

They didn't show the plus points like the fact that the amount of employment that it would bring into the area and stuff like that – I thought it was very, very one sided. The whole thing was very muddled, to make you remember all the stuff from the campaigners.

(N/AS)

citing instead reasons why farm land should be developed for other purposes:

... and we put so much land in set aside.[2] I mean are we – have we got more green land than we need? We're not growing corn, we're not growing produce on it. The farmers are putting hundreds and hundreds of acres set aside just to let the weeds and wild flowers grow.

(N/AS)

If you're going down the pan and somebody just said 'Well you know, here is half a million', would you say 'No, I'll keep my field' ... I mean how many farmers are doing alright?

(N/AS)

More environmental damage is being caused by farmers than has ever been caused by Tesco's supermarket. It's true. I'm sorry, but it's true. They ripped up hedgerows which have never been replaced. The chemicals that they pump into the field get into the waterways which end up in our drinking-water eventually. So, I mean I can't get upset about Tesco's building a supermarket.

(N/AS)

However, despite being less anxious about the loss of green space than the urban group, the Gloucestershire group were able to appreciate why city dwellers might react in the way they do to development on a greenfield site:

We have a job to understand like losing five acres of field. I mean if we lost a five-acre field we'd hardly ... we wouldn't notice it. We see so much of it every day.

(N/AS)

People who live in inner city areas have nowhere to go, have they?

(N/AS)

All these fights that they were talking about. They are city people, aren't they? The Bristol people didn't want it built on. But I mean that is city people that wanted to keep their green areas.

(N/AS)

At least we don't live sort of fifteen blocks up with two children under our feet and no ball games on the piece of green that is down there. I mean I think people in cities need much more help.

(N/AS)

... that piece of green is their countryside. That is all they've got and so you can understand why they're going to get so upset by one tree or – maybe not a lot of space but that's their countryside.

(N/AS)

In terms of general awareness of their immediate natural environment, the rural group in their discussion were much more likely than the urban group to make references to facts about the way 'the natural environment' works.

As the *Advice Shop* video was specifically about a very local issue it was perhaps to be expected that participants would limit their discussions almost exclusively to environmental issues which they confront on a day-to-day basis. In neither group was there any sustained discussion of wider global problems and indeed when the issue of deforestation was introduced by the moderator, using the loss of trees as a link to the question, participants in both groups were quick to point out that this was a very separate and distinctly different issue:

> But rainforests have a totally different effect on the environment, don't they . . . the way they're cutting the rainforest down will have a lot more effect on the environment than six trees here in Gloucestershire and four trees in Yorkshire and six trees in Manchester.
>
> (N/AS)

> If I were going to petition my local MP I'd much rather petition him about the hectares of Brazilian rainforest that's being cut down each year rather than these three trees in Bristol. You're talking about saving the planet. Then there we're really talking about a whole different ball game.
>
> (L/AS)

In contrast to the *Advice Shop* discussions, participants who watched the other UK film, *Growing Up*, spent very little time discussing local issues and most of the time discussing what they had seen as environmental problems in other parts of the world – reflecting the geographical diversity of the film. Initial comments on the film included:

> The video was a collection of world problems.

> It's a world-wide problem. It's not peculiar to any one country or any one part of the world. It's all over the world.
>
> (G/GU)

This said, participants – with one exception – still failed to make any connection during the ensuing discussion between problems in different parts of the world, despite the fact that the film's commentary made frequent references to global discrepancies between rich and poor nations. The one participant who did was commenting on the last case study in the film, which showed unequal land distribution and the resulting poverty and conflict under apartheid rule in South Africa:

> Seemed to be a kind of vignette for the whole kind of wider environmental thing . . . the little tail piece, the fact that there's the contrast between those two folks in the one environment was just a little reprise of the whole story. Between North and South having difficulties. But folks still can't share their resources in any meaningful way and have comparable lifestyles. For me it still goes back to the whole principle, again, the political decision. Not party-political decision but the whole . . . this political decision making process and the deficiencies in it.
>
> (G/GU)

This comment, however, failed to stimulate further discussion in the same vein. Other participants had not interpreted this part of the film in the same way and instead had seen it as being solely a comment on the political situation in South Africa:

I found that was the most political part of it really. You're getting back to the political situation in South Africa. The black living in poverty, the white living in luxury at the moment. But putting up these barricades and the gun with the fear of what is going to happen in the future. It touched very much on the situation in South Africa at the moment.

Whereas the rest of it was more general throughout the world without bringing politics or nationalism or anything into it at all.

(G/GU)

In the case of the two Indian films, participants' discussion was less centred on the issues tackled in the film and more related to their own general knowledge of several related issues. One of the reasons for this may be that, as they said themselves, participants found it difficult to follow the arguments of the Indian films:

Well, I don't know what the film was for.

(CS/IC)

What message were they trying to get across?

(N/MLD)

It was very unclear what the film was aiming to do.

(L/IC)

A second reason is undoubtedly linked to participants' level of knowledge of India and more specifically of India's environmental problems. Participants were unlikely to have had personal experience of the problems shown in the Indian films. Discussion was thus not specifically related to participants' knowledge of India. Instead it tended to draw on images from the film and to relate these to personal experiences of a similar problem or to other environmental topics closer to home.

In *Matter of Life and Death* the film looks at the problem of India's lack of clean drinking-water. In both rural and urban groups the discussions followed a similar pattern. After commenting on the difficulty they had in following the arguments of the film, participants then went on to relate their own experiences and knowledge of water pollution in the UK. This ranged from appreciation of the availability of unpolluted water sources in this country:

I think we're all lucky though because we've got a pure water supply in our homes.

(G/MLD)

We have jolly good water up here.

We're very lucky.

It's soft water.

(G/MLD)

... the water board – Severn and Trent isn't it? – they do come round and they do check. I know they checked the river down by us.

We're very lucky.

They always check your tap water as well. They come round every so many months or whatever.

Well it makes you realise how lucky we are over here.

(N/MLD)

You know over here like the fresh streams around here if you go out swimming and all that you don't think there's harm ... you just like swallow it and all that and you just don't think.

(N/MLD)

... We've got some of the cleanest parts of the world right over here. Beaches and that are far cleaner than somewhere out in the Mediterranean or whatever.

(N/MLD)

to a realisation that potable water is perhaps something that can no longer be taken for granted in the UK either:

But it's happening here as well.

Yes they're suing companies. Companies are suing water authorities.

(G/MLD)

There's all this problem here of the water metering. Now there are poor families who are not able now to pay for this water ... who cannot pay and they're having to – four of them use the toilet and just the last person flushes and they all use the same bath water.

That should not be in this country, in this day and age.

It shouldn't be in this country.

(G/MLD)

The very first time I ever went to London I was so thirsty I drank six glasses of water and the very next day I felt so unwell. I was really ill. We then went to visit friends ... and they said 'Have you drunk the water here? Did you drink any of the water?' And they said that was the problem.

(G/MLD)

When I was a young kid the whole of the Leadon was polluted and all the fish and that died. And that was just from one hop yard. It was polluted for about three or four years.

(N/MLD)

Blackpool. I mean the last time I went to Blackpool there was a big scandal up there. And then in the paper a few months ago the beaches were deserted.

Condemned as well.

It wasn't only just because people haven't got the money to go on a day trip or whatever to Blackpool. More likely because when I went up there there was all sorts of rubbish on the beach and crude oil and God knows ...

(N/MLD)

In the case of the other Indian film, *Insult to Civilisation*, again discussions in both rural and urban groups followed a similar pattern. After comments on the production of the film, participants attempted to focus their discussion on the main theme of the film – that of pollution in cities. However, in keeping with the comments which both groups made about the confusion they felt when they were watching the film – confusion over what point the film was trying to make – discussions in both groups soon deviated from the subject of the film itself and jumped from subject to subject throughout the discussion. Whilst participants may not have had any direct personal experience of the problems of pollution in an Indian city, their attempts to dissect and understand the meaning of the film led them to discuss a range of environmental problems including population, poverty, animal rights and public transport. One striking similarity between the discussions of the two groups was that in neither group was there any sustained discussion of the problem of inner city pollution in the UK. In the urban group the comment was made that the film could have been shot almost anywhere in the world:

You could make that film anywhere and just take images –

Latin America. Brazil. Rio.

Or Liverpool.

– of bad industrial pollution on the borders of any city you know. You could do it in London, you could do it in New York. You could do it in any city in Europe. You could have the same sort of imagery.

(L/IC)

but apart from this no one in the London group made any comment about city pollution. Interestingly, it was a participant from the rural group who made a comparison between urban and rural pollution in the UK:

You could argue that some of the worst pollution in the UK occurs out here in rural areas because of the use of fertilisers and DDTs and pesticides out on the farms and certainly in Anglia the drinking water is not ... is not to EEC-approved standards because of the use of pesticides and nitrates ... so I don't accept that just because you're a city you're the worst pollutant going.

(S/IC)

## India viewed from the UK

Given the importance of the subject as a factor in influencing the understanding and subsequent discussion of the film in a focus group, the choice of two Indian-made films here was obviously of significance in determining participants' interpretation of what they had been watching.

None of the participants had been told before meeting that they would be watching a film produced in India. None had any specialist knowledge or connection with the country, although two members of one of the London groups had been there as tourists. Of the four groups who watched the Indian-produced films, there is no strong correlation between group profiles and levels of either general knowledge or more specifically environmental knowledge about India. General comments made about India during the group discussions range from those of a very simplistic view of the Indian situation – often with reference to the somehow unfathomable factors of 'culture' and 'religion'- through to others which show a deeper understanding of the kinds of problems being faced in India today.

At the former end of the scale, were comments such as:

Could they not make a law where they didn't allow animals and people not to bathe in the same place?

That's their culture though, isn't it?

(G/MLD)

I think the culture is totally different. .... I mean, the animals for instance going in and out the Ganges for instance. People, when they die, are buried in the Ganges. Everything happens in the Ganges.

(G/MLD)

I think religion must have a lot to do with it . . . you can only like guess at what you read in the papers or what you see around you and there are some countries where religions are just extremely strong and maybe it follows along lines like that. But somewhere like India is slightly like sort of an enigma to most people and I think maybe they're more like us than we think.

(S/IC)

to other comments showing a greater understanding of development issues in India:

In the Sixties when India was a Third World country in terms of people starving to death they were importing food. And now they are a net exporter of food. So I mean they have moved on in the last twenty or so years.

(N/MLD)

The reason people left the land was because they were dying of poverty, not because they were looking for the bright lights particularly but to have jobs and trying to find a decent standard of living.

(L/IC)

It's a shame with countries like that though 'cos they're just trying to make a quid or two, aren't they? This is probably half the problem. They're just trying to develop as an economy.

(S/IC)

I imagine the cities are just getting bigger and bigger and then there is a depopulation from . . . well, there's a movement of people from rural to urban. And I assume it's to do with the population explosion and it's to do with land rights and how the families work.

(S/IC)

One issue on which several participants expressed an opinion was the issue of population growth in India and methods to control it. In two of the groups, participants recalled a birth-control programme in the 1970s which had been given wide media coverage in the UK – although one of the participant's recollection of events was rather confused:

Was this the time when they were doing enforced vasectomies and giving out transistor radios? . . . the government at the time were destroying people's homes, were enforcing sterilisation on the male population, were enforcing the caste system.

(L/IC)

Well I remember at one time . . . they had a scheme where I remember watching it or reading about it where after an Indian family had, say, two or three children they were giving the men vasectomies and the men didn't want to have these vasectomies. So they each get a transistor radio.

(G/MLD)

Other comments ranged from a sympathetic understanding of the situation:

I mean I think they do try hard with their health schemes. I mean you do see lots of these health visitors going through the villages, talking to the women about birth control and . . .

(G/MLD)

The adults die when they are very young, don't they?

And that's why they keep adding to their families because of the chance that they're going to lose some of them.

(G/MLD)

In undeveloped countries the way for a married couple to live ... to have a longer life is to have kids who then come in and bring money to the home and then keep them going when they've sort of worked their life. So, it's insurance to have a big family.

(S/IC)

to a slightly less sympathetic approach:

It has to control it's population before it can do anything ... I mean while it has to keep propping them up. The more civilised it becomes, the more people it has to keep alive, the ones you can't rely on to die. They have longer life spans, you have to feed them.

(S/IC)

Nowhere did participants attempt to identify or explain any link between poverty, overpopulation and environmental degradation. Nor did anyone mention changes in social or political structure as being prerequisites for tackling poverty and environmental degradation. Poverty was talked about as a reason for India's environmental problems but this was usually explained in terms of lack of ability to purchase new technologies:

And in a country like India they don't have the resources to upgrade their water systems.

(N/MLD)

It's just they don't have the resources to do it, to bring fresh water into the houses even if they had fresh water. I think the resources are not there. I think it must be an absolute nightmare.

(G/MLD)

But it's all back onto the money aspect. The technology is available for them to in fact put in all the equipment, drains, sewers, treatment works, reservoirs and what have you ... The water can certainly be used time and time again but it's the resources and these resources require money.

(G/MLD)

I get the impression that it's had very little development until recently and now it's trying to condense all the development to catch up with the rest of the world in, you know, within a very small time period and corners are being cut on a lot of environmental issues and things.

(S/IC)

The taxi drivers in India you know and the car is pumping out black smoke. But they're just trying to earn their living and that's probably the car that they can afford. They haven't got a catalytic converter in it or whatever.

(L/IC)

Another factor often cited as a reason for India's environmental problems was people's lack of education. There was a feeling amongst participants that if only Indians somehow knew more about safeguarding the environment they would be able to avoid further degradation:

I think it's education, not money.

But why can't they learn? I mean, they keep being ill.

(G/MLD)

But before you can teach them anything you'd have to make sure they realise the damage that they're causing ...

(N/MLD)

Surely by education and passing on the methods that we already know as a developed country about the reduction of pollution or whatever; surely if better education was passed on ... then the pollution problem can be sort of, well not eradicated, I mean you can never eradicate it, but it can be certainly less.

(N/MLD)

And even boiling the water. Do you think they could maybe not be educated to boil their water? Collect it and then boil it.

(G/MLD)

Perhaps there should be some sort of industrial incentive outside the cities ... perhaps ... perhaps ... people should go and teach them not only farming, but maybe also how to set up their own little workshop.

(S/IC)

I think I'd educate them to stop popping left, right and centre.

(S/IC)

Although transfer of technology and education of the masses were offered by participants as possible solutions to India's environmental problems, there was an overall feeling amongst all UK groups of powerlessness and slight despair in the face of global environmental problems. Having attempted to propose solutions to the problems identified in the Indian films, participants later tended to negate initial ideas with statements of realisation that on a global scale environmental problems were beyond their control.

**The environment as a global problem – with no solution**

There was a marked difference between UK participants' attitudes towards solving local and national environmental issues and those of a global nature.

Participants frequently commented on how bad environmental problems were in developing countries and how fortunate they were to be living in the UK:

I think we are fortunate in this country that we do not – it is not a country that suffers from these disasters like you get in – whether it's a flood, a typhoon, whatever it is. I mean we complain sufficiently about the weather, the conditions, but we don't have the extremities.

(G/GU)

We're all lucky though because we've got a pure water supply in our homes, which they don't have in India.

(G/MLD)

I think it's a lot to do with the climate as well ... The climate I think in a sense is kinder to us.

(G/MLD)

Factories are quite pollution-free at the moment aren't they? You take this country compared to the rest of the world. We're quite good.

(N/AS)

There was concern amongst participants about issues that can be identified as local or national 'environmental problems' – such as the planning issues talked about in the London and Gloucestershire groups and the lack of public transport identified by rural groups:

If you wanted to be more environmentally friendly, if you wanted to go shopping on public transport, you can't here. Because there isn't any. So where do you go from there?

(N/AS)

But I think you're right about public transport, I think that if we could move people by public transport – and the Green Party say that – and environmentally it would be better for everybody.

(S/IC)

What public transport? Out in the sticks here!

(S/IC)

However, for the majority of participants, 'environmental problems' are something which happen outside the UK – usually in developing countries but also in eastern Europe – and as such they are unable to prevent or solve them. While participants were able to discuss at length what might be termed as 'cosmetic' changes to their lifestyles in an effort to be more 'environmentally friendly' at home, such as buying green consumer products and recycling – but few would even go so far as to consider increasing their use of public transport – environmental issues such as global warming and ozone depletion were never discussed as problems which could either affect them or about which they should or could take substantial action. A comment by one participant reflected the attitude of many:

People will always get involved if it's something close at hand. If it affects them personally . . . . when they know their life will be changed because of it people are going to get much more organised. There's not only a feeling of your own interests being affected, but it's more effective to be involved at a local level. You can actually do something about it. Whereas, with the Amazon rainforest or whatever, it's very easy to feel that you're just a small drop in the ocean.

(L/AS)

## The role of the UK media

In all groups there was prompted and unprompted discussion of the media and its influence on people's environmental knowledge. From a 'straw poll' instigated by the moderator towards the end of each discussion when asked where they thought most of their information on the environment came from, participants' most frequent response was television, both news and documentaries. This response was qualified by participants saying that while they might be alerted to a particular problem by what they saw on television, to gain further information on the subject they would then turn to printed material – usually magazines rather than newspapers. Again there was a distinction in people's minds about what constitutes 'the environment'. When talking about the kinds of publications in which they read about the environment, participants mentioned trade journals and magazines such as *Time*, *Geographic* and the BBC *World Wildlife* magazine, saying that there was 'very little in daily newspapers'. Given that the types of environmental issues covered by these publications are not usually about the UK, this reflects the attitude that most participants have that the 'environment' is somehow happening 'out there' and not in their immediate surroundings. In terms of media consumption, older participants were more likely to mention printed materials and younger ones to talk about the relative importance of television as a source of information on the environment.

As well as being asked to comment specifically on their perceived sources of information on the environment, participants frequently made spontaneous references to the media during the course of the discussions. Analysis of the discourse shows that the most frequent reference to the media expressed a concern amongst participants that the media – particularly television – cannot be trusted:

> I think the media clearly is sensitive to all these issues that are going on but I think the feeling is that if you need any in-depth or balanced view you have to go somewhere else to back up what you're getting from the media.
>
> (G/GU)

> When I see dreadful things on television, whether it's even a small thing and it's an appeal for something I am inclined to believe it and accept that that's actually what's happening. I might be very naive.
>
> (G/GU)

> It's very manipulative, I feel, both papers and television ... particularly television. ... It's very manipulative ... and as I say they give people the high spot of the news that they want you to get and if they don't have enough news then they fill it in with all speculation which conditions people's minds.
>
> (G/GU)

> Sometimes I feel it's biased television. In the sense that maybe there is – the message it's trying – it's produced in such a way that it's trying to make you see it from their point of view whereas the whole point of a good documentary is that it should be well balanced and show the issues so that you can formulate your own idea.
>
> (L/IC)

> It's an environmentalist programme isn't it? And they can edit what they like as well to make it come out how they want it to.
>
> (N/AS)

This feeling was reflected in particular in comments from the groups who watched the *Advice Shop* video and who had had personal experience of the types of issues involved:

> They didn't show the plus points like the fact that the amount of employment that it would bring into the area and stuff like that – I thought it was very, very one-sided. The whole thing was very muddled, to make you remember all the stuff from the campaigners.
>
> (N/AS)

> If I was Tesco's I wouldn't have come along to that because it was definitely stacked against you. It was a kind of 'oh, Tesco's is a mean, horrible supermarket'.
>
> (N/AS)

This exaggeration factor is just the kind of thing we are seeing on the television. It makes a dull and boring programme if they actually state the case. It's in their interest to take a stand either one way or the other and to exaggerate rapacious developers with nasty bulldozers, smoking in dark rooms. It makes better television. So the media portrayal is also biased and it's in their interest to exaggerate and use hyperbole.

(L/AS)

To make it interesting on television and elsewhere we have to present it as the baddies versus the goodies . . . . You don't sell good news. . . . You only reacted there 'cos you saw the trees coming down. And I understand that. That was the most poignant part. I'm not arguing against that, that's what you have to see. If the media presents it that way, what other way is there to react?

(L/AS)

We really are at the mercy of the media because they present an argument on that programme and we all have sympathy with those people and want them to win their fight. Now what would have happened if they had decided to take the opposing view and say that the area needed a superstore? The trouble is that the media does have a lot of power in persuading us of anything.

(L/AS)

As well as this expression of mistrust, there was also a greater awareness amongst participants who had first-hand experience of the issues covered in *Advice Shop* of the role of the media in such campaigns:

With the Tesco's in Sheffield, as well as starting their battle earlier before the bull-dozers went in, possibly one of the best things that happened to them was that a television programme was made about it.

(L/AS)

I don't know why we did fail really.

It was because the opposition consisted of a few articles in the *Evening Standard*. I didn't see people marching in the streets to protest against it. Anna Ford on the London 6 o'clock programme and that was about it. I mean it wasn't a big major campaign.

(L/AS)

You're supposed to scan the local paper to see what planning permissions have been applied for.

(N/AS)

Relying on the papers isn't always a good thing because I know of a few instances that haven't been in the paper and they've gone through and nobody seemed to know anything about it and it's been built and nobody had put it in the paper.

(N/AS)

If you live in a big town they can't possibly put all these planning applications in the papers or if they did you wouldn't know what they all were. Perhaps there should be more publicity given to what is being applied for.

(N/AS)

In terms of the importance of the media in informing people about the environment, spontaneous comments made throughout discussions reinforced the responses from participants to the more specific question asked about perceived media sources of environmental information. Television was the most frequently mentioned medium, with newspapers second. Radio was rarely mentioned.

One television feature specifically mentioned by participants in two different groups – distinguishing it from any television programme – was an advertisement which was shown on UK television during 1993. The advertisement in question was part of a public information campaign run by the Department of the Environment to increase awareness of the need to reduce energy consumption in the home, aimed at curbing greenhouse gas emissions. Neither of the participants who mentioned the advert could remember who it had been made by and one was convinced it had been funded by an NGO:

> You get the odd advert on TV; a young girl's voice saying about. . . . But how often, I mean, that I really stop and listen to that. But there's not enough of them. The little girl. . . . on TV. She says about. . . . Saving energy.
>
> (N/MLD)

> One of the more difficult things I've had to turn off recently was an advert for that – I can't remember, it was Greenpeace probably. It was that wonderful children's cartoon with the little baby girl voice-over of 'we'll make the Greenhouse effect and then this ape here will die and we will all be sorry and then we'll cry' if I remember that. And it was just such a piece of emotional clap-trap. You know, it was as bad as that. It was one-sided, no discussion of the facts, no discussion this is what's happening. Bang! But it wasn't government paid for. It was a – it was a sort of 'now write your cheque in the amount of . . .'. I really can't remember who it was. Friends of the Earth probably and pop it in the post. See Freepost, blah, blah. And again it was a – it was bigoted but it wasn't produced by – oh sorry, biased. It wasn't produced by the state.
>
> (L/IC)

Whilst participants expressed a certain 'street-wise' mistrust of UK media, their awareness of the possible influence that the western media might have on viewers in the South was more limited. A few participants had a feeling for the sort of effects that western media might have on Indian viewers:

> If people [in India] go to get a video, usually they'll get the cheapest ones available on the market and which show sort of large cars, America, people smoking, you know a life of abundance which is what the western media is very often putting out to the Third World.
>
> (L/AS)

> I think . . . maybe, you know, like some of the, say the more educated, who get magazines and things and they see how the westerners live . . . they would love to be like us. But there is no way . . .
>
> (G/MLD)

> In India, they're not having the same video, aren't they? They must find it funny. To see people fighting over 'we don't want a shop'. If you think of the population of India . . . they could do with a supermarket.
>
> (N/AS)

but this did not generally extend to an acknowledgement that media messages of this kind were related to the environment. Given that for most participants the 'environment' means protection of a rare species, satellite photos of the ozone hole or saving the rainforests – i.e. the kind of issues covered by *Geographic* magazine and natural history documentaries – it seems likely that the sorts of implicit environmental messages contained in the western media remain just that – implicit rather than explicit. This does not deny, however, the effects that such implicit messages can have on people's attitudes and behaviour, despite the fact that they might remain unaware of them. It does, on the other hand, complicate attempts to correlate media consumption and environmental attitudes.

Participants expressed a belief that the media equip people with a certain amount of knowledge of environmental issues but that this level is insufficiently low for the majority of people to feel confident enough to discuss these issues in any depth. However, this lack of confidence amongst participants to enter into discourse about environmental issues betrays the level of knowledge which they possess. This was found to be particularly true of participants from rural areas who tended to be less articulate than those from urban areas. An analysis of their discussion reveals a much higher level of knowledge about the natural environment than among urban groups. Rural participants did not lack confidence in the knowledge that they have about the natural environment but rather their hesitancy in sharing this knowledge came from the supposition that the 'environment' concerns the global issues of the greenhouse effect and tropical rainforests.

Both urban and rural groups used media references much more frequently to justify statements made about the global environment than when they talked about local or national environmental issues.

## Summary of the UK responses

Where viewers of television programmes on an explicit environmental topic have no direct experience of the subject being tackled – for example, deforestation, lack of potable water, desertification – but have acquired knowledge of the subject through media exposure, they tend to conceptualise the issue and talk about it as an abstract concept, unrelated to their own experience.

Discussion of environmental problems about which they have direct experience – such as water pollution, vehicle pollution, loss of green space – is less frequently validated by reference to media sources. These problems tend to be treated as localised issues, the solutions of which require no more than mild changes in people's behaviour, easily achievable within the boundaries of present legislation.

In discussing environmental problems beyond their direct experience, most people generally do not have the knowledge to progress beyond discussion of the issue at a relatively superficial level. This is particularly true of older people living in an urban environment. People living in a rural environment are more likely to apply their knowledge of their own natural surroundings to environmental problems in a different geographical and cultural context. However, where a problem of the natural environment is compounded or complicated by other issues such as poverty or conflict, here again people have largely to rely on knowledge acquired from secondary sources, such as the media.

This leads to 'environmental problems' being categorised within people's minds as outside their everyday experience and thus well beyond their sphere of influence in terms of bringing about change.

## 6.5 THE RESPONSES IN INDIA

In this section we report on the focus group discussions held in Tamil Nadu and Maharashtra, two of the 22 states and 9 union territories of India. The chapter reports on the focus group discussions on the four documentaries in separate sections, but also attempts to bring together the views of participants on a range of subjects related to the environment.

Twenty-six focus groups were conducted in India from October 1993 to October 1994. Eighteen of these were conducted in the urban, rural and hilly regions of Tamil Nadu (South India); eight were held in the urban, rural and tribal regions of the western Indian state of Maharashtra. Tamil Nadu and Maharashtra are fairly large states, with a population of over 50 million each; that is, each with a population as large as the whole of Britain's. They are also fairly developed states compared to other states such as Bihar, Orissa, Andhra and Uttar Pradesh. In both states, there are campaigning non-government organisations (NGOs) which are attempting to raise the consciousness about the ecology of the region, and which are protesting against government developments and private industry which have an impact on the natural environment. The Narmada Bachao Andolan (see Chapter 4) is currently one of the best known of these.

Each focus group had six to eight participants and a moderator or animator. The participants came from different professional and demographic backgrounds. The focus group discussions were conducted in Tamil in the South, and in Marathi or Hindi or English or even a mix of these three languages in western India. The total number of participants in the 21 focus groups held was 136. None of the participants in the focus groups organised in India had seen the British documentaries earlier; nor is it likely that any participants had seen the Indian documentaries.

Each group watched one of the four video-taped documentaries (four groups in Maharashtra watched a British and an Indian documentary at one sitting). The two British films were dubbed in Tamil and Marathi, for focus group discussions in Tamil Nadu and Maharashtra. The viewing was immediately followed by a group discussion (in the local Indian language) with an assistant researcher acting as moderator/animator. The groups were kept informal, and the moderator's role was to keep the discussion moving. There was a minimum attempt to intervene or 'guide' the discussion along a certain track. No checklist of questions or instructions was given to the moderators.

All the discussions were taped, and later transcribed and translated into English. The English translations were then slightly edited (only where absolutely necessary), so that they made sense to an English reader. The translations were often literal translations (from Tamil, Marathi or Hindi) and did not always succeed in

*Table 6.6.1* Focus groups in India

| Film/Video | Region | Urban | Rural | Hills | Tribal | Total |
|---|---|---|---|---|---|---|
| *Advice Shop* | Tamil Nadu | 1 | 1 | 2 | 4 | 8 |
| | Maharashtra | 3 | - | - | 2 | 5 |
| *Growing Up* | Tamil Nadu | 1 | 1 | - | - | 2 |
| | Maharashtra | 2 | - | - | - | 2 |
| *Life and Death* | Tamil Nadu | 1 | 1 | 2 | - | 4 |
| | Maharashtra | - | - | - | - | – |
| *Insult to* | Tamil Nadu | 1 | 1 | 2 | - | 4 |
| *Civilisation* | Maharashtra | 3[1] | 1 | - | 2[1] | 1 |
| | Total | 9 | 5 | 6 | 6 | 26 |

*Note:* [1] These groups of Maharashtra were the same as for *Advice Shop*. Each of the groups watched *Advice Shop* and *Insult to Civilization* at one sitting.

bringing out the flavour of the discussions. However, they do offer a good summing up of the cut and thrust of the discussions that followed the viewing of each film.

### Advice Shop

Nine groups, four in Tamil Nadu and five in Maharashtra, watched one episode of the BBC2 series *Advice Shop* and later discussed the television documentary at one sitting.

One of the Tamil Nadu groups was from urban Tiruchirapalli (Trichy); the second from Mullal, a rural region of Tamil Nadu. The Trichy group was made up of five men and two women, while the Mullal group comprised three men and three women. The urban group was highly educated except for a housewife (primary education) and an attendant (high school), and included two professionals, a college lecturer and two postgraduate science students.

The rural group included three graduates, two young students and an illiterate coolie (porter). The average age of the urban group was 28 while that of the rural group was around 30. In terms of income, it could be concluded that the urban group was middle-class while the rural group was lower-class. One woman in the rural group was unemployed even though she was a science graduate.

Two Tamil Nadu groups were from the hill areas of Valparai and Mudumalai. Valparai is an urban hill area while Mudumalai is a rural hill area in Tamil Nadu. Two students, a housewife, a supervisor, a clerk and a coolie made up the Valparai group (average age 30), while two coolies, a teacher, a forest guide, a watchman and an ayah (maidservant) (average age 34) made up the Mudumalai group.

The Yavatmal group of Maharashtra was highly educated: the five men were graduates or postgraduates; the three women were all postgraduates. The average age of the group was 35, and the occupation was teaching or studying. The group also included a clerk, a factory technician and a housewife. Yavatmal is a district town in the Vidarbha region of Maharashtra. It is a semi-urban locality with a population of around 90,000. The nearest large city is Nagpur. Yavatmal is not an industrially developed region but is expanding rapidly.

The other groups from Maharashtra were from two tribal areas (Jawhar in Dahanu District, and Gadchiroli in Gadchiroli District), metropolitan Bombay, and urban Pune. Jawhar is a mere four-hour drive from Bombay, and borders the prosperous state of Gujarat. The village is electrified, has a college and a school (an 'ashram' school), and several tribal households possess their own radio/transistor set. (The focus group was conducted in the Educational Media Research sub-centre

*Table 6.6.2* Focus groups on *Advice Shop*

| Region | Type | Place | Number of participants |
|---|---|---|---|
| Tamil Nadu | Urban | Tiruchirapalli | 6 |
| | Rural | Mullal | 6 |
| | Hill | Mudumalai, Valparai | 6+6 |
| Maharashtra | Urban | Bombay, Poona, Yavatmal | 6+6+6 |
| | Rural/Tribal | Gadchiroli, Jawhar-Dahanu | 8+6 |
| | Total | Groups 9 | People 56 |

in Jawhar College.) The main occupation of the tribals is farming, though the younger generation is keen on taking up jobs elsewhere.

The tribal group at Jawhar comprised four men and two women, all but one of them barely literate or functionally literate. Three of the men were farmers (average age around 40), the fourth was an undergraduate student. One of the farmers was also a shop owner. Both the women in the group were housewives, though one was a farmer as well. The average household income of the farmers was about Rs. 1000 (or £20) per month.

The focus group held in Gadchiroli, a thickly forested tribal area in Chandrapur District, comprised eight participants, highly educated, except for a farmer and shop owner who were 'matriculates' (i.e. educated up to high school). The others included a bank employee, an insurance agent, a development officer, a social worker, a postgraduate student and a housewife. Three of the participants were distinctly upper middle class, and the others from the middle class.

The Bombay group consisted of six participants, five of whom were women. The moderator too was a woman. The participants were all graduates, two of them postgraduates, one in education, the other in law. As far as occupation was concerned, three were social workers, two students and one a schoolteacher. The income of the social workers was on average Rs. 6,000 per month, and that of the teacher Rs. 4,000 per month. The social workers were in their forties and fifties, the teacher in her late thirties, and the students in their mid-twenties. The Pune city group comprised four men and two women. Two of the men ran their own business, the third was a school-teacher and the fourth a young college student. Of the two women, one was a house-keeper in a hotel and the other a housewife. All the participants were highly educated: either graduates or postgraduates. The average age of the group was around 35, and the average income around Rs. 6,000 per month.

## Themes in participants' responses

### Admiration for the Golden Fields (anti-Tesco) Movement

The nine groups of Tamil Nadu were immensely impressed by the people's movement (discussed in *Advice Shop*) targeted at halting the establishment of the Tesco supermarket in Golden Fields. Invariably, they compared the unity and persistence of the people of the community to what they found in their own communities in India. The highly educated participants of Trichy, Bombay and Pune were aware of the Chipko movement in India against the clearing of the forests in the Himalayan region, the Tehri Dam project, the Narmada Bachao Andolan (Save the Narmada Movement) as well as the people's movement in Kerala (against the Silent Valley project), in Maharashtra (against the Enron Power project), and in Gujarat and Karnataka against the multinational, Cargill. However, they wished that the kind of unity the residents of Golden Fields displayed in the video was present in India. The video turned the attention of the participants to talking about their own local problems related to the environment. They realised the Indian context was 'different'. Said a Pune businessman, 'Our country does not need supermarkets; hence cutting down trees for it is of no use. Further, our problems and their problems are very different.' He found the episode to be biased, since 'Only bright points about Britain are shown ... I think they have an extreme bias about everything!' (MA/Pun/AS-IC).

But a young Bombay undergraduate student took a different view: 'Whether a supermarket is necessary or not is not the question. The motive of the film is to show how people are able to fight a big organisation if they have the will, the methodical approach and motivation as a driving force.' Even though Tesco has a large infrastructure, the people's movement was strong and powerful, she added (MA/Bom/AS).

A young postgraduate science student of Trichy observed, 'Apart from the wonderful co-operation of the community, these people had very good support and guidance from the authorities and politicians which I do not think we can expect in our country.'

An engineer in the same group endorsed the view, adding that 'Here our politicians and even people will lose their motivation if given money'. As an example he cited the case of agricultural land being used for mining by the Salem Steel Plant which was 'initially protested by the local people and politicians, things were brought to the negotiation table, but the politicians and the organisers received heavy amounts from the industrial authorities. As a result, the uprising lost its momentum.'

Another participant, a young science lecturer, remarked that 'The same thing happened in the Kodambakkam Atomic Power Plant Case: there was not enough resistance from the people' (TN/Tri/AS). [The power station should have been named Kalpakkam.]

The Vice-President of the village *panchayat* of Mullal too cited the case of Kothari Sugars, which started a distillery unit. The people initially objected, but they failed because of lack of co-operation (TN/Mul/AS). A middle-aged science teacher of Yavatmal cited the case of the Nippon Enro factory near Nagpur which released the waste and by-products from its galvanised sheet factory into a nearby canal without any kind of prior processing. The pollution was so heavy that the water from nearby wells became contaminated. Though questions were raised in the state legislative assembly about the factory, nothing was done to stop the pollution. A young science graduate of Yavatmal observed, 'It is not that we are inert towards the environment; we just take a little more time to come together, that's all' (MA/Yav/AS).

Several of the participants in the nine focus groups understood that the struggle was targeted at Tesco's attempts to set up a supermarket, in the process destroying an open space. They understood the implications of having a supermarket that would entail making space available for parking lots and other facilities. They were full of admiration for the leaders of the struggle. Several wished and hoped that a similar kind of awakening could take place in India: 'The methods the people use', observed a 40-year-old city engineer of Trichy, 'and the way the people are united is really excellent. That same people-power is lacking in our country.' Another (a 19-year-old science student of the same group) exclaimed that

> the steps taken by the people is amazing. They have to face the police, chain themselves to the heavy vehicles, conduct opinion surveys, and spend a lot of time and money. I wish that type of motivation and awareness should reach our people.

A 25-year-old housewife of Trichy remarked that it was 'wonderful' to see 'a lot of people fighting for something'(TN/Tri/AS). A 23-year-old university student of Yavatmal was full of admiration for the woman leader of the Golden Fields struggle: 'Look at the lady in the film, she spent an hour every day for the movement. How many ladies with that kind of dedication can you come across in India?'

But a fellow student, also a member of the Yavatmal group, disagreed.

No, there are quite a few dedicated people in our country too. What happened to Cargill [the multinational firm that tried to set up a salt manufacturing plant in Gujarat, but ultimately had to withdraw because of the active resistance of the local people] was the result of a mass movement. So also in Dhabol of Ratnagiri District. And of course the Narmada Bachao Movement. . . . It is not that we are inert towards the environment; we just take a little more time to come together and organise ourselves.

(MA/Yav/AS)

A 24-year-old housekeeper in a Pune hotel blamed 'dirty politics' for coming in the way of 'the power of the people', adding that 'We also in India try to oppose government policies but our efforts are fruitless because we don't unite' (MA/Pun/AS-IC). A highly educated 30-year-old clerk from the tea-estate of Valparai remarked: 'If we could have good co-operation among us like what we have seen in the cassette, I think we can definitely stop the destruction of the forests.' 'One reason why they do not co-operate and unite with one another,' argued a young college-going woman, was that 'everybody has a fear of the tea-estate employer' since all of them work on the estate and are dependent on the tea industry. The role of government is vital in the film, but, observed the clerk, 'I don't think in India our government would support the people' (TN/Val/AS).

However, the majority of participants in the Gadchiroli group felt that such a 'movement' could not take place in their tribal area. The watchman in the Mudumalai group declared: 'We should only bother about our own lives. . . . Revolution against employer, and strike and roadblocks are only for the city folk.' The 37-year-old coolie in the same group said: 'Our life is more important than the environment,' though he was impressed that 'the people of Golden Fields did achieve something to stop the cutting of trees'.

The forest guard in the group chipped in with, 'There [in UK] everything is done by machines . . . Nobody does any hard work. What they do is switch on some buttons.' This was in response to the maidservant's exclamation: 'How they cut trees, there! Here in Mudumalai men spend one full day to cut a tree, but there a single man does it in minutes' (TN/Mud/AS).

When the educated in the group pointed out that the felling of trees would reduce the amount of rain, the coolies came up with a totally unexpected riposte: 'The amount of rain we receive is more than enough. . . . I personally hate working in the rain!' Another coolie in the group asked pertinently: 'What do we do for fire-wood? We have to cut down trees for firewood' (TN/Madu/AS).

The members of the group were well able to pin down the responsibility for the destruction of forests. Asserted a forest guard of Mudumalai: 'The Forest Department is responsible for most of the felling of trees. They allow contractors to cut a huge number of trees' (TN/Mud/AS).

### Was the Golden Fields 'Movement' a success? Is such a movement possible in India?

While one view is that the struggle was a great success, another strongly expressed view was that the movement proved to be only a 'partial' success. As one young science student of Trichy observed, 'I accept it is only a partial success for the people of Golden Fields, because so many trees were destroyed.' Another remarked that some trees were destroyed because 'it usually takes some time for the people to organise themselves; it is a part of the game' (TN/Tri/AS).

Observed a Valparai coolie: 'The type of revolt we have seen on the cassette cannot be practised here. Mainly because of our economic situation. For example, when my employer asks me to cut a tree, I cannot say no to him.' He stressed, 'We are all coolies depending on this tea-estate. Even if we know that we have to protect the forests, I don't think we can do it' (TN/Val/AS). A watchman and a coolie from Mudumalai endorsed this approach. The watchman, said

What will we do if our officer drives us out of this estate? Remember always, we should always think about our lives first. This tea-estate is our life; we stay here, we work here and we should be grateful to our employer. Whatever they ask us to do, we should do. For, cutting a tree is not all bad. In my lifetime I have cut down more than a thousand trees. I have also planted quite a few trees. Now you are all happily working on the estate only because I have destroyed a few acres of forest land.

The coolie supporting this line of argument, affirmed: 'What uncle [the watchman] says is correct. We should obey our masters. Why should we bother about trees?' When a young teacher in the group warned that 'in a few years there won't be any forest, trees left in our area; nor will there be any wild life', the coolie responded: 'How is the wild life useful to us? The elephants destroy our plantation and the panther attacks our people.' A maidservant in the same group declared: 'I wish we can cut all the trees and convert these areas either into tea or coffee plantations' (TN/Mud/AS).

The rural group of Mullal had a different perspective. A young unemployed science graduate asserted that he understood that Tesco, a leading company, purchased a plot to develop it into a supermarket, but he believed that the people should not have agitated against the company. If the people want to admire the trees and the forest, let them go somewhere else. Why should they prevent development which is very much essential now?

However, an agriculturist and the Vice-President of the village *panchayat* took a different view. He stressed that

The people are united. They consult with many officers and they fight with the police to stop the cutting down of trees. They failed initially but later on they got the majority vote in the council and were able to stop the building. To me what the people have done is correct, because we have to have lots of trees and forests since they bring rain.

A retired teacher and now an agriculturist in the same village countered: 'The forest will bring rain whereas these types of supermarkets and industries will bring socioeconomic development. It provides job opportunities. So I strongly disapprove the action of the people in preventing the construction' (TN/Mul/AS).

Some participants in the Valparai and Mudumalai groups expressed similar down-to-earth sentiments about the greater importance of preserving and promoting jobs than that of the conservation of forests.

Could a Golden Fields Movement happen in India? Could a 'movement' or 'campaign' like that in Golden Fields take place in a country like India? A highly educated middle-class bank employee asserted:

I don't think so. ... If there were a proposal tomorrow for a big industry in Gadchiroli, which might endanger the environment, it will hardly face any opposition. This is mainly because the local people want jobs, some stable employment. That is why the environment will take a back seat.

Snapped a 62-year-old local farmer:

That's not true . . . Don't forget the Tultuli project. The central government has ordered that the project be stopped. Some forest land will be submerged, they say. Baba Amte and his followers have opposed the project and the central government has put a brake on the progress of the project. There has been no progress since 1982.

MODERATOR:   What is the project about?
YOUNG ARTS STUDENT:   It's an irrigation project.
BANK EMPLOYEE:   The central government stopped the work not only because of the threat to the environment. There are other reasons too. There are no funds with the central government. Secondly, the concern about the environment in this case has been put forward by the forest department and not by the people. I believe some rules regarding forest land have been flouted.

The discussion then veered to the region of Vidarbha around Gadchiroli. The elderly farmer asked desperately:

But why is the Forest Department after Vidarbha only? There are big dams and projects in other states, but there is hardly any word of protest from the department. True, our district [Chandrapur] has a dense forest cover, but does that mean we will be deprived of projects and developmental work for ever?

WOMAN SOCIAL WORKER:   In Western Maharashtra and other parts, the local leaders and councillors have done nothing, and the present councillor is active in the environment movement only to serve his political interests.
FARMER:   But that is a recent thing. The work was stopped in 1982. Crores of rupees have already been spent. And what about the Karwaka dam project? It neither affects the forests nor the people; then why has it been stopped?
SHOPOWNER AND INSURANCE AGENT (aged 42):   I think there is a need for proper guidance. We are not fully aware of the environment problems.
FARMER:   Look at the double standards of the central government. When Baba Amte fought against the Tultuli project, the government acceded to his demands and stopped the project, but now when he is fighting against the Narmada project on similar grounds, the government is not ready to listen to him.

(MA/Gad/AS)

### An Insult to Civilisation

Eight focus groups discussed the Films Division's documentary *Insult to Civilisation*. Four groups were held in Tamil Nadu and four in Maharashtra. The Tamil Nadu groups were held in Tiruchirapalli (urban), Sembarai (rural), and the two hill areas of Valparai (urban) and Karumalai. The Maharashtra groups were held in the urban areas of metropolitan Bombay and Pune city, and the semi-urban Yavatmal near Nagpur. The fourth group was from the rural and tribal parts of Dahanu. It must be noted that the Maharashtra groups also watched and discussed Advice Shop at the same sitting as they watched and discussed *Insult to Civilisation*.

### Themes in participants' responses

The Pune group was, like the others, critical of the film.

*Table 6.6.3* Focus groups on *An Insult to Civilisation*

| Region | Type | Place | Number of participants |
|---|---|---|---|
| | Urban | Trichy | 6 |
| Tamil Nadu | Rural | Sembarai | 6 |
| | Hills | Valparai, Karumalai | 6+6 |
| | Urban | Bombay, Pune, Yavatmal | 6+6+6 |
| Maharashtra | Rural | Gadchiroli | 6 |
| | Tribal | - | |
| | Total | Groups 8 | People 48 |

In all the 20 minutes, they presented so many points that it was confusing. No one knew what exactly they were trying to say.

(Teacher (42), MA/Pune/IC)

Whether it is industrial pollution, auto pollution, nuclear energy or the consumption of electricity, each concept is half dealt with.

(Businessman (42), MA/Pune/IC)

It looks as if it is just a government propaganda film. There is no participation of the people. It only gives you information, and provides no guidance at all.

(Teacher (42), MA/Pune/IC)

They should make it more understandable. The voice should be more friendly.

(Housewife (40), MA/Pune/IC)

Others pointed out that

It is not only environment we should stress, for it is the outcome of many other issues. Take, for example, population, traffic problems, deterioration in health care, corruption, water shortage and illiteracy – all these are interrelated. These are the main causes of environmental degradation. But we seem to stress trivial issues like tree-cutting and industrialisation, and overlook the major issues.

(Teacher (42), MA/Pune/IC)

### Factors responsible for environmental pollution

The Trichy group was dominated by two young teachers of Environmental Science and a 31-year-old college lecturer in Commerce. The other participants included a young businessman, a 38-year-old typist, and a young attendant. These last three had to be coaxed by the animator to contribute their views to the discussion. The discussion revolved round 'pollution at various places', 'types of pollution'. The group was very appreciative of the title *Insult to Civilisation*, but felt that some solutions to the problems should have been suggested at the end. As one young Environmental Science lecturer put it: 'The documentary talks about various types of pollution, how they are caused, what the effects of pollution. But nothing is said about control and preventive methods. Moreover, the documentary ends abruptly without giving any message to the viewers.' He also pointed out that little importance is given to 'recycling of waste products' as a method of pollution prevention. The film shows only 'solid waste recycling' as a solution (TN/Tri/IC).

The factors that were seen as responsible for pollution can be grouped under several heads.

*(i) Overpopulation.* The film points out that 'overpopulation' is a contributing factor to pollution. A woman participant bemoaned that the 'government has not been able to do very much about it'. The businessman was emphatic that 'Any act which leads to economic development will definitely lead to certain ill effects which is usually due to human error or negligence' (TN/Tri/IC).

From another participant, a woman student:

We know that the population explosion is such a big problem in our country. But our politicians are still not prepared to make family planning compulsory, irrespective of religion. A Common Civil Code should be introduced in the country. But the political leaders would not dare do so. They fear losing the Muslim vote bank.

(MA/Yav/IC)

But a 27-year-old upper middle-class housewife of Gadchiroli perceived:

The environment is the biggest and most crucial problem. I have been taught right from my childhood that population is the biggest problem; now the time has come for everybody to be taught – right from his or her childhood – that environment is the biggest and most crucial problem facing the world.

(MA/Gad/IC-AS)

A Development Officer (30) in an insurance company, upper middle-class and not a native of the village, declared: 'I do think that it is largely educated people like us who are responsible for the environmental degradation' (MA/Gad/IC-AS).

The Valparai group was also an urban group, but from the hills. It comprised four men (average age 45) and two young women, one a student (18) and the other unemployed (21). Two of the men were educators, and of the other two (both illiterate), one was a tailor (48) and the other a peon (35).

The two other groups were from rural parts of Tamil Nadu, one from Sembarai on the plains, and the second from Karumalai on the hills. The Sembarai group consisted of four men (a young mechanic, two farmers and a coolie), and two women (a coolie and a middle-aged housewife). The Karumalai group comprised two coolies (a 52-year-old man and an illiterate 40-year-old woman), a middle-aged primary school teacher, a young student, a 45-year-old woman cook, and a young woman whose occupation is organising noon meals (lunch) for schoolchildren.

The town of Valparai has no big industries and therefore the kind of pollution graphically depicted in the documentary has not been experienced by locals. However, as the illiterate tailor of the group pointed out, 'Look at the drainage at our bus-stand and in the market; all these places are dirty.' And the peon observed, 'But comparing to what we have seen on the cassette this is nothing.' Averred a 52-year-old teacher of Valparai: 'Overpopulation is responsible for all the problems associated with environmental pollution. Family planning is the only way by which we can check pollution.' An Assistant Education Officer, a government employee, held: 'The government is responsible for all these problems, they should take necessary action against the people who are responsible for pollution.' But the tailor of the group believed that 'If the government and the people are united we can solve the problem. The high number of vehicles which we use in towns and cities contribute to the pollution' (TN/Val/IC).

The Trichy group that discussed *Growing Up* observed that 'one main reason for environmental pollution is population'. A 25-year-old typist noted: 'We have to produce more to feed these people; so the amount of pollution is also high.' A science lecturer endorsed this view: 'Whenever the population increases, the pollution level also increases.' A science teacher, however, had a different view on the matter: 'Population may be high, but if the people were given proper facilities, the pollution level could be kept under control.'

ATTENDANT (31): Yes, at least the basic facilities like toilets, drainage, roads and street lights.

SCIENCE LECTURER: I think poverty too is closely related to pollution. I believe the children are exposed to pollution mainly because of poverty. For instance, the Kenyan lady [in the documentary] had to cut a tree to feed her three children.

*(ii) Urbanisation* 'Urbanisation is another factor responsible for industrial pollution,' affirmed an educated but unemployed man of Valparai. 'Illiteracy is yet another factor.' He also believed that 'inequality' was an equally important contributory factor in India. When challenged to show how inequality influences pollution, he responded:

Since the rich people are becoming richer they want to make more money. So they make money by exploiting poor people. The poor people live in filth, defecate in the open air, and throw their garbage out in the open. If they were provided basic facilities, I am certain they will pollute the environment much less.

(TN/Val/IC)

In the Yavatmal group of Maharashtra, there was much discussion of urbanisation leading to pollution:

TEACHER (30): The 'flat system' culture in the cities is equally bad.

YOUNG SCIENCE STUDENT: But it saves space. Instead of haphazard expansion it is good to restrict the cities by introducing the 'flat system' culture.

HOUSEWIFE (54): But then it should take place with appropriate and adequate facilities. . . . You must ask: Is there a proper waste disposal system? Or parking space?

YOUNG ARTS STUDENT: Even educated people like us are not aware of the proper waste disposal systems.

TEACHER: If you have a small house, you generally plant at least a tree or two in the backyard. How can you do this if you live in a flat? Now, about waste disposal . . . we can make good quality compost from the household waste.

YOUNG EDUCATED TECHNICIAN: The day is not far off when we will need an 'oxygen bar' in the city square.

(MA/Yav/IC)

*(iii) Lack of public sanitary facilities* As one of the Valparai participants pointed out, there was only one public toilet in the whole town, and this was unusable because of the lack of water. The need for the construction of public toilets and bathrooms was stressed by all the participants. 'Public toilets and bathrooms would stop people from polluting the environment' (TN/Val/IC). The Karumalai group also dwelt at length on the need for toilets and bathrooms. When a high school student asserted that 'we don't harm the environment', a senior teacher (55) said this was not so at all.

Think about what most of us do in the morning or night: we go to the nearby open land to deposit our waste. And we wash in the stream close by. This is also considered to be polluting the environment.

(TN/Kar/IC)

But he was immediately challenged by a coolie: 'What shall we do? We don't have toilets. Only the officers in our estates are provided with toilets. All others are forced to use the open air. So there is no point in blaming us poor people.'

The senior teacher proffered another instance of environmental pollution in the hilly estates of Karumalai village. He spelt out how the placenta of calves was packed in dry leaves in the belief that this would help cattle give more milk. 'We should stop this. There is no meaning in it. But people do it unknowingly. What happens is the bad smell remains for days together, and there are maggots all over the place' (TN/Kar/IC).

Economic status is closely associated with the pollution problems. But an Education Officer (TN/Val/IC) saw economic status as a partial excuse. He believed that 'The only thing we require is people awareness and government action'. 'We should stop harming nature. We should educate people to stop polluting the environment. But the economic status of our [Karumalai] people has to improve for this' (TN/Kar/IC).

People's participation' and 'people's awareness' were stressed again and again in the focus group discussions.

You cannot have a housewife dumping the domestic waste on the roadside.

(TN/Val/IC)

No, no. He's not saying that only women are polluting the environment. All of us are doing it, either knowingly or unknowingly. So all of us should realise that we should not pollute the environment.

(TN/Val/IC)

*(iv) Religious practices* The Yavatmal group (of Maharashtra) and the Permulai group (of Tamil Nadu) referred to the pollution of the sacred river, the Ganga.

RETIRED HEADMASTER:   It is true that some of our religious practices are responsible for the pollution of the Ganga. But don't forget that the haphazard development of industries along the Ganga is the main cause of pollution. It is in the last 30 to 40 years that Mother Ganga has become polluted. Mahatma Gandhi was really a seer. He had warned against the superficial development long ago. But his warnings were ignored in the atmosphere charged with 'development mania'.

CLERK (24):   This has been the tragedy of almost all great leaders. Their followers have always betrayed their ideology. These followers play up only those aspects of the ideology which suits their interests.

(MA/Yav/IC)

The Permulai focus group was as critical:

ARMY JAWAN (30):   Have you ever looked at the water of the Ganga [Ganges]?

COOLIE (37):   You mean Ganga, the holy river?

ARMY JAWAN:   Yes. If you have looked at the Ganga you would not say that it is a holy river.

WOMAN COOLIE (37):   Why do you say that? Some of my relatives went for a pilgrimage to the Ganga.

ARMY JAWAN:   I believe that the Ganga is the dirtiest river I have ever seen. It more or less resembles the river shown in the film. It seems that they have shot the film in the Ganga. You can see dead animals, human beings, waste water from industries and a lot of rubbish in the river. It is highly polluted.

WOMAN GRADUATE STUDENT (20): I have heard that they are cleaning up the Ganga river.

ARMY JAWAN: Yes, they are cleaning up the river right from the Himalayas to West Bengal. But the people who live on the banks of the river should stop polluting it. Otherwise, the cleaning-up operations will have to be done every year.

(Tn/Peru/MLD)

## Pollution: is it an urban problem?

The Karumalai group was of the general view that the pollution depicted in the film was an urban problem, the result of the use of soap and detergents, the high density of traffic, the high level of inequality in cities, and the 'ease of life' (as evidenced in the urban people's use of washing machines, pressure cookers, electric mixers and grinders, etc.). A 52-year-old coolie asserted: 'Here we don't see that type of dirty water, and there are plenty of vehicles and big industries. I don't understand why they show all this.'

In Maharashtra, four focus groups discussed *Insult to Civilisation*, together with an episode from *Advice Shop*. These four groups did not discuss the Indian film at a separate sitting, as was the case with the Tamil Nadu focus groups. The Yavatmal group was reminded of the industrial pollution caused by the Nippon Deenro factory near Nagpur, which manufactures galvanised sheets. As a young science teacher put it: 'The waste and the by-products from the factory are released into a nearby *nullah* [ditch] without any processing. The pollution was so heavy that the water from nearby wells got contaminated. Questions were raised in the state legislative assembly.'

HOUSEWIFE (54): How can the local authorities give a 'No Objection Certificate' (NOC) to such factories?

LADY TEACHER (42): Perhaps the company did not know about the possible consequences.

YOUNG ARTS STUDENT: No, no ... these companies know very well of the possible hazards of the waste and by-products ... but they just don't care.

(MA/Yav/IC)

The Yavatmal group was also reminded of the Bhopal disaster:

RETIRED HEADMASTER (62): What happened in Bhopal is a similar case. We don't pay attention to these problems. Take for example these cement concrete houses, the tarred roads ... these are all comfortable for us but these comforts have an adverse effect on the environment.

(MA/Yav/IC)

YOUNG ARTS STUDENT: But automation saves time, you see.

HEADMASTER: It's of no use. You save time and you start indulging in the things you shouldn't. Remember: 'An idle mind is the devil's workshop!' Have you read Minoo Masani's *Our India*? He had cautioned Indian society about increasing mechanisation. Mahatma Gandhi put forward a whole philosophy of living with nature.

YOUNG CLERK: The earth is round! We will again return to the same place from where we started.

(MA/Yav/IC)

*Rural groups' perception of urban life*

It is interesting to see how the rural groups compared their own lifestyle with that of the urban people:

> Our hill status is fine comparing to what I have seen in the film. We usually walk 10 to 15 km a day. Even if we happen to travel by bus or train we usually chat with our neighbour. We make friends during our travel, but in the film a lady passenger is reading a book and another passenger is listening to the radio. I feel that city people don't even know how to relate to others.
>
> (Coolie (52), TN/Kar/IC)

> City people don't have time.
>
> (Teacher, TN/Kar/IC)

> Another important thing is that we sit together and talk. The town and city people consider this a waste of time. We have unity, a sense of oneness; city people lack this.
>
> (Teacher (55), TN/Kar/IC)

> Even though the people in cities are rich and more educated and own a lot of property, they live a mechanical life. They are so busy that they don't have good family relationships like us in villages. . . . They live in a complex world. But our life in the hills is calm, quiet and we fully depend on other people for some activities.
>
> (Teacher (55), TN/Kar/IC)

The focus group from the tribal area of Dahanu in Maharashtra expressed similar sentiments, but more aggressively:

> In the film [*Insult to Civilisation*], vehicles, factories, railways are shown. Do you think we don't understand that? We live a simple lifestyle. According to me, if city people were to gradually change their lifestyles – towards simplicity – I think we will be able to change the environment for better.
>
> (Functionally literate farmer (30), MA/Jaw/IC)

> We live up to 80–90 years, but they [city folk] start ageing at 45 or 50.
>
> (Illiterate farmer (42), MA/Jaw/IC)

> Every time they [the urban folk] tell us to use solar energy, smokeless *chulhas* [domestic stoves burning coal or wood], but why can't they themselves use them. They want better lifestyles at our cost.
>
> (Illiterate housewife (40), MA/Jaw/IC)

> They will use cars and planes, but we must walk. ... And if we want some comforts they say that we are destroying our culture.
>
> (Undergraduate student (22), MA/Jaw/IC)

A Gadchiroli shop-owner (also a farmer and money-lender) (40):

> People in villages live in the close company of animals and trees. They always maintain a balance between humans and animals, and thus preserve the environment. The only thing is that they speak about the environment using a different terminology. If you go away from this village you will find a *nullah* [ditch]. But you will not find one in a small village, because there the sewage-water goes directly to the trees and plants. They give waste vegetables and other unused eatables to cattle, or compost it. Plastic is used nowadays but earlier the villagers used to do with paper bags. They used to recycle paper.

Other suggestions made for the reduction of pollution included: the small family norm, the working out of a long-term plan to correct environmental pollution problems, and rigorous punishments for industries responsible for pollution.

## A Matter of Life and Death

The environment documentary, *A Matter of Life and Death*, begins with the declaration that water is a source of life, but that this source is being polluted. We are shown in graphic detail (generally in top-angle shots) Bombay's slums where people (and buffaloes) bathe, wash their clothes, draw drinking water, wash dishes, often using the same water source. As in the documentary *An Insult to Civilisation*, this documentary too has a background commentary, with hardly any points of view expressed by the people shown in it. We are told that there are 130 river basins in India. We are shown that the major polluters of these river basins are: tanning of leather, industrial wastes, dyes and pharmaceutical industries. The commentary warns that these industrial and other practices of dumping of waste return the wastes to the sources of life and thus contaminate the very sources of our food and drink. This gives rise to many epidemics. The documentary, while touching on the various sources of pollution, offers no solutions, except to state that fighting pollution is 'a matter of life and death'.

Four focus groups on *A Matter of Life and Death* were held in Tamil Nadu (Table 6.6.4). No focus group on this film was conducted in Maharashtra. Two of the groups were from the urban areas and the other two from the rural areas of the state. The urban areas included Trichy on the plains and Valparai on the hills, while the rural parts included Sembarai on the plains and Perumalai on the hills.

While viewers in British focus groups were 'shocked' by the sight of people in urban slums of India bathing, washing clothes and even drinking the polluted water from the same sources, Indian viewers considered this 'very pathetic'. As one 40-year-old coolie of Perumalai expressed it:

> Water is important, but it is very pathetic to see how people are spoiling it. There are dead animals, the drainage mixes directly with the river water. But the people do not seem to be worried about it. They simply take a bath and wash their clothes in the same dirty water.

(TN/Per/MLD)

*Table 6.6.4* Focus groups on *A Matter Of Life And Death*

| Region | Type | Place | Number of participants |
| --- | --- | --- | --- |
| Tamil Nadu | Urban | Tiruchirapalli | 6 |
| | Rural | Sembarai | 6 |
| | Hills | Perumalai, Valparai | 6+6 |
| | Total | Groups 4 | People 24 |

**Themes in participants' responses**

*Industry as a source of pollution*

The consensus was that industries are largely responsible for the water pollution, but a number of participants hold the people themselves and the lax attitude of the government equally responsible for the tragic situation. Most participants in the four focus groups could relate the issues raised in the film to the environmental pollution caused by industries located in their own neighbourhoods or regions. For instance, the Sembarai group could identify with the issues raised, and cited the pollution caused by the Kothari Sugars and Chemicals factory in their own locality.

RETIRED ACCOUNTS OFFICER: The film clearly indicates that industries contribute to environmental pollution, especially to water pollution.
(TN/Tri/MLD)

GRADUATE STUDENT: Industries are mainly responsible for polluting water. As shown in the film, the leather industry and the paper industry are largely responsible.
(TN/Per/MLD)

But this was contested by an army jawan (private) in the same group: 'There are other chemical industries and dyeing industries which cause pollution to a much larger extent.'

The Sembarai group discussed industrial pollution in a similar vein:

COOLIE: But what can the industries do with the polluted water? They have to send it somewhere.

GRADUATE STUDENT: There are a few treatment methods by which they can purify the water before sending it outside. One such method was shown in the film. Scientists even say that the water thus purified is suitable for drinking.

ARMY JAWAN: Yes, I have visited an industry in Rajasthan where they treat the polluted water and use the water for agriculture just outside the industrial campus.

COOLIE: How was the yield?

ARMY JAWAN: I believe there is a good yield of crops, and also they are not found to be affected by the chemicals of the original effluent.

WOMAN COOLIE: Then why don't all the factories follow their example?

STUDENT: It involves money. They have to spend some money for the purification of water. So, instead of spending that amount of money, the industries usually drain off the untreated polluted water.

ELECTRICIAN (26): ... at the same time, big industries pollute the [drinking] water by mixing their effluent with the water of rivers. Further, they pollute the seas though this can be prevented. ... This is associated with development.

BRICKLAYER/MASON (24): In our region, the effluents released by Kothari Sugars and Chemicals cause a lot of damage to the paddy fields, and also to the people. It is more or less like what we have seen in the film.
(TN/Sem/MLD)

*The responsibility and failure of government in pollution abatement*

The Government and other regulating bodies were blamed by participants for being too lenient with industries:

The industries are responsible for the pollution of water. But the government is not taking enough steps to control the pollution of water.

(Army jawan (soldier) (30), TN/Per/MLD)

Why doesn't the government take necessary steps to prevent mixing of polluted and unpolluted water? The animals should be washed in a different place.

(Graduate student)

Yes, here also some people wash their domestic animals in the stream from where we take water for drinking.

(Housewife (45), TN/Per/MLD)

Even though all of us, including the government, are aware that industries are polluting, by dumping their wastes in places close to human habitation, no action is taken against them [The film too does not show any action taken against industries]. ... I have visited an industry [Fenner India Ltd, Madurai] where they don't send the effluent outside, but store it in various tanks (like perforated tanks) and the resultant treated water is used for gardening and other purposes within the campus itself. If every industry could follow this example, water pollution could be prevented to a great extent.

(Businessman (33))

A science lecturer in a local college pointed out that 'The government is taking some initiative to prevent or stop pollution: I remember the Taj Mahal was given a chemical wash to remove the chemical sediments deposited by nearby chemical industries' (TN/Tri/MLD).
Other responses included:

Water pollution can be controlled to a large extent if government enforces environmental protection acts, properly and efficiently.

(Science teacher (40), TN/Val/MLD).

Why is the government not doing anything about it? I know that most of the industries are owned by either politicians or their relatives ... that is the reason why the government is keeping quiet.

(Schoolteacher (45))

Yes, I agree with you. Government is responsible for most environmental disasters.

(Science teacher in college, TN/Val/MLD)

## Role of the public in water pollution

Several participants held the people themselves responsible for pollution of the sources of water in urban and rural parts:

RETIRED ACCOUNTS OFFICER: Slum dwellers contribute to pollution by using the same water for washing their animals and for dumping dead animals, solid wastes, etc.

SCIENCE STUDENT: I don't think we should blame only the people; they do it because of their poor socio-economic status.

COLLEGE LECTURER: Yes, this is shown very clearly in the film: the slum people live close to the polluted areas, draw water and drink from the pond, even from a ditch; bathe in the polluted water where animals also bathe. The people risk being affected by various diseases.

(TA/Tri/MLD)

YOUNG HOUSEWIFE:   I have not seen anything like this before. Now I am very worried about the way we pollute water – without which we cannot survive. I don't know whether people are doing it knowingly or unknowingly.

FIELD OFFICER (37):   If people are doing it unknowingly we can educate them to prevent environmental disaster; but it is difficult to tackle the people who do it knowingly. Only Acts and laws can stop them from doing so. Punishment should be given to them.

SCIENCE TEACHER (40):   The question is not whether people are doing it knowingly or unknowingly; they should not do it. Any one who pollutes the water is a criminal.

BUSINESSMAN (37):   But it is not always possible to punish everyone to stop water pollution. People should realise the importance of water to their lives.

FIELD OFFICER (37):   It is pathetic to see even developed countries like the US and Japan being involved in polluting water by dumping wastes into the sea.

SCIENCE TEACHER:   The major disaster related to water pollution was the Gulf War where we saw, on television, oil dumped into the sea. Many birds and animals and fish died as a result.

PRIMARY SCHOOL TEACHER (45):   I too saw that on television. I also saw people cleaning up the polluted beach.

BUSINESSMAN:   I don't understand why people pollute water even though they know how important water is to their lives. All major rivers in India are polluted.

SCIENCE TEACHER:   I feel the film was shot in Delhi or Bombay. The rivers in north India are more polluted than rivers in south India. The Cauvery, Vaigai and Aliyar rivers, for example, are comparatively cleaner than Ganga and other rivers.

FIELD OFFICER:   The rivers are less polluted than the lakes, ponds and other stagnant waters, since in the rivers there is a constant flow of water.

### The solutions to environmental pollution

FIELD OFFICER (37):   Why don't the people do something about it like the agitation against the Narmada project in north India? They don't want to disturb the environment. So they have fought a battle with the government, with the politicians. Now the World Bank has decided not to give any grant for the project.

UNIVERSITY STUDENT (21):   A similar example [in south India] is the Koodangulam Atomic Power Project [should be Kalpakkam]. But we should remember that we have to feed more and more people. So environmental degradation is inevitable.

SCIENCE TEACHER (40):   Yes, that is true, but let us think about well-developed countries like US, UK and Japan. Think about the amount of pollution they have; I believe it is comparatively less.

FIELD OFFICER:   I do not agree with you. If you see the incidence of cancer and other related physical problems, it is higher in the developed countries. Here in India, we usually see the dirt and the filth but there it is concealed or they dump it into the sea. So pollution is there every where.

(TN/Val/MLD)

BUSINESSMAN:   I believe the main theme of this documentary is to show how water is polluted at various levels.

FIRST SCIENCE LECTURER:   Yes, it argues that industrialisation and urbanisation contribute to pollution.

EDUCATED HOUSEWIFE (a science graduate): Urbanisation results in formation of slums where people live in poor-quality housing, drink dirty water and
dump waste everywhere.
RETIRED ACCOUNTS OFFICER: I think the only way to stop this is to educate
the people, especially the poor.
SECOND SCIENCE LECTURER: Why only the poor? We should educate
everyone. We can use this type of documentary for the purpose.

(TA/Tri/MLD)

In three of the four focus groups the need for boiling water for drinking was
emphasised. As one young housewife observed: 'After seeing the film I am going
to boil water before I give it to my children.' A businessman in the same group
endorsed this, saying: 'Yes, this is a very simple way by which we can prevent
various water-borne diseases' (TN/Val/MLD).

PETTY SHOPKEEPER (25): The only way to prevent the spread of disease is
to drink boiled water. Because I don't think we can get pure drinking water
anywhere.
ILLITERATE COOLIE (48): We should have separate canal systems for river
water and effluent.
ELECTRICIAN (26): It is not practical. Even if we did it, the ground-water
would certainly be affected. We could dispose of the polluted water in our waste-
lands. For all these problems associated with water pollution, government is
responsible because they provide licences to a large number of factories.

(TN/Sem/MLD)

## Growing Up: children in the developed and developing world

The documentary is in three parts, the first lasting for 25 minutes, the second for
12 minutes and the last for 15 minutes. The three parts were shown at a stretch,
without a break or interval. The documentary traces the growth of a child from
conception, through birth and development. This explicitly and implicitly makes
the point that we made in the Foreword, that all aspects of development from physical conception to political future are linked.

Four focus groups were held on the British TV documentary, *Growing Up*: two
in Tamil Nadu and two in Maharashtra (Table 6.6.5). The Tamil Nadu groups were
held in Trichy and Pullambadi, the first an urban area, and the second a rural area.
The Trichy group comprised four men and two women; the Pullambadi group three
men and three women. Two housewives, a teacher, a lecturer, a typist and an attendant made up the first group, while a housewife, a teacher, a physiotherapist, a
driver and a coolie made up the second. The driver and the coolie were illiterate,
but the others in both the Trichy and the Pullambadi groups were well educated.

*Table 6.6.5* Focus groups on *Growing Up*

| Region | Type | Place | Number of participants |
|---|---|---|---|
| Tamil Nadu | Urban | Tiruchirapalli | 6 |
| | Rural | Pullambadi | 6 |
| Maharashtra | Urban | Pune | 8+8 |
| | Total | Groups 4 | People 28 |

The two focus groups of Maharashtra were held in the city of Pune, 150 km from Bombay. Both the groups had eight participants. The first group comprised two housewives, two students, a businessman, an engineering consultant and a retired senior citizen. The second group brought together two 'government servants', an industrial worker, a salesman, a household worker (domestic), a manager, a student and a retired senior citizen. Except for the household worker and a housewife, the other participants were all well-educated.

## Themes in participants' responses

### *The Rio Summit: expectations and disappointments*

Most participants 'enjoyed' watching the documentary, but found it 'very long', 'rather boring', though 'the message was very clear'. Some found the introductory section (the birth of a child) irrelevant to the main theme. However, they interpreted the main theme/thrust of the film differently.

According to a science student in Pune, 'It is about the degradation of our ecology and how it is affecting the human race.' But a senior citizen of the same city believed the documentary was about 'the problem of development, and how civilisation is affected by the degeneration of the environment. It is also about the problem of overpopulation.' A Pune salesman observed, 'I liked the film because it compares the situation in the developed and underdeveloped countries' (MA/Pun1/GU). One young housewife felt that 'the film was mainly about child rearing, and how men should also help women in child rearing' (TN/Pull/GU).

A sugarcane inspector saw a problem with the documentary: 'The idea is to tell people about poor environmental conditions in which children live, but what the documentary does focus on is children feeding, bathing, playing, etc.' (TN/Pull/GU). An illiterate driver summed up the 'contradictions' and 'confusions' in the documentary: 'They show children, then industry, then cooking, some animals, people going to work, people cutting down trees . . . there is no smooth flow of information like what we see in Tamil pictures' (TN/Pull/GU).

The commentary by Roger Moore made an impression on the Pullambadi folk. 'See how a famous actor has involved himself in environmental education,' remarked a participant. 'Everyone should take up the challenge to educate others [about the environment]' (Tn/Pull/GU).

The consensus among the Indian focus group participants of *Growing Up* was that the resolutions of the Rio Summit did not make any difference to the ground realities of the environment situation in the country or the world. The world leaders were seen to be no better than the leaders of India. Stated a 42-year-old science teacher: 'There is no point in our leaders signing a treaty; it is useless.' He was also suspicious of the 'loans' being given by the rich to the poor countries to help them improve their environment. He said:

> They will give loans, but in turn they will ask for something else. It may be for a military base or permission to carry on trade . . . the rich countries are not at all helpful. They are indirectly swindling the wealth of poor countries.
>
> (TN/Tri/GU)

A sugarcane inspector (50) of Pullambadi village, Tamil Nadu, felt let down. 'Two years ago I read about the Earth Summit at Rio. I was very happy to know that so many countries came together to save the environment, but now I realise it achieved nothing.' A physiotherapist (40) of the same village despaired of Rio making any difference: 'I think we cannot eradicate pollution. We may sign agreements and treaties, but nothing is going to work out. Our children are going to suffer. . . . Our children may die of hunger but not owing to pollution.' A science student commented that the film 'exposed the hypocrisy in the actions of all governments, against the backdrop of the Rio Earth Summit'. A mechanical engineer, seconding the comment, pointed out the 'dumping of waste by developed countries in the waters of the developing countries, the cutting of rainforests in Brazil and Malaysia by developed countries for their own use'(MA/Pun1/GU).

Comparisons were drawn between ways of nursing and rearing children in the affluent industrialised West and the poorer countries of Asia and Africa. A young housewife of Trichy commented: 'I was happy to see how children are nursed in different countries. I noticed that in most countries men also care for the children, but in our country men do not seem to help the women.' Another young housewife of the same group was 'amazed to see how the rich white people spend so much money to nurse their children. They provide children with separate rooms, clothes, toys, etc., while the poor black people have nothing; they even carry their children to their places of work.' A young science lecturer believed that this difference was 'basically due to the economic status of individuals' (TN/Tri/GU).

The presence of the husband in the operating theatre while a woman was giving birth to her baby was a strange experience for some participants; they wondered what connection it had with environmental protection! One young housewife wondered 'how can they allow a husband during the time of delivery? I thought that man was also a doctor!' Another housewife observed that 'if we could have the same practice in India, there could be a reduction in the rate of child births'.

## The portrayal of India

Most participants were disturbed (even angry) at the way India was portrayed in the documentary. (India was one of the case studies shown in the documentary.) A Pune businessman said he liked the film, but objected to the Indian scenes: 'They have only shown the poverty of our country. I could not understand why they should show beggars in India, when the film is about the environment. Doesn't the developed world have poverty?' (MA/Pun1/GU). However, a government employee (46) countered that 'the film has shown the two sides of a coin very well'.

Participants were particularly disturbed by the portrayal of child labour in India:

ATTENDANT (31): I was not at all happy with the way they had shown India. There are other rich and beautiful places [in India] and there are better ways of child rearing in the country, but these are not shown in the documentary.
SCIENCE LECTURER (30): The idea is not to degrade the beauty of our country, but rather to show how our children are exposed to risk at work.
SCIENCE TEACHER (42): Why did they talk about child labour in India? I don't think there was any need for it, as they do not even hint at child labour in other countries. It also appears that there is child labour only in India.

A debate ensued on the need for child labour at Sivakasi's fireworks-manufacturing company.

SCIENCE LECTURER: Three weeks ago a fire in the factory killed many children. Who cares about them? Not the government, or the parents or the employer.

TYPIST (25):  We have to think about the economic conditions of the people who live in Sivakasi. They don't have any other job, so naturally they have to work and send their children to work knowing full well that this would lead to health problems.

YOUNG HOUSEWIFE:  How can a mother send her own children to work in a place where there is so much risk? Look at how parents in other countries, especially in the West, care for their children. I have heard that in those countries they don't have proper families. The husband and wife don't live together, and they never bother about their children. But in the film, it is very different: they do care for their children.

(TN/Tri/GU)

When questioned about what could be done for the children of Sivakasi, a physiotherapist (40) averred that 'it was their fate', though this was contested by a schoolteacher (27): 'How can we say it is their fate. We should do something about it' (TN/Pull/GU).

A housewife was also struck that no scenes of pollution in the film related to the United States or to Britain.

How can you say that in the US and UK pollution is a problem? I have not seen anything in the film. Everybody seems to have a good house, beautiful roads, and everywhere it is very clean. The children are healthy and beautiful. I don't believe there is pollution in these wealthy countries.

Science lecturer:

Even in those places which appear to be clean there is pollution. They have air pollution, radiation pollution because they are developed nations. They use much power, especially atomic energy. They also use chemicals. They use power for cooking and for production.

(TN/Tri/GU)

'Power' means specifically electric power: power stations are associated with pollution – but in India most people cook on their own solid fuel or kerosene stoves, therefore they are 'not contributing to pollution'.

### Themes cutting across all four documentaries

#### Indian government policy on the environment

The Indian Government's over-centralised top-down policy on environment came in for a lot of flak: 'The government in Bombay or New Delhi should not decide for Gadchiroli. Local people must be taken into consideration' (rich shop-owner of Gadchiroli, MA/Gad/AS-IC).

DEVELOPMENT OFFICER:  Now, the Government has some policy on the environment, but this is policy on the national level. A district or local level policy is needed. Almost every town has its own police station. In the same way, why can't there be an environmental station in every town? Bombay, Delhi, London, should not decide for the rest of the world. The decision makers and enforcement machinery should be decentralised. See, we have a Regional Transport Office (RTO) here in Gadchiroli; hence one thinks twice before breaking any traffic rules, but in Kunughada [a small village nearby] you are a free bird! The important point is the decentralisation of the process. If it takes place some government officials, field-workers will come to the towns and villages and would act as agents of environmental awareness.

INSURANCE AGENT (42): Every time it seems that we are dependent on the government machinery: the government will do this, and the government will do that.

WOMAN SOCIAL WORKER: You are right. People must come forward. Voluntary organisations should take up this cause. But how are they to raise the huge funds required for the implementation of the policies? Who will pay the officials and workers?

(MA/Gad/AS-IC)

### Environment and development

The apparent conflict between environment and development was touched on by several participants during the group discussions on the four documentaries. A 20-year-old science student of Yavatmal believed that to strike a balance between environment and development is really difficult: 'Now we want big dams for electricity and irrigation but at the same time we would not like any damage to be done to the environment' (MA/Yav/AS). A 54-year-old housewife of Yavatmal declared: 'We can't put the clock back; it is not possible.' But a retired headmaster of the same city observed that 'one should not blindly follow the western model: that would make us dull and dependent'.

Most Yavatmal participants endorsed the Gandhian philosophy of 'living with nature'. Gandhi had cautioned against 'increasing mechanisation'.

An aspect of the environment debate in India that is often not properly considered is the political factor. Participants in the Indian focus group discussions showed awareness of the political role in environmental movements. A Yavatmal science teacher asserted:

The political factor is also important as far as the environment is concerned. But unfortunately in our country, the environment is not on the political agenda of our political parties. The environment has not yet become a crucial social and political issue [in India] ... as yet. Family planning and anti-dowry movement could really become a people's movement. We really need an environment awareness campaign.

(Young science teacher, MA/Yav/AS-IC)

A 19-year-old arts student from Gadchiroli:

Economic development must incorporate, among other factors, a deep concern for the environment. We must stop the exploitation of nature. Let's not become a third party in the process. Instead we should become a part of the larger process. We must restrict our needs and stop the haphazard use of natural resources like coal, petrol, minerals, etc. Nature provides all the answers; the only thing is that we must be ready to make some compromises. See, we can minimise quite easily the unnecessary use of vehicles.

(MA/Gad/AS-IC)

In the Trichy group that discussed *Growing Up*, questions were raised about why the developed countries were obsessed with 'environmental protection':

SCIENCE TEACHER: It is difficult to stop cutting trees. We need wood to build a house; the poor have to cut trees to get firewood to cook their food. It is very difficult indeed.

TYPIST (25): Many poor countries of Africa depend on forests. They have to cut down the forests to export wood to other countries in order to earn foreign exchange.

SCIENCE LECTURER (30): You are right. There is a lot of criticism about it. All the developed countries like the USA, UK and Europe who have money and power are very interested now in environmental protection, whereas the poor countries are not at all bothered about it. It is also being argued that the rich countries have already converted their forests to industrial and production areas; thereby they have become rich countries. So they do not want the poorer countries to follow that path of development. That is why the rich countries are talking about environmental protection. Of course, this may not be entirely true. But there is a need to protect our global environment, otherwise future generations will suffer.

(TN/Tri/GU)

### Environmental awareness and the mass media

Several suggestions for improving the state of the environment in India were offered by the participants in the various focus groups during their discussion of Advice Shop. A common theme was the need for 'movements' and 'campaigns', especially 'awareness campaigns'. As one Yavatmal science teacher put it, 'Family planning and anti-dowry campaigns could really become a people's movement. We really need an environment awareness campaign' (MA/Yav/AS). A highly educated housewife (40): 'Everyone should start becoming environmentally conscious. Only then can we think of acting like people who have done something against the establishment' (MA/Pune/IC).

A 29-year-old Development Officer in an insurance company:

> In the beginning, people used to laugh at the family planning programme, but now even the most illiterate person is aware about the need and concept of family planning. The problems of the environment cannot be solved in a day or two. It will take several years. We must launch our efforts today. Maybe we could hasten the process of environmental awareness with the help of the media.
>
> (MA/Gad/AS-IC)

A retired headmaster of Yavatmal observed that there has already been a 'movement' in his school for the past few years, and 'It has really yielded results. The students of the school planted about 300 trees on the school campus and from the income generated by it two classrooms could be built' (MA/Yav/AS). He therefore suggested that it should be a legal requirement for all factories in India to plant a certain number of trees.

A 30-year-old science lecturer in a Yavatmal college suggested, 'We can make good-quality compost from our household waste' (MA/Yav/AS). He believed that the practice, started by some people to request relatives in their wills to plant a tree or two in their names, would become a 'mass movement' very soon.

A 42-year-old schoolteacher from the same city suggested that 'the media should be put to work to stimulate thinking about the environment'. He added that 'some advertisements on environment are shown on TV but they are very few as compared to the advertisements for consumer products'.

But a 54-year-old housewife from Yavatmal believed that 'the young people are more enthusiastic than the older generation. So they should come forward and start a movement in schools and college' (MA/Yav/AS).

Members of the Valparai group suggested that the local people should be educated about the importance of forests so that they could also organise themselves to fight against the authorities. They suggested the use of films, wall posters, loudspeakers fitted in vans, and slides in theatres.

One young person, a 20-year-old science student of Trichy, urged people to say 'No' to polythene bags, and to start carrying cloth bags. 'Even that will be a considerable contribution towards environment protection.' He also suggested that there should be 'at least one electric crematorium in each district and that . . . other environment friendly practices should be made compulsory' (TN/Tri/AS).

A 40-year-old shop-owner and farmer of Gadchiroli:

> The environment problem is complex; so are the answers. But one thing I can tell you. Every answer starts with a tree at home. It is not just an environment-friendly activity, but rather a value of life. I have six mango trees at home, and my father says he has six grandsons to look after him. I believe that is the concept we should start with.
>
> (MA/Gad/AS-IC)

This suggestion made by a 27-year-old highly educated upper middle-class housewife was greeted with laughter and applause:

> The government should make it compulsory for everyone seeking a job – whether in a government, semi-government or private organisation – to plant at least one sapling. But it should not stop there. The employed person should look after it and tend it till it grows into a tree. When he or she retires from the organisation, pension should be disbursed only after due inspection of the state of the tree.
>
> (MA/Gad/AS-IC)

A Development Officer in an insurance company endorsed the suggestion: 'It's like insurance instalments! Unless the company deducts the instalments from the salary, no one would bother to pay them up voluntarily' (MA/Gad/AS-IC).

> RICH GADCHIROLI SHOP-OWNER: Some people have recently floated the idea of investments in 'commercial' trees like teak, saal, etc. This is a good scheme. You can earn money and at the same time help to preserve the environment. The government should encourage such schemes. But what is needed above all this is the national spirit and the virtue of honesty!
>
> YOUNG ARTS STUDENT: No, the environment is not the problem of a single country. It is a global problem. So, while developing the national spirit and awareness, similar efforts should also be made at the global level.
>
> INSURANCE AGENT: People should be motivated to take up the environment issue.
>
> UPPER MIDDLE-CLASS HOUSEWIFE: I think many times we tend to talk too much and do very little. Of course, there are other priorities. But then, among other things, the environment should become a priority in our daily lives.
>
> (MA/Gad/AS-IC)

## Summary of the Indian responses

Contemporary trends in television audience studies ('reception analysis') point to the need to regard audiences as 'active' readers of the media, as 'makers of meaning' in terms of the 'interpretative communities' they belong to (cf. Lindlof, 1988). While some readers accept the intended messages of the media for what they are worth, there are several others who reject them outright, seeing through the bias and the prejudice. Other readers negotiate the meanings of the intended messages, and still others impose their own meanings on the given media texts. What is evident from much of recent media research (especially that conducted in the 'culturalist' and 'critical' traditions) is that 'reception' of media texts is unpredictable across a range of mass audiences. Media producers cannot know for certain on what kind of ground (to use a biblical image) the 'seeds' of their messages fall, how they might be interpreted and what fruit they might bear.

Take, for instance, the Indian viewers' interpretation of the two Films Division documentaries. Several immediately saw them as government propaganda films; many blamed the government itself for the pollution; others blamed the people, the industries, urbanisation, overpopulation, and even religious practices. From the cut and thrust of the discussions it is not possible to conclude that there was either widespread agreement with or rejection of an intended message. In almost every one of the 21 focus groups covering all four films, there were subtleties in the arguments, and a critique of the embedded messages. Even in their admiration of the Golden Fields Movement, the nine focus groups which discussed Advice Shop were not unanimous even as they recognised that such a movement would be difficult to replicate in India. Yet, examples were proffered of the environmental movements in India (targeted at Narmada, Tehri, Dhabol, Chilka and other projects) that have had some measure of success.

The focus group discussions also point to a second conclusion: that audiences in small groups interpret environment documentaries in terms of their own cultural and demographic backgrounds and their experiences with, and awareness of, environmental issues. There were obvious differences (and some similarities too) in the understandings and interpretations of the four documentaries among focus groups from urban, rural and tribal backgrounds. Rural and tribal region focus groups, for instance, were concerned about the preservation of forests but not at the cost of development and jobs. They were highly suspicious of developed countries and environmental organisations that made a fetish of conservation in the developing countries and yet exploited the natural resources of the poorer countries.

What is the nature of people's talk about the environment and environment documentaries? In the first place, it is evident that people in focus groups do not talk as much about the documentaries shown as about their own experiences and their own problems. They use the documentaries as merely a peg to talk about their own experiences. The Golden Fields Movement led Indian focus groups to talk about their own struggles against the government, industries and other vested interests in protecting their own regions against pollution. Second, there is a tendency among focus group participants to make personal references and even to turn autobiographical. It takes some effort for the animator to intervene and bring the participants back to the theme of the documentary under discussion.

Third, 'turn-taking' in the discussions appears to be natural and easy, even when participants are hostile in their disagreements. The presence of an animator and other participants can help focus the discussions, though there are occasions when the animator seems to 'lose control'. However, there were also several occasions in the Indian focus groups when the animators had to elicit the opinions of men and women who were shy and reluctant to join in the discussion.

Most participants, even when they did not like the environmental documentaries and rejected the arguments of the films, were happy to have been given an opportunity to express their views. They liked the use of the visual medium to disseminate the message of environment protection. They were conscious that people's awareness and people's active participation were imperative for the protection of the environment in India and the world. They were of the view that the mass media had an important role in creating public awareness and supporting public participation in the cause of the environment.

## 6.6 CONCLUSIONS

In both countries at least some members of these groups express clear reservations about whether or not these videos in particular, and to some extent the media more generally, can bring them objective information about other places. In the UK one of the reservations is about the diktat resulting from the format – that there have to be 'goodies' and 'baddies' – exactly as Nigel Wade pointed out in Chapter 3. This does not necessarily mean that the story is untrue, just that the viewer is aware of the roles that must be played out. The other reason, and the commoner one in India, why the video might be suspect is that it does not confirm the viewer's world-view – particularly where (s)he can verify statements that are made about a known locality. There are also clear asymmetries – accepted on both sides. The greater wealth and technology per capita in Britain is acknowledged by both, though there is confusion in India whether the result is that Britain is more badly polluted as a result, or in fact cleaner. This is an area where the power of the current world-view is clearly important in making a judgement one way or the other. Another asymmetry seen from India's side is obviously the power of the North and the suspicion that any international treaties, environmental ones included, will be used to disadvantage India.

In India the tension between development and environment is clearly played out. Many of the participants are aware of the need for better water, drainage, and some even accept the clearing of forests to provide estate land. In benign Britain, where all dangerous bears and wolves have long been vanquished, we might wish to preserve biodiversity. In India, Project Tiger has been so successful (but possibly blunted by increased poaching rates over the last few years) that the number of humans killed by tigers has been growing quite fast again. In Himachal Pradesh panthers prowl the urban fringes, and children in particular are likely prey. In these exchanges one labourer disliked the forest because it harbours dangerous animals. None of this is inconsistent with the almost universal approval of the planting of trees – for this does not also include the breeding of wild animals. The arboreal landscape which Indians seem to have in mind is clearly a tamed landscape, neatly tied to pensions.

Participants in both nations obviously have images of the other country. It may not be a contradiction to say that these are simultaneously vague but well formed – vague in the sense that participants do not express specific and detailed knowledge about other places, but well formed in that they are often quite sure about what does *not* apply. Thus participants give the impression of having criteria for testing their knowledge of distant places but we are left uncertain about the kinds of specific knowledge they might or might not hold. To the extent that members of the public hold images of other countries, there are clearly many sources which contribute to their formation, from deep but diffuse cultural memories – of the British Raj, for example – to contemporary holiday adverts. In Chapter 5 it was clear that the media in India and Britain actually say very little about each other's country, and we *presume* in consequence that they are not a very strong component of the image-forming process. In addition, the media clearly have to mediate:

Melissa Ballard of CNN commented in Chapter 3 that she did not think that their edited material might make much sense in the Third World and in this chapter a labourer has complained of just that – too many abrupt jumps. Since the media may not be strong components of the image-forming process, the images that they do carry may be rejected by the audience if they do not conform with a stronger current world-view, or else they have to accommodate some of the audience's existing preconceptions. Despite this, some of the Indian participants specifically said that the media ought to play a bigger role in raising environmental awareness (but implicitly and explicitly in India and about India – not at the global level). We will touch on these problems again in the concluding chapter.

## NOTES

1    Abbreviations used in the text:

### The UK transcripts

The moderator for these groups was Caroline Fraser. The following abbreviations have been used to indicate from which of the UK focus groups quotations originate:

| | |
|---|---|
| L/IC | London group viewing *Insult to Civilisation* |
| L/AS | London group viewing *Advice Shop* |
| G/MLD | Glasgow group viewing *Matter of Life and Death* |
| G/GU | Glasgow group viewing *Growing Up* |
| N/MLD | Newent (Glos) group viewing *Matter of Life and Death* |
| N/AS | Newent (Glos) group viewing *Advice Shop* |
| S/IC | Church Stretton (Shrops.) group viewing *Insult to Civilisation* |

### The Indian transcripts

The focus groups were 'animated' by Professor A. Relton and his assistants in Tamil Nadu, and by Seema Ginde, Vishram Dhole and Shekar Nagarkar in Maharashtra. They also assisted in dubbing the documentaries into the local languages, and in transcribing and translating the discussions.

### Abbreviations

The following abbreviations have been used throughout this chapter, to indicate the names/titles of documentaries, and also the regions and places where the focus groups were held:

| | |
|---|---|
| AS | *Advice Shop* |
| GU | *Growing Up* |
| IC | *An Insult to Civilisation* |
| MLD | *A Matter of Life and Death* |
| TN | Tamil Nadu |
| MA | Maharashtra |
| Tri | Trichy/Tiruchirapalli |
| Per | Perumalai |
| Val | Valparai |
| Sem | Sembarai |
| Bom | Bombay |
| Pun | Pune |

Pun1   Pune, 1st focus group on *Growing Up*
Pun2   Pune, 2nd focus group on *Growing Up*
Jaw    Jawhar-Dahanu
Gad    Gadchiroli
Pul    Pullambadi
Mud    Mudumalai
Kar    Karumalai
Yav    Yavatmal
Mul    Mullal.

2   This refers to a scheme in Europe to stop overproduction, by which farmers are paid to leave land uncultivated.

## REFERENCES

Goldman, E. (1962) 'The group depth interview', *Journal of Marketing*, 26: 61–8.
Jensen, Klaus Bruhn and Jankowski, Nicholas W. (eds) (1991) *A Handbook of Qualitative Methodologies for Mass Communication Research*, London: Routledge.
Lindlof, Thomas A. (ed.) (1987) *Natural Audiences: Qualitative Research of Media Uses and Effects*, Norwood, NJ: Ablex.
—— (1988): 'Media audiences as interpretative communities', in Anderson, James A. (ed.) *Communication Yearbook* 11: 81–107.
Livingstone, Sonia and Hunt, Peter (1994) *Talk on Television: Audience Participation and Public Debate*, London: Routledge.
Merton, R.K., Fiske, M. and Curtis, A. (1946) *Mass Persuasion*, New York: Harper & Row.
Schlesinger, Philip *et al.* (1992) *Women Viewing Violence*, London: British Film Institute.
Stewart, David W. and Shamdasani, Prem N. (1990) *Focus Groups: Theory and Practice*, London: Sage.
White, Robert A. (1994): 'Audience interpretation of media: emerging perspectives', *Communication Research Trends*, 14(3) (Saint Louis University, Missouri).

# 7  Public understanding of the environment in India and the UK

## 7.1 INTRODUCTION

Knowledge about, and attitudes towards, the environment are ultimately the property of individual human beings, however we group them: into the categories of editors or reporters, Joe Public or Scientific Expert, Briton or Indian. As reported in earlier chapters, we have tried to find out something of the attitudes of the 'gatekeepers' in the media, we have tried to make some kind of semi-objective statement about what is actually fed via the gatekeepers to the public consumers of the mass media in Britain and India, and we have looked at how focus group representatives of the public respond to specimen TV programmes on the environment. The next logical step in our project was to find out to what extent the public think about any of these issues, what they know about some of these issues, and to try – in so far as we could – to relate this to their consumption of media sources. Our next step is not the definitive last step that we could take – the production of knowledge and images in the public mind reiterates over a never-ending cycle of ideas circulating between the expert community, the journalists and the public, with none of them starting or ending the cycle, but each projecting feedback onto the others. This is a perspective that some of the expert community might find hard to swallow, but those who fight on their behalf for the funds to further research would accept the reality of this claim. The next step of our research upon which we report is not therefore the last step; it is simply the last empirical step of the current project that we report on in this volume. We were concerned to make some attempt at assessing the concerns of the public in India and Britain on at least a roughly comparable basis. To do so we constructed a set of questions to use in a public opinion survey. The different cultural contexts of the two countries meant that there had to be some slight variation in approach, and the different agencies we used in the two countries also meant that there were some divergencies in both the data collected and the processing of the results.

The surveys were conducted in both countries in November 1993. In Britain we used the established Social Surveys (Gallup Poll) company to put our own questions to a sample of 1,000 people in Britain as an addendum to it normal weekly survey. In India we used our own appointed representatives drawn from the research community by Keval Kumar to implement a similar questionnaire (see Appendix IV) to samples of the population from three different parts of India. Given the diversity of development in India, of culture and of language, we are not aware of how we could attempt to draw a 1,000 person sample which claimed to be representative of India's population in its totality. The three survey groups amounted to a total poll of 1,623 people but each group should be taken to reflect its own regional and social biases. The surveys were made in and around Bhopal, the capital of Madhya Pradesh and part of the Hindi belt (North survey, sample 362); near Kodaikanal in Tamil Nadu, a less developed and more rural area of the South (716); and in and around Pune, one of the most developed and dynamic cities of India (West: 545).

In Britain the answers were tabulated for us by Gallup, with tabulation for individual questions being by sex, age, social class, and by three regions (the South, the Midlands and Wales, and the North and Scotland). In India all respondents were identified by sex, and asked about their age, caste, class (not an easy concept in many areas) and religion (except in the case of Pune), and they had already been identified by region. Quite a few respondents (but a minority) did not answer questions about caste, class, religion and income. To a very large extent it seems as if the breakdowns by caste, class and income replicate each other, which is what most observers would expect: further, Muslims were mostly in the lower income and class brackets, but their numbers in the surveys were in any event small. We have therefore reduced the size of most of the tables reproduced here by omitting breakdowns by caste, class and religion. The Gallup survey did not give breakdowns by media consumption pattern – but to some extent that can be surmised from the social class breakdowns. Since we have conducted our own analysis of the Indian returns, we have been able to produce breakdowns by media access, and they are an important part of what follows. What we did not do, which clearly in retrospect would have added some other insights, was to identify respondents by urban/rural location. The samples in the North and West are almost certainly more urban than local state figures would suggest for a representative stratified sample.

We have grouped the Indian respondents into three income classes: high, meaning an income of more than Rs. 5,000 per month; middle income for the range Rs. 1,000 to Rs. 4,999 per month; and low for those with less than Rs. 1,000. The distribution of respondents by income and region is then as shown in Table 7.1.1

The absence of low-income respondents in the West is either the result of not finding a representative sample, or is the result of using the same income classes for a dynamic metropolitan economy as for the rural South, or a compound of both factors. There is no doubt that costs of living and wage levels do vary regionally by large amounts – however, we have not used any regional deflators. We indicate when and where this may have some bearing on the results, but given the divergent nature of the sample for the three regions, for most of the time we show the breakdowns of the answers for each region separately.

Many of the questions were open-ended – as, for example, 'What are the most important issues facing the world today?' The responses have been classed for us in the case of the British survey by Gallup. For the Indian surveys the questionnaires were reproduced in the appropriate regional language but the answers were translated into English. This obviously leaves a margin for error – the extent to which a very specific answer may have been translated generally – but since for

*Table 7.1.1*  Distribution of survey respondents in India by region and income

|  | All | North | South | West |
|---|---|---|---|---|
| *Sample* | *1623* | *362* | *716* | *545* |
|  | | *Percentages* | | |
| No answer | 29 | 37 | 35 | 12 |
| High (> Rs. 5,000) | 13 | 1 | 6 | 32 |
| Middle | 35 | 25 | 19 | 56 |
| Low (< Rs. 1,000) | 23 | 27 | 39 | 0 |

*Table 7.1.2*  The scaling of answers to the question 'What do you think are the most important issues affecting the world today?'

| | % distribution | |
|---|---|---|
| | *Before scaling* | *After scaling* |
| Sample size | 1623 | 1623 |
| Don't know/no response | 31 | 31 |
| Communalism | 9 | 6 |
| Corruption | 3 | 2 |
| Crime | 1 | 1 |
| Disasters | 0 | 0 |
| Economic problems | 12 | 8 |
| Environmental problems | 19 | 13 |
| Health | 4 | 3 |
| Northern hegemony | 3 | 2 |
| Political problems | 2 | 1 |
| Population | 11 | 7 |
| Poverty | 7 | 5 |
| Civil violence | 15 | 10 |
| War | 13 | 9 |
| Other | 4 | 3 |
| Total | 134 | 100 |

most answers most of the translated responses are specific, this seems not to have been too large a problem. We have then made our own classification of the answers to produce the aggregate tables shown here. To do this we have used the same approach as for the content analysis of Chapter 5. A dictionary is produced of all the answers used for a question, and the entries in this dictionary are then allocated to higher level words. The aggregation of individual records to these higher level words is then done automatically – which means that aggregation is consistent: if two or more people have given the same answer, it will always lead to the same higher level aggregation. But, also following the methodology used before, if an answer has more than one element, it may be aggregated to more than one higher level word. For example, 'Poor roads and hospitals' aggregates into both Transport and Health. As before, therefore, the output can initially lead to a percentage distribution that can sum to more than 100 per cent. To make the pattern more consistent with the Gallup survey in which percentage distributions do sum to 100 per cent, the Indian results have then been scaled so that they too sum to 100 per cent. Table 7.1.2 shows an example. The 69 per cent of those giving an answer has been redistributed in the second column according to the relative frequency of the answers in the first column.

The last part of this introduction is to remind ourselves of what was happening 'in the world' in November 1993. This is a fairly bizarre idea, because it all depends on whose 'world' we are thinking about. Here is a random but quite representative sample of the November broadcast headlines:

**In India**

Bomb Blast At Dimapur Railway Station Assam
Iran Extends Full Support To Ind On Kashmir Issue

Assembly Elections Polling In Harayana Tomorrow
Militants Will Not Bend Until Demands Are Met: Hazratbal Crisis (Kashmir)
Rupee Will Be Fully Convertible Within A Short Time Finance Minister
One Village Adopted By Coal India In Earthquake Hit Area (Latur, Maharashtra)
Indo US Relations Not As Good As Expected Commerce Minister
Tamil Tigers Attack Sri Lankan Military: More Than 300 Soldiers Killed
Black Jawar [a millet food grain] To Be Purchased By Govt At 190 Rs Per Quintal
Kerala, Andhra Pradesh: Cyclone 36 Killed
Mr Ahmed Lenghar Selected As Pakistan's President
Well Known Gandhian Leader Manibhai Desai Dead
India's First Prime Minister Nehru:Birth Centenary Celebrated
Himachal-Pradesh Polls Tomorrow
Efforts Continue To Resolve Hazratbal Shrine Issue
First Ever Multi Party Elections In Jordan
Over 60 per cent Polling In Himachal Pradesh
Major Fire Breaks Out In Pak Nat Assembly Bldg In Islamabad:
Heavy Rains Claim Many Deaths
Repolling Ordered In 81 Polling Stations In Rajasthan
26 People Killed In W Bengal In Trawler Capsize
Iran Has Appreciated India's Policy Of Restraint
30 Persons Feared Killed In Landslides In Tamil Nadu
Govt Has Permitted Indian Companies To Raise Money From Global Markets
PM Says New Delhi Favours Free & Equal Trade For Benefit Of All Countries

**In the UK**

Princess Of Wales Says She Will Sue Over Sneaked Photos
Miners See All Hope Fade As Closures Loom Again
Row Over Seat Belts After 10 Die In Coach Crash
Trade Improves But Unsold Nissans Tell Another Story
Mystery Over Popstar Michael Jackson's Whereabouts After He Cancels His World
    Tour
Iraq: Saddam Hussein Resumes Executions In Baghdad
Britain Prepares Compromise With China Over Hong Kong:
President Menem Tells C4 News Falkland Islands Will Be Argentina's By 2000
PLO Chairman Condemns Killings Of Jewish Settler
Head Of RAF Accuses Some Govt Ministers Of Campaign To Undermine Services
Head Of RAF Apologises For Speech In Which He Made An Attack On Chancellor
    & Chief Secretary
First Evidence Of Widespread Massacre In Burundi
UN Security Council Voted To Enforce Tougher Sanctions On Libya
UN Troops Help Mentally Ill Abandoned In Bosnian Hospital
Back To Basics Speech By John Major & Home Secretary
Should Education Aim For Excellence
Single Mothers Persecuted
Eurodisney £4 Million Loss A Week

This ought to dispel any idea that there is any agreement on the significant news
items, even supposing that the channels in each country were presented with the
same list from which to make a selection in the first place.

Table 7.2.1  British consumption of television, November 1993

| | Total | Sex | | Age | | | | | Class | | | | Region | | | |
| | | Male | Female | 16-24 | 25-34 | 35-44 | 45-64 | 65+ | AB | C1 | C2 | DE | South | Wales | Midlands | North Scotland |
|---|---|---|---|---|---|---|---|---|---|---|---|---|---|---|---|---|
| Base | 940 | 453 | 487 | 159 | 200 | 168 | 246 | 166 | 169 | 224 | 266 | 281 | 375 | 236 | 236 | 329 |

*In the last week how many hours a day on average would you say you spent watching TV?*

| | Total | Male | Female | 16-24 | 25-34 | 35-44 | 45-64 | 65+ | AB | C1 | C2 | DE | South | Wales | Midlands | North Scotland |
|---|---|---|---|---|---|---|---|---|---|---|---|---|---|---|---|---|
| Less than 1 hour | 6 | 7 | 4 | 8 | 6 | 7 | 6 | 1 | 9 | 7 | 5 | 4 | 8 | 4 | 4 | 5 |
| 1-3 hours | 36 | 39 | 34 | 39 | 40 | 42 | 32 | 30 | 46 | 37 | 37 | 30 | 40 | 34 | 36 | 34 |
| 3-5 hours | 32 | 31 | 33 | 27 | 29 | 30 | 34 | 40 | 31 | 32 | 36 | 30 | 30 | 36 | 30 | 32 |
| More than 5 hours | 24 | 20 | 27 | 26 | 22 | 18 | 24 | 28 | 13 | 21 | 22 | 34 | 19 | 24 | 24 | 28 |
| None/never watch TV | 2 | 2 | 1 | 0 | 3 | 3 | 1 | 0 | 1 | 3 | 0 | 2 | 2 | 2 | 2 | 1 |
| Don't know | 0 | 0 | 1 | 0 | 0 | 0 | 1 | 1 | 1 | 1 | 0 | 0 | 1 | 0 | 0 | 0 |

*Thinking of all the channels you can watch on TV, which one do you watch the most?*

| | Total | Male | Female | 16-24 | 25-34 | 35-44 | 45-64 | 65+ | AB | C1 | C2 | DE | South | Wales | Midlands | North Scotland |
|---|---|---|---|---|---|---|---|---|---|---|---|---|---|---|---|---|
| BBC1 | 30 | 31 | 29 | 26 | 21 | 32 | 33 | 38 | 47 | 35 | 25 | 20 | 30 | 28 | 31 | 31 |
| BBC2 | 4 | 5 | 3 | 2 | 6 | 3 | 3 | 5 | 4 | 4 | 2 | 4 | 4 | 2 | 4 | 4 |
| ITV | 39 | 32 | 46 | 48 | 40 | 37 | 35 | 40 | 26 | 31 | 45 | 49 | 38 | 40 | 40 | 40 |
| Channel 4 | 5 | 7 | 2 | 5 | 7 | 6 | 4 | 4 | 6 | 6 | 6 | 3 | 5 | 4 | 5 | 5 |
| Satellite/cable TV | 7 | 10 | 4 | 10 | 10 | 9 | 4 | 2 | 7 | 7 | 7 | 8 | 8 | 5 | 7 | 7 |
| Video | 0 | 0 | 0 | 0 | 0 | 1 | 0 | 0 | 0 | 0 | 0 | 0 | 0 | 0 | 0 | 0 |
| Other | 0 | 0 | 1 | 1 | 0 | 1 | 1 | 1 | 1 | 0 | 0 | 1 | 1 | 0 | 0 | 0 |
| Don't know | 8 | 7 | 9 | 2 | 10 | 8 | 13 | 5 | 0 | 9 | 8 | 9 | 9 | 10 | 10 | 6 |
| None | 6 | 7 | 6 | 6 | 6 | 5 | 8 | 6 | 0 | 9 | 6 | 6 | 4 | 10 | 10 | 6 |

Table 7.2.1 Continued

*Which television news programme do you watch most regularly?*

| | Sex | | | Age | | | | | Class | | | | Region | | |
|---|---|---|---|---|---|---|---|---|---|---|---|---|---|---|---|
| | Total | Male | Female | 16-24 | 25-34 | 35-44 | 45-64 | 65+ | AB | C1 | C2 | DE | South | Midlands Wales | North Scotland |
| Base | 940 | 453 | 487 | 159 | 200 | 168 | 246 | 166 | 169 | 224 | 266 | 281 | 375 | 236 | 329 |
| BBC 1 o'clock lunchtime news | 5 | 5 | 6 | 3 | 3 | 5 | 7 | 8 | 4 | 7 | 5 | 5 | 6 | 6 | 4 |
| BBC 6 o'clock news | 19 | 21 | 17 | 18 | 15 | 11 | 20 | 30 | 17 | 20 | 20 | 17 | 17 | 18 | 20 |
| BBC 9 o'clock news | 16 | 17 | 16 | 10 | 19 | 20 | 16 | 17 | 30 | 20 | 13 | 8 | 18 | 17 | 14 |
| ITN lunchtime news | 4 | 1 | 6 | 7 | 4 | 3 | 3 | 4 | 1 | 5 | 4 | 5 | 2 | 5 | 6 |
| ITN early evening news | 13 | 12 | 14 | 15 | 13 | 9 | 12 | 17 | 8 | 9 | 14 | 19 | 13 | 10 | 15 |
| ITV News at Ten | 24 | 23 | 26 | 22 | 28 | 30 | 24 | 17 | 20 | 19 | 26 | 30 | 24 | 28 | 22 |
| Newsnight | 1 | 1 | 0 | 0 | 2 | 1 | 0 | 1 | 1 | 1 | 1 | 0 | 1 | 0 | 2 |
| Chan 4 7 o'clock news | 2 | 3 | 1 | 3 | 1 | 3 | 2 | 1 | 2 | 3 | 2 | 1 | 2 | 2 | 1 |
| BBC1 Breakfast news | 1 | 1 | 1 | 1 | 0 | 1 | 2 | 1 | 1 | 1 | 1 | 0 | 1 | 1 | 1 |
| BBC2 Breakfast news | 0 | 0 | 0 | 0 | 0 | 0 | 0 | 0 | 0 | 0 | 0 | 0 | 0 | 0 | 0 |
| Big Breakfast news on Chan 4 | 1 | 1 | 1 | 3 | 0 | 0 | 0 | 0 | 1 | 1 | 0 | 1 | 1 | 1 | 0 |
| GMTV regional news on ITV | 1 | 0 | 1 | 2 | 1 | 0 | 0 | 0 | 0 | 2 | 0 | 0 | 1 | 2 | 0 |
| Newsround | 0 | 0 | 0 | 0 | 0 | 0 | 0 | 0 | 0 | 1 | 0 | 0 | 0 | 0 | 0 |
| Other | 2 | 3 | 1 | 0 | 3 | 3 | 1 | 0 | 1 | 1 | 2 | 2 | 2 | 1 | 2 |
| Don't know | 3 | 2 | 4 | 4 | 1 | 3 | 5 | 2 | 3 | 3 | 3 | 3 | 3 | 2 | 4 |
| None | 9 | 11 | 8 | 15 | 10 | 11 | 7 | 4 | 9 | 9 | 9 | 9 | 11 | 8 | 8 |

*Table 7.2.2*  British radio listening and readership of newspapers, November 1993

*In the last week which radio station did you listen to most regularly?*

| | Sex | | | Age | | | | | Class | | | | Region | | |
| | Total | Male | Female | 16–24 | 25–34 | 35–44 | 45–64 | 65+ | AB | C1 | C2 | DE | South | Midlands/Wales | North/Scotland |
| Base | 940 | 453 | 487 | 159 | 200 | 168 | 246 | 166 | 169 | 224 | 266 | 281 | 375 | 236 | 329 |
| Radio 1 | 15 | 15 | 15 | 39 | 23 | 7 | 5 | 2 | 11 | 21 | 15 | 12 | 14 | 16 | 14 |
| Radio 2 | 7 | 8 | 7 | 0 | 4 | 4 | 10 | 18 | 7 | 8 | 8 | 7 | 8 | 9 | 6 |
| Radio 3 | 1 | 2 | 1 | 0 | 1 | 0 | 3 | 2 | 3 | 1 | 0 | 1 | 2 | 1 | 0 |
| Radio 4 | 11 | 10 | 12 | 1 | 10 | 17 | 10 | 17 | 26 | 13 | 5 | 6 | 13 | 12 | 8 |
| Radio 5 | 1 | 3 | 0 | 4 | 1 | 0 | 2 | 1 | 2 | 1 | 2 | 1 | 2 | 2 | 1 |
| Classic FM | 3 | 5 | 2 | 0 | 3 | 2 | 5 | 6 | 8 | 3 | 3 | 2 | 2 | 5 | 4 |
| Virgin 1215 | 2 | 3 | 1 | 5 | 4 | 2 | 0 | 0 | 3 | 2 | 3 | 2 | 2 | 3 | 1 |
| BBC local radio | 7 | 8 | 6 | 2 | 3 | 6 | 9 | 15 | 5 | 7 | 8 | 7 | 7 | 11 | 5 |
| BBC World Service | 0 | 0 | 0 | 0 | 0 | 0 | 0 | 0 | 0 | 0 | 0 | 0 | 0 | 0 | 0 |
| Commercial/independent local radio | 21 | 19 | 22 | 23 | 23 | 30 | 19 | 10 | 16 | 18 | 26 | 21 | 21 | 10 | 29 |
| Other | 8 | 8 | 8 | 12 | 9 | 11 | 6 | 2 | 5 | 7 | 8 | 10 | 9 | 8 | 7 |
| None | 21 | 18 | 23 | 14 | 18 | 19 | 26 | 25 | 13 | 19 | 19 | 29 | 18 | 23 | 23 |
| Don't know | 2 | 1 | 2 | 0 | 0 | 1 | 5 | 1 | 1 | 1 | 2 | 3 | 1 | 1 | 3 |

Table 7.2.2  Continued

| | Total | Sex | | Age | | | | | Class | | | | Region | | |
|---|---|---|---|---|---|---|---|---|---|---|---|---|---|---|---|
| | | Male | Female | 16-24 | 25-34 | 35-44 | 45-64 | 65+ | AB | C1 | C2 | DE | South | Midlands Wales | North Scotland |
| Base | 940 | 453 | 487 | 159 | 200 | 168 | 246 | 166 | 169 | 224 | 266 | 281 | 375 | 236 | 329 |

*Which national daily newspaper do you read most frequently?*

| | Total | Male | Female | 16-24 | 25-34 | 35-44 | 45-64 | 65+ | AB | C1 | C2 | DE | South | Midlands Wales | North Scotland |
|---|---|---|---|---|---|---|---|---|---|---|---|---|---|---|---|
| Daily Express | 6 | 7 | 4 | 6 | 4 | 3 | 8 | 6 | 7 | 8 | 6 | 2 | 6 | 8 | 4 |
| Daily Mail | 9 | 8 | 9 | 7 | 6 | 10 | 8 | 12 | 13 | 10 | 8 | 6 | 12 | 8 | 5 |
| Daily Mirror | 15 | 17 | 13 | 18 | 12 | 13 | 15 | 15 | 7 | 12 | 16 | 19 | 13 | 14 | 17 |
| Sun | 16 | 17 | 15 | 24 | 16 | 15 | 14 | 12 | 2 | 11 | 18 | 26 | 18 | 16 | 14 |
| Today | 4 | 4 | 3 | 2 | 5 | 2 | 5 | 3 | 3 | 7 | 3 | 2 | 4 | 6 | 2 |
| Daily Star | 3 | 4 | 2 | 4 | 2 | 4 | 2 | 2 | 0 | 1 | 4 | 4 | 0 | 4 | 5 |
| Guardian | 4 | 5 | 4 | 4 | 4 | 8 | 2 | 5 | 13 | 5 | 2 | 1 | 6 | 1 | 3 |
| The Independent | 3 | 3 | 2 | 4 | 3 | 3 | 2 | 2 | 6 | 4 | 1 | 1 | 4 | 4 | 2 |
| Daily Telegraph | 5 | 5 | 5 | 1 | 3 | 6 | 6 | 7 | 13 | 6 | 2 | 2 | 8 | 4 | 2 |
| The Times | 1 | 2 | 0 | 2 | 1 | 2 | 1 | 0 | 4 | 1 | 2 | 1 | 1 | 2 | 1 |
| Financial Times | 0 | 1 | 0 | 1 | 0 | 1 | 0 | 0 | 1 | 0 | 0 | 0 | 1 | 0 | 0 |
| Other | 7 | 6 | 8 | 4 | 8 | 5 | 7 | 11 | 5 | 8 | 7 | 8 | 3 | 4 | 14 |
| None | 28 | 22 | 35 | 23 | 36 | 28 | 30 | 22 | 25 | 27 | 32 | 28 | 23 | 32 | 32 |

*Table 7.2.3*  Sample media access in India, November 1993

|  | | Sex | | | Age | | | |
|---|---|---|---|---|---|---|---|---|
|  | Total | Male | Female | 16–24 | 25–34 | 35–44 | 45–64 | 65+ |
| NORTH | | | | | | | | |
| Respondents | 362 | 69 | 82 | 46 | 39 | 54 | 14 | 3 |
| Radio | 88 | 91 | 87 | 85 | 90 | 94 | 57 | 100 |
| No radio | 12 | 9 | 13 | 15 | 10 | 6 | 43 | 0 |
| TV | 83 | 91 | 70 | 74 | 90 | 87 | 64 | 33 |
| No TV | 17 | 9 | 30 | 26 | 10 | 13 | 36 | 67 |
| English paper | 19 | 17 | 15 | 24 | 8 | 13 | 21 | 33 |
| Language paper | 57 | 51 | 65 | 65 | 72 | 44 | 50 | 0 |
| No paper | 27 | 36 | 24 | 11 | 23 | 50 | 36 | 67 |
| SOUTH | | | | | | | | |
| Respondents | 716 | 340 | 323 | 130 | 192 | 160 | 194 | 34 |
| Radio | 72 | 71 | 73 | 67 | 70 | 81 | 74 | 53 |
| No radio | 28 | 29 | 27 | 33 | 30 | 19 | 26 | 47 |
| TV | 63 | 64 | 62 | 69 | 60 | 66 | 64 | 26 |
| No TV | 37 | 36 | 38 | 31 | 40 | 34 | 36 | 74 |
| English paper | 25 | 24 | 26 | 33 | 21 | 31 | 23 | 9 |
| Language paper | 20 | 19 | 20 | 20 | 21 | 21 | 20 | 6 |
| No paper | 55 | 57 | 54 | 47 | 58 | 48 | 58 | 85 |
| WEST | | | | | | | | |
| Respondents | 545 | 332 | 200 | 84 | 152 | 184 | 89 | 19 |
| Radio | 75 | 77 | 75 | 65 | 73 | 83 | 76 | 63 |
| No radio | 25 | 23 | 26 | 35 | 27 | 17 | 24 | 37 |
| TV | 75 | 73 | 77 | 83 | 64 | 77 | 79 | 68 |
| No TV | 25 | 27 | 24 | 17 | 36 | 23 | 21 | 32 |
| English paper | 33 | 31 | 36 | 48 | 34 | 24 | 36 | 37 |
| Language paper | 52 | 53 | 52 | 49 | 44 | 60 | 52 | 42 |
| No paper | 26 | 26 | 26 | 14 | 33 | 26 | 25 | 32 |

*Table 7.2.3* Continued

| | Income | | | Radio and TV | | | | | Newspaper | | |
|---|---|---|---|---|---|---|---|---|---|---|---|
| | High | Middle | Low | No resp | Radio | No rad. | TV | No TV | English | Vernac. | None |
| **NORTH** | | | | | | | | | | | |
| Respondents | 2 | 128 | 97 | 135 | 320 | 42 | 300 | 62 | 69 | 207 | 94 |
| Radio | 100 | 91 | 92 | 84 | 100 | 0 | 90 | 81 | 86 | 89 | 89 |
| No radio | 0 | 9 | 8 | 16 | 0 | 100 | 10 | 19 | 14 | 11 | 11 |
| TV | 100 | 92 | 57 | 93 | 84 | 71 | 100 | 0 | 94 | 86 | 68 |
| No TV | 0 | 8 | 43 | 7 | 16 | 29 | 0 | 100 | 6 | 14 | 32 |
| English paper | 100 | 23 | 2 | 27 | 18 | 24 | 22 | 6 | 100 | 4 | 0 |
| Language paper | 100 | 60 | 38 | 67 | 58 | 52 | 60 | 45 | 12 | 100 | 0 |
| No paper | 0 | 23 | 62 | 6 | 28 | 24 | 23 | 48 | 0 | 2 | 100 |
| **SOUTH** | | | | | | | | | | | |
| Respondents | 44 | 136 | 279 | 257 | 516 | 200 | 450 | 266 | 182 | 141 | 396 |
| Radio | 68 | 74 | 76 | 67 | 100 | 0 | 79 | 60 | 77 | 80 | 67 |
| No radio | 32 | 26 | 24 | 33 | 0 | 100 | 21 | 40 | 23 | 20 | 33 |
| TV | 98 | 86 | 41 | 68 | 69 | 47 | 100 | 0 | 87 | 77 | 47 |
| No TV | 2 | 14 | 59 | 32 | 31 | 54 | 0 | 100 | 13 | 23 | 53 |
| English paper | 48 | 54 | 14 | 19 | 27 | 21 | 35 | 9 | 100 | 2 | 0 |
| Language paper | 20 | 29 | 16 | 18 | 22 | 14 | 24 | 12 | 2 | 100 | 0 |
| No paper | 32 | 18 | 69 | 64 | 51 | 66 | 41 | 79 | 0 | 0 | 100 |
| **WEST** | | | | | | | | | | | |
| Respondents | 173 | 305 | 0 | 67 | 410 | 135 | 407 | 138 | 179 | 285 | 132 |
| Radio | 80 | 76 | 0 | 60 | 100 | 0 | 78 | 68 | 69 | 82 | 70 |
| No radio | 20 | 24 | 0 | 40 | 0 | 100 | 22 | 32 | 31 | 18 | 30 |
| TV | 98 | 59 | 0 | 87 | 77 | 67 | 100 | 0 | 94 | 96 | 12 |
| No TV | 2 | 41 | 0 | 13 | 23 | 33 | 0 | 100 | 6 | 4 | 88 |
| English paper | 67 | 14 | 0 | 31 | 30 | 41 | 42 | 7 | 100 | 18 | 0 |
| Language paper | 54 | 50 | 0 | 57 | 57 | 38 | 67 | 9 | 28 | 100 | 0 |
| No paper | 4 | 41 | 0 | 15 | 25 | 29 | 6 | 84 | 2 | 2 | 100 |

## 7.2 ACCESS TO THE MEDIA

Some aspects of media access in Britain are revealed by Tables 7.2.1 and 7.2.2. Relatively speaking, most of the patterns do not vary much by regional breakdown, by class breakdown, or by age. There are some exceptions to this: in broadcasting the balance between viewing BBC1 and ITV changes with social class, with the commercial channel being more attractive to the working class and BBC1 to the professional classes, and the young are more attracted to Radio 1 than other radio channels. The *Daily Telegraph* and *The Times* attract an older readership, which will not surprise those who know the UK media.

The public use of broadcasting is fairly intensive: 2 per cent of this sample profess never watching TV, while 92 per cent watch for more than 1 hour everyday, and 86 per cent watch a TV news programme. Satellite and cable seem to have made only small inroads into the audience so far; most of the viewing is of the established terrestrial channels. Newspaper readership is dominated by the tabloids, and our 'sample' tabloid (the *Daily Mirror*) of Chapter 5 vies with the *Sun* at about 15 per cent of the population. (It is sobering to realise that 15 per cent of the Indian population aged over 16 would mean a readership of about 80 million). Our sample broadsheet paper, *The Independent*, has the third largest share of the 'quality' press, at about 3 per cent, behind the *Telegraph* and the *Guardian* at 5 and 4 per cent each; 28 per cent of the sample claim not to read newspapers.

Table 7.2.3 is derived from the Indian questionnaire. In many of the tables we have kept the three regions separate, and we have cross-tabulated the types of media access, as well as age and income class. The urban bias of the sample is evident from the fact that TV penetration is shown as being as high as 83 per cent for the North sample, and 63 per cent for the South sample – whereas DD is claimed (Chapter 1) to have a penetration that may be as low as 25 per cent. Nevertheless, there are quite large representations of the groups with no TV, and some interesting associations emerge: 84 per cent of those in the West and 79 per cent of those in the South who do not have a TV do not read a newspaper either. In all three regions those who do not have a TV set are overwhelmingly likely to belong to the lowest income groups. (A demand for a TV set has become a fairly routine dowry demand.) It also almost looks as if the level of newspaper readership has reached British proportions (but this is an urban-biased sample), only 26 per cent saying they do not have access to newspapers in the West and North, though 55 per cent in the South. But these percentages do vary considerably with income. In terms of language, the smallest English readership is in the North, part of the Hindi belt; the largest is in and near metropolitan Pune in the West, though it is still less than the vernacular readership and dominated overwhelmingly by the highest income group. The English readership in the South is also smallish, but it is actually bigger than the vernacular readership in this sample. A representative sample would not show it as more important than local languages such as Tamil, but it is true that the use of English in the South is more prevalent than in the North, and its use as a lingua franca rather than Hindi is stoutly defended in the South.

## 7.3 ISSUES OF PUBLIC CONCERN AT WORLD, NATIONAL AND LOCAL LEVELS

### The most important issues affecting the world today

The first thing that distinguishes the British and Indian responses to this question, reported in Table 7.3.1, is the pattern of 'Don't know/no response'. In aggregate it is higher in India – somewhere over 30 per cent perhaps as opposed to the really rather small 7 per cent in Britain. But, as with other statistics, the 7 per cent in Britain does not vary much when broken down by age or class or sex. In India, however, the variation according to newspaper readership is extreme: in all three cases the percentage of 'Don't know/no response' increases first by a small degree from the English readership to the language readership, then by a vast amount among the 'No readership' group. Similar variations show up in the 'Don't know/no response' figures for many of the following tables too.

In Britain War and Unemployment figure highly – though to what extent the latter is seen as a world problem is not clear. The Environment is a middle-ranking item. In India many of the responses are either not really about world issues, or are seen through Indian equivalents. Civil Violence (terrorism in particular) outdoes War, though War is seen as important, if not as important as in Britain, which was involved in the former Yugoslavian conflict. In the North 'Communalism' – mainly meaning the tension between Hindu and Muslim communities, heightened in recent years over the problem of the mosque at Ayodhya – is seen as an important world problem, and much of the conflict elsewhere in the world is seen in communalistic terms. (There are quite a few western commentators who see the clash of religion and culture as important on the world stage, rather than the clash of nations. The British public however may see things more in terms of nations.) In two of the regions of India, the Environment comes out as an issue. To find out why, the dictionary entries which have been aggregated into the category Environment have been printed out. A very few of these are specifically identifiable as world issues – there are some references to ozone holes and global warming. There are also some which just generally say Environmental Problems. However, the majority are about pollution, and many are also tied to other themes like Population: Unemployment and Pollution. But in all three regions there is something of a fault-line which differentiates the TV viewers and newspaper readers – of either press – on the one side, concerned about the environment, from the non-viewers and non-readers on the other side, who are seemingly unconcerned.

### National issues

The results for this question are reported in Table 7.3.2. In Britain there is one clear leader as a national issue – Unemployment. Law and Order and 'Other economic issues' follow, and the Environment has become a non-issue – just as the focus groups also suggested. In India the number of 'don't knows' is substantially reduced from the former question, though still heavily biased towards those who have poor media access. In India Communalism and Economic issues (including unemployment) dominate – and if Communalism is in some sense related to Law and Order, then the results from the two countries are not so dissimilar thus far. India has another major concern though, and that is with Population. Here the disaggregations show that in the North and South of India the poorer sections of the populace have a greater concern over population – perhaps contrary to some

## 7.3 ISSUES OF PUBLIC CONCERN AT WORLD, NATIONAL AND LOCAL LEVELS

### The most important issues affecting the world today

The data items that distinguish the British and Indian responses to the questions reported in Table 7.3.1 is the pattern of 'Don't know' responses, which is higher in India — somewhere over 20 per cent — whereas in the rather earlier stage. 'War' even in Britain, but it is with other similar types within a British class and sex subdivision broken down by age or class or sex. In these cases, the proportion of 'Don't know' responses increases progressively from the 16–24 readership to the language readership: the by a very unequal among the less educated, and similar values. Table 7.3.1 gives the response figures for many of the following tables too.

In Britain, War and Crime, however, figure less in Britain as what people later is seen as a world problem, not clearly. The environmental issues form one item. In India many of the responses are, rather, not really about world issues but are seen through Indian eyes. Even the 'Civil Wars' are generally concerned. War, though, War is seen as important, if not as important as it should, and is involved in the former Yugoslavian conflict, in the North Communalism — thus meaning the tension between different communities in India. Tension in Ayodhya over the problem of the mosque at Ayodhya — is seen as an important world problem, and against the conflict elsewhere in the world is seen in communalist terms. (There are quite a few British correspondents who take a rather different role and rather less important on the world stage, rather than the clash of nations. The British public, however, takes the issues as rather more important in their own regions of India. The Environment comes as an issue of concern in all these ordinary entries which have been aggregated into the category 'environment issues' — most carried out. A very large these aggregated 'environment as world issue' — there are more references to ozone holes and global warming. There are some when not generally any Environmental problems. However, the majority — world problems, and many indeed have different 'category populations', environment and pollution, but in all these regions there is something of a gap over which differentiates the TV. However, in a question it is chiefly of other kinds, so in the sense of a marginal gap the most general, more persuasively very mostly, extended for the other areas, who are decidedly more even.

The British public, less precisely aggregated form, is also concerned, and proportionately more; and relative to the Environment more clearly a separate. First, in specific terms, while the Environment more recently a more usefully first, particularly from the bare prominence of direct concern in more responses referred more to, for just discussions which on global warming and pollution, more aggregate matters, for pollution, in their conditions, while anxious or global more clearly a separate. Clearly the more than half the smaller gap over concern in the North and South of India the particular section of the population have a greater concern over population — perhaps coherent to wider.

Table 7.3.1  Answers to 'What do you think is the most important issue affecting the world at the present time?'

**BRITAIN**

| | Total | Sex | | Age | | | | | Class | | | | Region | | |
| | | Male | Female | 16–24 | 25–34 | 35–44 | 45–64 | 65+ | AB | C1 | C2 | DE | South | Midlands/Wales | North/Scotland |
| Base | 940 | 453 | 487 | 159 | 200 | 168 | 246 | 166 | 169 | 224 | 266 | 281 | 375 | 236 | 329 |
| Unemployment | 13 | 12 | 14 | 17 | 12 | 12 | 13 | 12 | 8 | 13 | 15 | 15 | 13 | 12 | 14 |
| War | 28 | 23 | 32 | 23 | 27 | 30 | 31 | 26 | 32 | 28 | 25 | 27 | 26 | 29 | 29 |
| Disease/health issues | 3 | 4 | 2 | 3 | 3 | 4 | 4 | 7 | 3 | 2 | 2 | 4 | 4 | 1 | 4 |
| Overpopulation | 4 | 6 | 3 | 4 | 4 | 2 | 7 | 5 | 7 | 4 | 3 | 3 | 6 | 5 | 1 |
| Environmental issues | 9 | 12 | 7 | 13 | 13 | 7 | 8 | 5 | 10 | 9 | 11 | 7 | 9 | 9 | 9 |
| Homelessness | 2 | 2 | 1 | 3 | 1 | 2 | 2 | 1 | 1 | 1 | 3 | 2 | 1 | 1 | 1 |
| Crime | 7 | 5 | 9 | 7 | 4 | 6 | 12 | 12 | 3 | 6 | 9 | 9 | 7 | 7 | 8 |
| Famine/hunger | 10 | 12 | 9 | 9 | 7 | 12 | 11 | 13 | 9 | 13 | 9 | 9 | 7 | 14 | 8 |
| Racial/ethnic/religious prejudice or discrimination | 3 | 4 | 3 | 1 | 3 | 4 | 5 | 1 | 7 | 3 | 4 | 1 | 4 | 2 | 3 |
| Other | 14 | 15 | 12 | 11 | 15 | 16 | 14 | 12 | 17 | 14 | 11 | 13 | 15 | 11 | 13 |
| Don't know | 7 | 5 | 9 | 10 | 9 | 6 | 6 | 6 | 4 | 6 | 8 | 9 | 5 | 9 | 9 |

Table 7.3.1 Continued

INDIA

NORTH

| | Total | Sex | | Age | | | | | Income | | | | Radio and TV | | | | Newspaper | | |
|---|---|---|---|---|---|---|---|---|---|---|---|---|---|---|---|---|---|---|---|
| | | Male | Female | 16–24 | 25–34 | 35–44 | 45–64 | 65+ | High | Middle | Low | Noresp | Radio | Norad. | TV | No TV | Eng | Vern. | None |
| Base | 362 | 69 | 82 | 46 | 39 | 54 | 14 | 3 | 2 | 128 | 97 | 135 | 320 | 42 | 300 | 62 | 69 | 207 | 94 |
| Don't know/no response | 17 | 30 | 6 | 7 | 15 | 37 | 7 | 0 | 0 | 19 | 29 | 8 | 19 | 5 | 20 | 6 | 4 | 7 | 49 |
| Communalism | 11 | 10 | 9 | 7 | 9 | 9 | 16 | 0 | 0 | 13 | 3 | 14 | 12 | 4 | 12 | 5 | 15 | 13 | 2 |
| Corruption | 0 | 0 | 1 | 0 | 2 | 0 | 0 | 0 | 0 | 0 | 2 | 0 | 0 | 0 | 0 | 3 | 0 | 1 | 0 |
| Economic | 5 | 5 | 4 | 2 | 4 | 9 | 0 | 0 | 0 | 4 | 3 | 8 | 5 | 9 | 7 | 0 | 9 | 5 | 4 |
| Environment | 20 | 24 | 16 | 38 | 15 | 11 | 5 | 0 | 0 | 22 | 7 | 28 | 20 | 17 | 22 | 10 | 25 | 27 | 0 |
| Health | 2 | 2 | 2 | 3 | 4 | 0 | 0 | 0 | 0 | 0 | 3 | 3 | 2 | 0 | 2 | 3 | 2 | 1 | 2 |
| Northern hegemony | 1 | 1 | 1 | 0 | 0 | 3 | 0 | 0 | 0 | 1 | 0 | 1 | 1 | 0 | 1 | 0 | 0 | 1 | 2 |
| Population | 6 | 3 | 9 | 9 | 8 | 3 | 11 | 0 | 0 | 8 | 7 | 3 | 6 | 0 | 5 | 10 | 6 | 6 | 5 |
| Poverty | 4 | 2 | 6 | 0 | 6 | 5 | 11 | 0 | 0 | 3 | 5 | 5 | 3 | 13 | 5 | 3 | 3 | 5 | 4 |
| Civil violence | 14 | 2 | 26 | 9 | 23 | 9 | 38 | 25 | 0 | 15 | 12 | 14 | 13 | 26 | 14 | 15 | 13 | 14 | 15 |
| War | 15 | 14 | 19 | 17 | 13 | 12 | 11 | 75 | 100 | 13 | 28 | 6 | 15 | 17 | 9 | 43 | 17 | 14 | 18 |
| Other | 4 | 5 | 1 | 9 | 2 | 2 | 0 | 0 | 0 | 1 | 0 | 11 | 4 | 9 | 4 | 3 | 5 | 6 | 0 |

Table 7.3.1 Continued

INDIA

SOUTH

| | Total | Sex | | Age | | | | | Income | | | | Radio | | Radio and TV | | Newspaper | | |
|---|---|---|---|---|---|---|---|---|---|---|---|---|---|---|---|---|---|---|---|
| | | Male | Female | 16-24 | 25-34 | 35-44 | 45-64 | 65+ | High | Middle | Low | Noresp | Radio | Norad. | TV | No TV | Eng | Vern. | None |
| Base | 716 | 340 | 323 | 130 | 192 | 160 | 194 | 34 | 44 | 136 | 279 | 257 | 516 | 200 | 450 | 266 | 182 | 141 | 396 |
| Don't know/no response | 40 | 34 | 47 | 31 | 41 | 36 | 44 | 59 | 9 | 13 | 50 | 48 | 32 | 59 | 28 | 60 | 17 | 23 | 56 |
| Communalism | 5 | 6 | 4 | 4 | 4 | 6 | 6 | 12 | 6 | 11 | 4 | 3 | 6 | 4 | 6 | 4 | 6 | 10 | 3 |
| Corruption | 1 | 2 | 1 | 1 | 1 | 2 | 2 | 0 | 4 | 2 | 1 | 1 | 1 | 1 | 1 | 1 | 2 | 2 | 1 |
| Crime | 1 | 2 | 1 | 2 | 2 | 1 | 1 | 0 | 4 | 3 | 1 | 0 | 1 | 1 | 2 | 0 | 2 | 2 | 1 |
| Disaster | 1 | 1 | 1 | 1 | 0 | 1 | 0 | 0 | 0 | 1 | 0 | 1 | 1 | 1 | 1 | 1 | 1 | 1 | 1 |
| Economic | 5 | 4 | 6 | 6 | 5 | 6 | 4 | 0 | 2 | 6 | 6 | 4 | 6 | 2 | 5 | 5 | 7 | 7 | 3 |
| Environment | 6 | 6 | 4 | 6 | 6 | 7 | 5 | 3 | 21 | 12 | 1 | 4 | 5 | 6 | 8 | 2 | 12 | 4 | 3 |
| Health | 5 | 6 | 4 | 11 | 5 | 4 | 3 | 0 | 11 | 9 | 3 | 4 | 5 | 6 | 7 | 2 | 7 | 7 | 3 |
| Northern hegemony | 1 | 1 | 1 | 2 | 1 | 2 | 0 | 0 | 1 | 3 | 1 | 1 | 2 | 1 | 1 | 0 | 2 | 1 | 3 |
| Politics | 2 | 2 | 2 | 1 | 2 | 4 | 2 | 0 | 4 | 4 | 2 | 1 | 2 | 2 | 3 | 1 | 3 | 3 | 2 |
| Population | 4 | 3 | 5 | 6 | 4 | 5 | 3 | 3 | 8 | 4 | 6 | 2 | 6 | 1 | 4 | 6 | 6 | 6 | 3 |
| Poverty | 7 | 10 | 5 | 8 | 4 | 5 | 11 | 6 | 0 | 6 | 5 | 11 | 9 | 1 | 8 | 5 | 5 | 7 | 8 |
| Civil violence | 14 | 14 | 13 | 14 | 14 | 18 | 13 | 3 | 15 | 17 | 15 | 11 | 16 | 8 | 16 | 10 | 20 | 16 | 10 |
| War | 5 | 6 | 4 | 6 | 7 | 6 | 3 | 3 | 8 | 8 | 4 | 4 | 5 | 4 | 7 | 2 | 7 | 7 | 4 |
| Other | 2 | 4 | 1 | 2 | 3 | 1 | 1 | 12 | 8 | 3 | 2 | 2 | 2 | 3 | 3 | 1 | 3 | 3 | 2 |

Table 7.3.1  Continued

INDIA

WEST

| | Total | Sex | | Age | | | | | Income | | | | Radio and TV | | | | Newspaper | | |
|---|---|---|---|---|---|---|---|---|---|---|---|---|---|---|---|---|---|---|---|
| | | Male | Female | 16–24 | 25–34 | 35–44 | 45–64 | 65+ | High | Middle | Low | Noresp | Radio | Norad. | TV | No TV | Eng | Vern. | None |
| Base | 545 | 332 | 200 | 84 | 152 | 184 | 89 | 19 | 173 | 305 | 0 | 67 | 410 | 135 | 407 | 138 | 179 | 285 | 132 |
| Don't know/no response | 27 | 27 | 28 | 18 | 36 | 28 | 22 | 26 | 2 | 44 | 0 | 15 | 26 | 30 | 9 | 80 | 3 | 9 | 92 |
| Communalism | 5 | 6 | 5 | 4 | 5 | 5 | 9 | 4 | 13 | 2 | 0 | 2 | 6 | 3 | 7 | 0 | 12 | 5 | 0 |
| Corruption | 3 | 3 | 2 | 4 | 1 | 3 | 2 | 2 | 2 | 2 | 0 | 7 | 2 | 4 | 4 | 0 | 1 | 4 | 1 |
| Crime | 0 | 0 | 0 | 1 | 0 | 0 | 0 | 0 | 0 | 0 | 0 | 1 | 0 | 0 | 0 | 0 | 0 | 0 | 0 |
| Disaster | 0 | 0 | 0 | 0 | 0 | 0 | 1 | 0 | 1 | 0 | 0 | 0 | 0 | 0 | 0 | 0 | 0 | 0 | 0 |
| Economic | 12 | 12 | 12 | 10 | 11 | 13 | 13 | 11 | 21 | 8 | 0 | 8 | 13 | 8 | 15 | 4 | 16 | 15 | 2 |
| Environment | 16 | 17 | 14 | 20 | 14 | 17 | 13 | 11 | 13 | 17 | 0 | 22 | 16 | 17 | 20 | 4 | 15 | 23 | 2 |
| Health | 1 | 1 | 2 | 3 | 3 | 1 | 0 | 0 | 2 | 1 | 0 | 2 | 1 | 2 | 2 | 0 | 3 | 1 | 0 |
| Northern hegemony | 4 | 4 | 3 | 3 | 3 | 4 | 6 | 7 | 9 | 1 | 0 | 2 | 4 | 2 | 5 | 1 | 8 | 3 | 0 |
| Politics | 1 | 0 | 2 | 1 | 1 | 1 | 1 | 0 | 1 | 0 | 0 | 2 | 1 | 2 | 1 | 0 | 2 | 1 | 0 |
| Population | 11 | 11 | 11 | 11 | 10 | 11 | 12 | 13 | 7 | 12 | 0 | 15 | 11 | 10 | 14 | 3 | 8 | 17 | 2 |
| Poverty | 4 | 3 | 5 | 5 | 4 | 3 | 5 | 2 | 4 | 3 | 0 | 4 | 3 | 5 | 4 | 3 | 5 | 4 | 1 |
| Civil violence | 5 | 5 | 6 | 10 | 4 | 5 | 2 | 4 | 5 | 4 | 0 | 11 | 5 | 6 | 7 | 0 | 7 | 7 | 0 |
| War | 8 | 9 | 8 | 7 | 6 | 8 | 13 | 16 | 19 | 3 | 0 | 3 | 9 | 7 | 11 | 2 | 18 | 8 | 1 |
| Other | 2 | 1 | 2 | 4 | 1 | 1 | 1 | 2 | 1 | 1 | 0 | 4 | 2 | 2 | 2 | 1 | 3 | 1 | 0 |

Table 7.3.2 National issues: 'What do you think is the most important issue affecting this country at the present time?'

BRITAIN

| | Total | Sex | | Age | | | | | Class | | | | Region | | |
|---|---|---|---|---|---|---|---|---|---|---|---|---|---|---|---|
| | | Male | Female | 16–24 | 25–34 | 35–44 | 45–64 | 65+ | AB | C1 | C2 | DE | South | Midlands Wales | North Scotland |
| Base | 940 | 453 | 487 | 159 | 200 | 168 | 246 | 166 | 169 | 224 | 266 | 281 | 375 | 236 | 329 |
| Unemployment | 45 | 48 | 42 | 50 | 42 | 49 | 43 | 41 | 46 | 37 | 45 | 50 | 46 | 39 | 48 |
| Law and order | 14 | 11 | 17 | 10 | 10 | 12 | 16 | 23 | 10 | 16 | 15 | 15 | 14 | 13 | 16 |
| Cost of living | 3 | 3 | 4 | 5 | 4 | 2 | 3 | 2 | 2 | 4 | 5 | 2 | 3 | 3 | 3 |
| Other economic issues | 10 | 11 | 8 | 9 | 14 | 10 | 9 | 6 | 13 | 13 | 8 | 6 | 10 | 13 | 7 |
| Health | 3 | 3 | 3 | 2 | 2 | 3 | 4 | 2 | 2 | 4 | 3 | 3 | 4 | 3 | 3 |
| Housing | 2 | 2 | 1 | 3 | 0 | 3 | 1 | 1 | 3 | 2 | 1 | 1 | 1 | 3 | 2 |
| Education | 1 | 1 | 1 | 2 | 2 | 2 | 0 | 0 | 2 | 1 | 2 | 0 | 1 | 1 | 0 |
| Pensions | 1 | 1 | 1 | 0 | 0 | 0 | 1 | 3 | 0 | 1 | 1 | 1 | 1 | 1 | 0 |
| Immigrants | 1 | 1 | 0 | 0 | 0 | 1 | 1 | 1 | 0 | 1 | 0 | 0 | 1 | 1 | 0 |
| Europe | 1 | 1 | 1 | 0 | 0 | 0 | 2 | 2 | 1 | 1 | 1 | 1 | 1 | 1 | 0 |
| Environment | 1 | 2 | 1 | 2 | 3 | 2 | 0 | 1 | 3 | 1 | 1 | 1 | 2 | 2 | 1 |
| International affairs | 0 | 1 | 0 | 0 | 0 | 1 | 0 | 0 | 0 | 1 | 0 | 0 | 0 | 0 | 0 |
| Defence | 0 | 0 | 0 | 0 | 0 | 0 | 0 | 0 | 0 | 0 | 0 | 0 | 0 | 0 | 0 |
| Council tax | 0 | 0 | 0 | 0 | 0 | 0 | 0 | 0 | 0 | 0 | 0 | 0 | 0 | 0 | 0 |
| Strikes | 0 | 0 | 0 | 0 | 0 | 0 | 0 | 0 | 0 | 0 | 0 | 0 | 0 | 0 | 0 |
| Conservative Party | 4 | 4 | 4 | 2 | 8 | 1 | 3 | 5 | 3 | 3 | 2 | 7 | 3 | 3 | 6 |
| VAT on fuel | 2 | 1 | 2 | 0 | 3 | 2 | 1 | 2 | 1 | 1 | 3 | 2 | 1 | 2 | 3 |
| Other | 8 | 8 | 7 | 7 | 6 | 9 | 9 | 7 | 11 | 8 | 9 | 5 | 8 | 10 | 6 |
| Don't know | 5 | 2 | 7 | 8 | 4 | 2 | 5 | 5 | 3 | 4 | 5 | 6 | 5 | 4 | 5 |

Table 7.3.2  Continued

INDIA

NORTH

| | | Sex | | | Age | | | | | Income | | | | Radio and TV | | | | Newspaper | | |
|---|---|---|---|---|---|---|---|---|---|---|---|---|---|---|---|---|---|---|---|---|---|
| Base | Total 362 | Male 69 | Female 82 | 16-24 46 | 25-34 39 | 35-44 54 | 45-64 14 | 65+ 3 | High 2 | Middle 128 | Low 97 | Noresp 135 | Radio 320 | Norad. 42 | TV 300 | No TV 62 | Eng 69 | Vern. 207 | None 94 |
| Don't know/no response | 12 | 23 | 2 | 0 | 10 | 26 | 14 | 0 | 0 | 16 | 19 | 4 | 13 | 5 | 14 | 3 | 3 | 6 | 34 |
| Communalism | 15 | 15 | 14 | 15 | 17 | 13 | 0 | 0 | 0 | 17 | 12 | 17 | 15 | 15 | 15 | 14 | 12 | 19 | 7 |
| Corruption | 2 | 2 | 3 | 0 | 2 | 3 | 6 | 33 | 0 | 5 | 2 | 1 | 2 | 4 | 3 | 0 | 8 | 1 | 0 |
| Economic | 17 | 19 | 14 | 19 | 17 | 13 | 11 | 67 | 100 | 9 | 17 | 24 | 17 | 19 | 17 | 17 | 21 | 19 | 11 |
| Education | 9 | 9 | 10 | 7 | 10 | 7 | 11 | 0 | 0 | 8 | 6 | 13 | 10 | 0 | 10 | 6 | 13 | 12 | 0 |
| Environment | 7 | 7 | 5 | 19 | 3 | 4 | 0 | 0 | 0 | 5 | 3 | 11 | 6 | 15 | 7 | 6 | 7 | 10 | 0 |
| Health | 0 | 0 | 1 | 0 | 0 | 0 | 6 | 0 | 0 | 1 | 1 | 0 | 0 | 4 | 0 | 0 | 2 | 0 | 0 |
| Northern hegemony | 0 | 0 | 1 | 0 | 0 | 1 | 0 | 0 | 0 | 1 | 0 | 0 | 0 | 4 | 0 | 0 | 0 | 0 | 2 |
| Politics | 2 | 0 | 5 | 0 | 3 | 1 | 11 | 0 | 0 | 3 | 3 | 0 | 1 | 8 | 1 | 6 | 0 | 2 | 4 |
| Population | 19 | 13 | 26 | 29 | 16 | 15 | 17 | 0 | 0 | 15 | 25 | 18 | 20 | 8 | 15 | 39 | 19 | 16 | 25 |
| Poverty | 9 | 6 | 10 | 5 | 10 | 9 | 17 | 0 | 0 | 13 | 8 | 5 | 9 | 8 | 9 | 6 | 6 | 8 | 12 |
| Civil violence | 5 | 2 | 8 | 5 | 7 | 4 | 6 | 0 | 0 | 7 | 3 | 5 | 4 | 11 | 5 | 6 | 8 | 5 | 4 |
| War | 0 | 0 | 1 | 0 | 2 | 0 | 0 | 0 | 0 | 0 | 2 | 0 | 0 | 0 | 0 | 0 | 0 | 0 | 2 |
| Other | 2 | 3 | 1 | 2 | 2 | 3 | 0 | 0 | 0 | 1 | 2 | 2 | 2 | 0 | 2 | 0 | 2 | 3 | 0 |

Table 7.3.2 Continued

INDIA

SOUTH

| Base | Total | Sex | | Age | | | | | Income | | | | Radio and TV | | | | Newspaper | | |
|---|---|---|---|---|---|---|---|---|---|---|---|---|---|---|---|---|---|---|---|
| | | Male | Female | 16–24 | 25–34 | 35–44 | 45–64 | 65+ | High | Middle | Low | No resp | Radio | Norad. | TV | No TV | Eng | Vern. | None |
| | 716 | 340 | 323 | 130 | 192 | 160 | 194 | 34 | 44 | 136 | 279 | 257 | 516 | 200 | 450 | 266 | 182 | 141 | 396 |
| Don't know/no response | 30 | 24 | 36 | 21 | 31 | 25 | 37 | 41 | 2 | 10 | 38 | 38 | 24 | 47 | 20 | 48 | 8 | 16 | 45 |
| Communalism | 15 | 17 | 12 | 13 | 19 | 17 | 13 | 9 | 15 | 23 | 12 | 14 | 16 | 12 | 18 | 9 | 22 | 17 | 11 |
| Corruption | 2 | 4 | 2 | 2 | 2 | 4 | 2 | 3 | 2 | 3 | 2 | 2 | 2 | 3 | 3 | 1 | 6 | 1 | 1 |
| Disaster | 0 | 0 | 0 | 1 | 0 | 0 | 0 | 0 | 0 | 0 | 0 | 0 | 0 | 0 | 0 | 0 | 0 | 1 | 0 |
| Economic | 14 | 10 | 17 | 19 | 14 | 14 | 12 | 3 | 19 | 16 | 15 | 12 | 15 | 12 | 16 | 11 | 17 | 22 | 10 |
| Education | 1 | 2 | 1 | 1 | 2 | 2 | 1 | 0 | 0 | 1 | 1 | 2 | 2 | 1 | 2 | 1 | 2 | 1 | 1 |
| Environment | 1 | 1 | 2 | 2 | 1 | 1 | 1 | 0 | 2 | 4 | 0 | 1 | 1 | 2 | 2 | 1 | 3 | 1 | 0 |
| Health | 2 | 2 | 2 | 3 | 2 | 1 | 1 | 0 | 0 | 2 | 1 | 0 | 2 | 0 | 2 | 0 | 2 | 1 | 2 |
| Northern hegemony | 0 | 0 | 0 | 0 | 0 | 0 | 0 | 0 | 0 | 0 | 0 | 0 | 0 | 0 | 0 | 0 | 0 | 0 | 0 |
| Politics | 9 | 11 | 6 | 6 | 11 | 10 | 8 | 12 | 24 | 13 | 7 | 6 | 9 | 8 | 10 | 7 | 11 | 11 | 7 |
| Population | 9 | 9 | 10 | 12 | 8 | 10 | 8 | 12 | 15 | 10 | 10 | 8 | 12 | 4 | 11 | 7 | 12 | 12 | 8 |
| Poverty | 7 | 9 | 6 | 8 | 5 | 7 | 9 | 9 | 8 | 7 | 5 | 10 | 8 | 6 | 7 | 7 | 6 | 8 | 8 |
| Civil violence | 6 | 7 | 5 | 9 | 4 | 6 | 6 | 9 | 6 | 8 | 7 | 4 | 7 | 4 | 6 | 6 | 6 | 7 | 6 |
| War | 0 | 0 | 0 | 1 | 0 | 0 | 0 | 0 | 0 | 0 | 0 | 0 | 0 | 0 | 0 | 0 | 0 | 0 | 0 |
| Other | 2 | 3 | 1 | 3 | 2 | 2 | 2 | 3 | 8 | 1 | 2 | 2 | 2 | 1 | 2 | 2 | 3 | 1 | 2 |

Table 7.3.2  Continued

INDIA

WEST

| | Total | Sex | | Age | | | | | Income | | | | Radio and TV | | | | Newspaper | | |
|---|---|---|---|---|---|---|---|---|---|---|---|---|---|---|---|---|---|---|---|
| | | Male | Female | 16–24 | 25–34 | 35–44 | 45–64 | 65+ | High | Middle | Low | Noresp | Radio | Norad. | TV | No TV | Eng | Vern. | None |
| Base | 545 | 332 | 200 | 84 | 152 | 184 | 89 | 19 | 173 | 305 | 0 | 67 | 410 | 135 | 407 | 138 | 179 | 285 | 132 |
| Don't know/no response | 8 | 7 | 9 | 6 | 11 | 7 | 6 | 16 | 0 | 12 | 0 | 7 | 5 | 16 | 1 | 27 | 1 | 0 | 31 |
| Communalism | 8 | 9 | 7 | 7 | 7 | 9 | 9 | 8 | 11 | 7 | 0 | 5 | 8 | 9 | 9 | 7 | 11 | 8 | 6 |
| Corruption | 5 | 5 | 5 | 9 | 5 | 5 | 4 | 4 | 4 | 5 | 0 | 12 | 6 | 5 | 6 | 2 | 6 | 6 | 3 |
| Economic | 20 | 21 | 19 | 15 | 20 | 25 | 18 | 15 | 21 | 21 | 0 | 17 | 22 | 15 | 22 | 15 | 18 | 25 | 13 |
| Education | 6 | 6 | 5 | 6 | 5 | 4 | 8 | 6 | 8 | 4 | 0 | 6 | 6 | 4 | 6 | 5 | 7 | 6 | 5 |
| Environment | 7 | 7 | 7 | 9 | 4 | 8 | 8 | 2 | 3 | 8 | 0 | 13 | 6 | 10 | 8 | 3 | 4 | 9 | 2 |
| Health | 1 | 1 | 0 | 1 | 1 | 1 | 1 | 2 | 1 | 1 | 0 | 1 | 1 | 1 | 1 | 1 | 1 | 1 | 1 |
| Northern hegemony | 5 | 5 | 4 | 3 | 4 | 5 | 6 | 8 | 10 | 1 | 0 | 1 | 5 | 3 | 5 | 2 | 8 | 4 | 1 |
| Politics | 4 | 3 | 4 | 5 | 5 | 3 | 4 | 2 | 6 | 2 | 0 | 4 | 3 | 6 | 5 | 1 | 7 | 3 | 0 |
| Population | 28 | 27 | 28 | 21 | 28 | 30 | 28 | 25 | 26 | 31 | 0 | 21 | 30 | 21 | 28 | 31 | 23 | 30 | 36 |
| Poverty | 5 | 5 | 7 | 10 | 7 | 2 | 5 | 8 | 5 | 5 | 0 | 6 | 5 | 6 | 6 | 5 | 7 | 5 | 3 |
| Civil violence | 1 | 1 | 1 | 2 | 2 | 1 | 0 | 0 | 1 | 1 | 0 | 2 | 1 | 1 | 2 | 0 | 2 | 1 | 0 |
| War | 0 | 0 | 0 | 1 | 0 | 0 | 0 | 0 | 0 | 0 | 0 | 0 | 0 | 0 | 0 | 0 | 0 | 0 | 0 |
| Other | 2 | 2 | 2 | 4 | 3 | 0 | 2 | 2 | 3 | 1 | 0 | 3 | 2 | 3 | 2 | 0 | 4 | 2 | 0 |

*Table* 7.3.3   Local issues: 'What do you think is the most important issue affecting your town or village at the present time?'

BRITAIN

| | Total | Sex | | Age | | | | | Class | | | | Region | | |
|---|---|---|---|---|---|---|---|---|---|---|---|---|---|---|---|
| | | Male | Female | 16–24 | 25–34 | 35–44 | 45–64 | 65+ | AB | C1 | C2 | DE | South | Midlands Wales | North Scotland |
| Base | 940 | 453 | 487 | 159 | 200 | 168 | 246 | 166 | 169 | 224 | 266 | 281 | 375 | 236 | 329 |
| Unemployment | 38 | 37 | 38 | 41 | 40 | 39 | 40 | 26 | 28 | 37 | 37 | 45 | 34 | 32 | 46 |
| Law and order | 14 | 13 | 15 | 14 | 10 | 15 | 13 | 18 | 10 | 15 | 13 | 16 | 13 | 14 | 15 |
| Cost of living | 2 | 1 | 3 | 4 | 1 | 2 | 2 | 1 | 2 | 2 | 3 | 1 | 3 | 1 | 2 |
| Other economic issues | 3 | 4 | 2 | 1 | 6 | 2 | 3 | 3 | 4 | 4 | 2 | 2 | 2 | 5 | 2 |
| Health | 1 | 1 | 2 | 0 | 1 | 0 | 3 | 3 | 0 | 1 | 2 | 1 | 3 | 0 | 1 |
| Housing | 2 | 1 | 3 | 2 | 3 | 4 | 3 | 2 | 1 | 1 | 3 | 3 | 2 | 3 | 2 |
| Education | 1 | 1 | 1 | 1 | 2 | 2 | 1 | 1 | 2 | 1 | 1 | 0 | 1 | 1 | 1 |
| Pensions | 0 | 1 | 0 | 0 | 0 | 0 | 0 | 0 | 0 | 0 | 0 | 1 | 0 | 0 | 1 |
| Immigrants | 0 | 1 | 0 | 0 | 0 | 0 | 1 | 2 | 0 | 0 | 1 | 0 | 0 | 1 | 0 |
| Europe | 0 | 0 | 0 | 0 | 0 | 1 | 0 | 0 | 0 | 0 | 1 | 0 | 0 | 0 | 0 |
| Environment | 2 | 2 | 2 | 0 | 0 | 4 | 3 | 2 | 3 | 2 | 1 | 1 | 3 | 1 | 1 |
| International affairs | 0 | 0 | 0 | 0 | 0 | 0 | 0 | 0 | 0 | 0 | 0 | 0 | 0 | 0 | 0 |
| Defence | 0 | 0 | 0 | 0 | 0 | 0 | 0 | 0 | 0 | 0 | 0 | 0 | 0 | 0 | 0 |
| Council tax | 1 | 2 | 0 | 2 | 1 | 0 | 0 | 2 | 0 | 1 | 2 | 1 | 1 | 2 | 0 |
| Strikes | 0 | 0 | 0 | 0 | 0 | 0 | 0 | 0 | 0 | 0 | 0 | 0 | 0 | 0 | 0 |
| Traffic congestion | 2 | 3 | 1 | 0 | 1 | 2 | 3 | 5 | 5 | 1 | 2 | 1 | 3 | 1 | 3 |
| Other | 17 | 19 | 15 | 12 | 17 | 17 | 17 | 20 | 22 | 16 | 17 | 13 | 18 | 18 | 14 |
| Don't know | 17 | 19 | 18 | 22 | 18 | 12 | 13 | 20 | 21 | 18 | 14 | 15 | 17 | 21 | 13 |

Table 7.3.3 Continued

INDIA

NORTH

| | Total | Sex | | Age | | | | | Income | | | Noresp | Radio and TV | | | | Newspaper | | |
|---|---|---|---|---|---|---|---|---|---|---|---|---|---|---|---|---|---|---|---|
| | | Male | Female | 16-24 | 25-34 | 35-44 | 45-64 | 65+ | High | Middle | Low | | Radio | Norad. | TV | No TV | Eng | Vern. | None |
| Base | 362 | 69 | 82 | 46 | 39 | 54 | 14 | 3 | 2 | 128 | 97 | 135 | 320 | 42 | 300 | 62 | 69 | 207 | 94 |
| Don't know/no response | 20 | 19 | 23 | 15 | 21 | 33 | 14 | 0 | 0 | 16 | 41 | 10 | 21 | 14 | 17 | 39 | 6 | 11 | 51 |
| Communalism | 3 | 2 | 2 | 2 | 3 | 1 | 0 | 0 | 0 | 2 | 2 | 5 | 3 | 4 | 3 | 3 | 5 | 3 | 0 |
| Corruption | 1 | 0 | 2 | 0 | 0 | 2 | 0 | 0 | 0 | 2 | 0 | 0 | 1 | 0 | 1 | 0 | 0 | 1 | 0 |
| Disorganisation | 2 | 3 | 1 | 0 | 3 | 1 | 0 | 33 | 100 | 1 | 2 | 1 | 2 | 0 | 2 | 0 | 4 | 2 | 0 |
| Drainage, garbage, sanitation | 12 | 9 | 11 | 17 | 12 | 5 | 15 | 0 | 0 | 10 | 2 | 22 | 12 | 14 | 14 | 0 | 15 | 15 | 2 |
| Drinking water | 4 | 1 | 7 | 3 | 2 | 2 | 20 | 0 | 0 | 2 | 3 | 6 | 2 | 18 | 4 | 3 | 0 | 4 | 5 |
| Economic | 9 | 10 | 7 | 8 | 3 | 10 | 5 | 67 | 0 | 6 | 11 | 12 | 10 | 4 | 9 | 12 | 8 | 11 | 7 |
| Education | 16 | 22 | 15 | 24 | 16 | 14 | 10 | 0 | 0 | 21 | 8 | 17 | 18 | 4 | 18 | 9 | 23 | 20 | 2 |
| Electricity | 0 | 0 | 1 | 0 | 0 | 0 | 5 | 0 | 0 | 1 | 0 | 0 | 0 | 4 | 0 | 0 | 0 | 0 | 2 |
| Environment | 8 | 8 | 10 | 8 | 10 | 6 | 10 | 0 | 0 | 9 | 8 | 6 | 7 | 11 | 8 | 6 | 15 | 9 | 2 |
| Health | 2 | 1 | 3 | 3 | 3 | 0 | 0 | 0 | 0 | 2 | 2 | 1 | 1 | 4 | 1 | 6 | 2 | 2 | 2 |
| Politics | 0 | 1 | 0 | 0 | 0 | 1 | 0 | 0 | 0 | 1 | 0 | 0 | 0 | 0 | 0 | 0 | 0 | 0 | 2 |
| Pollution | 4 | 3 | 4 | 5 | 3 | 2 | 5 | 0 | 0 | 7 | 0 | 2 | 4 | 0 | 3 | 6 | 0 | 5 | 2 |
| Population | 2 | 5 | 0 | 0 | 2 | 6 | 0 | 0 | 0 | 3 | 5 | 3 | 3 | 0 | 3 | 0 | 0 | 0 | 11 |
| Poverty | 3 | 5 | 1 | 2 | 7 | 4 | 0 | 0 | 0 | 3 | 5 | 2 | 4 | 0 | 4 | 0 | 0 | 2 | 9 |
| Transportation | 10 | 4 | 14 | 7 | 14 | 9 | 10 | 0 | 0 | 13 | 8 | 10 | 9 | 18 | 10 | 12 | 18 | 10 | 4 |
| Trees | 2 | 4 | 1 | 3 | 2 | 2 | 0 | 0 | 0 | 2 | 2 | 2 | 2 | 4 | 3 | 3 | 0 | 3 | 0 |
| Civil violence | 1 | 1 | 0 | 0 | 0 | 0 | 5 | 0 | 0 | 0 | 2 | 1 | 0 | 4 | 0 | 3 | 1 | 1 | 0 |
| Other | 0 | 1 | 0 | 2 | 0 | 0 | 0 | 0 | 0 | 0 | 0 | 1 | 0 | 0 | 0 | 0 | 2 | 0 | 0 |

Table 7.3.3  Continued

INDIA

SOUTH

| | Base | Sex | | Age | | | | | Income | | | | Radio and TV | | | | Newspaper | | |
|---|---|---|---|---|---|---|---|---|---|---|---|---|---|---|---|---|---|---|---|
| | Total | Male | Female | 16–24 | 25–34 | 35–44 | 45–64 | 65+ | High | Middle | Low | Noresp | Radio | Norad. | TV | No TV | Eng | Vern. | None |
| Base | 716 | 340 | 323 | 130 | 192 | 160 | 194 | 34 | 44 | 136 | 279 | 257 | 516 | 200 | 450 | 266 | 182 | 141 | 396 |
| Don't know/no response | 18 | 15 | 20 | 18 | 16 | 16 | 21 | 12 | 7 | 14 | 15 | 25 | 14 | 28 | 14 | 24 | 9 | 10 | 24 |
| Communalism | 6 | 8 | 5 | 6 | 8 | 7 | 4 | 13 | 0 | 4 | 10 | 5 | 6 | 6 | 6 | 7 | 4 | 6 | 8 |
| Corruption | 1 | 2 | 1 | 1 | 0 | 2 | 1 | 3 | 0 | 3 | 0 | 1 | 1 | 1 | 2 | 0 | 2 | 1 | 1 |
| Disorganisation | 0 | 0 | 0 | 1 | 0 | 1 | 0 | 0 | 0 | 1 | 0 | 0 | 0 | 0 | 0 | 0 | 0 | 0 | 0 |
| Drainage, garbage, sanitation | 7 | 3 | 10 | 8 | 7 | 9 | 5 | 3 | 11 | 8 | 4 | 7 | 6 | 9 | 8 | 4 | 12 | 10 | 3 |
| Drinking water | 8 | 6 | 11 | 15 | 8 | 7 | 7 | 5 | 11 | 8 | 11 | 5 | 9 | 7 | 7 | 10 | 8 | 10 | 7 |
| Economic | 12 | 18 | 7 | 7 | 15 | 10 | 12 | 32 | 17 | 13 | 14 | 8 | 12 | 12 | 12 | 13 | 10 | 12 | 13 |
| Education | 3 | 3 | 4 | 3 | 4 | 5 | 2 | 0 | 2 | 4 | 3 | 3 | 4 | 3 | 4 | 3 | 5 | 4 | 2 |
| Electricity | 4 | 2 | 5 | 4 | 3 | 5 | 3 | 0 | 2 | 3 | 4 | 3 | 4 | 3 | 4 | 3 | 4 | 2 | 1 |
| Environment | 2 | 4 | 1 | 0 | 4 | 3 | 2 | 0 | 2 | 3 | 3 | 1 | 3 | 0 | 3 | 1 | 3 | 4 | 1 |
| Health | 5 | 7 | 4 | 3 | 4 | 5 | 7 | 5 | 6 | 5 | 3 | 7 | 6 | 3 | 6 | 3 | 4 | 5 | 5 |
| Politics | 2 | 2 | 1 | 1 | 3 | 2 | 0 | 0 | 0 | 5 | 0 | 2 | 1 | 2 | 2 | 0 | 4 | 3 | 0 |
| Pollution | 2 | 3 | 1 | 3 | 1 | 2 | 3 | 0 | 13 | 3 | 0 | 2 | 2 | 2 | 3 | 0 | 6 | 1 | 1 |
| Population | 2 | 2 | 1 | 1 | 3 | 1 | 1 | 3 | 8 | 3 | 1 | 1 | 2 | 2 | 2 | 1 | 3 | 1 | 1 |
| Poverty | 5 | 9 | 1 | 2 | 4 | 6 | 8 | 11 | 4 | 2 | 8 | 5 | 5 | 6 | 4 | 7 | 2 | 3 | 8 |
| Transportation | 16 | 11 | 20 | 22 | 14 | 17 | 15 | 5 | 10 | 11 | 18 | 17 | 19 | 7 | 16 | 16 | 15 | 23 | 14 |
| Trees | 1 | 1 | 0 | 2 | 0 | 0 | 1 | 0 | 0 | 1 | 0 | 1 | 0 | 2 | 1 | 0 | 1 | 0 | 1 |
| Civil violence | 1 | 1 | 0 | 0 | 0 | 2 | 0 | 0 | 2 | 2 | 0 | 0 | 1 | 0 | 1 | 0 | 0 | 2 | 0 |
| Other | 6 | 5 | 7 | 6 | 5 | 5 | 7 | 8 | 6 | 8 | 4 | 7 | 5 | 9 | 6 | 5 | 7 | 4 | 6 |

*Table 7.3.3* Continued

INDIA

WEST

| Base | Total 545 | Sex Male 332 | Sex Female 200 | Age 16–24 84 | Age 25–34 152 | Age 35–44 184 | Age 45–64 89 | Age 65+ 19 | Income High 173 | Income Middle 305 | Income Low 0 | Radio and TV Noresp 67 | Radio and TV Radio 410 | Radio and TV Norad. 135 | Radio and TV TV 407 | Radio and TV No TV 138 | Newspaper Eng 179 | Newspaper Vern. 285 | Newspaper None 132 |
|---|---|---|---|---|---|---|---|---|---|---|---|---|---|---|---|---|---|---|---|
| Don't know/no | 4 | 4 | 4 | 2 | 5 | 4 | 3 | 0 | 2 | 4 | 0 | 6 | 3 | 7 | 3 | 6 | 7 | 2 | 3 |
| Communalism | 1 | 1 | 1 | 2 | 0 | 0 | 1 | 0 | 0 | 1 | 0 | 1 | 1 | 1 | 1 | 1 | 1 | 0 | 1 |
| Corruption | 1 | 1 | 2 | 2 | 0 | 0 | 1 | 3 | 0 | 0 | 0 | 4 | 1 | 3 | 2 | 0 | 3 | 1 | 0 |
| Crime | 0 | 0 | 0 | 0 | 0 | 0 | 0 | 0 | 0 | 0 | 0 | 0 | 0 | 0 | 0 | 0 | 0 | 0 | 0 |
| Disorganisation | 6 | 4 | 7 | 8 | 6 | 4 | 5 | 3 | 9 | 3 | 0 | 6 | 6 | 6 | 6 | 2 | 13 | 3 | 2 |
| Drainage, garbage, sanitation | 2 | 2 | 1 | 2 | 2 | 1 | 1 | 3 | 0 | 2 | 0 | 3 | 1 | 1 | 2 | 0 | 1 | 3 | 0 |
| Drinking water | 16 | 17 | 14 | 12 | 17 | 20 | 13 | 9 | 6 | 22 | 0 | 16 | 16 | 16 | 15 | 19 | 5 | 20 | 20 |
| Economic | 8 | 8 | 8 | 6 | 9 | 8 | 7 | 12 | 4 | 11 | 0 | 7 | 8 | 9 | 6 | 17 | 3 | 6 | 19 |
| Education | 2 | 2 | 1 | 0 | 2 | 2 | 1 | 3 | 1 | 2 | 0 | 7 | 3 | 3 | 2 | 2 | 1 | 3 | 1 |
| Electricity | 1 | 2 | 2 | 0 | 2 | 2 | 1 | 0 | 0 | 2 | 0 | 1 | 2 | 0 | 2 | 1 | 0 | 1 | 1 |
| Environment | 6 | 5 | 6 | 7 | 4 | 7 | 6 | 0 | 7 | 5 | 0 | 6 | 6 | 6 | 7 | 5 | 6 | 8 | 1 |
| Health | 1 | 1 | 0 | 2 | 0 | 0 | 0 | 0 | 0 | 0 | 0 | 2 | 1 | 0 | 1 | 1 | 0 | 0 | 1 |
| Northern hegemony | 0 | 0 | 0 | 1 | 0 | 0 | 0 | 0 | 0 | 0 | 0 | 0 | 0 | 0 | 0 | 0 | 0 | 0 | 0 |
| Politics | 1 | 1 | 0 | 0 | 0 | 0 | 1 | 3 | 0 | 1 | 0 | 2 | 2 | 2 | 1 | 0 | 0 | 0 | 0 |
| Pollution | 10 | 11 | 8 | 8 | 12 | 12 | 7 | 18 | 20 | 4 | 0 | 6 | 11 | 5 | 12 | 3 | 17 | 8 | 2 |
| Population | 11 | 11 | 13 | 9 | 12 | 9 | 16 | 21 | 17 | 10 | 0 | 5 | 11 | 12 | 10 | 17 | 14 | 9 | 19 |
| Poverty | 6 | 5 | 6 | 5 | 8 | 3 | 9 | 9 | 5 | 7 | 0 | 2 | 6 | 6 | 3 | 14 | 3 | 3 | 15 |
| Transportation | 22 | 21 | 22 | 19 | 19 | 23 | 27 | 18 | 24 | 21 | 0 | 20 | 23 | 17 | 23 | 17 | 24 | 23 | 17 |
| Trees | 2 | 2 | 3 | 4 | 2 | 2 | 1 | 0 | 0 | 3 | 0 | 3 | 2 | 2 | 3 | 0 | 1 | 4 | 0 |
| Civil violence | 0 | 0 | 0 | 0 | 0 | 0 | 0 | 0 | 0 | 0 | 0 | 0 | 1 | 1 | 0 | 0 | 0 | 0 | 0 |
| Other | 1 | 1 | 1 | 3 | 0 | 1 | 1 | 0 | 0 | 1 | 0 | 3 | 2 | 2 | 1 | 0 | 2 | 0 | 0 |

expectations. In the North, but not the other two regions, there is greater concern amongst the young about population. The environment *qua* 'environment' has also become a much lower ranking issue – in the South it is not a real issue, and for the poorer sections of society it is a non-issue. There is no obvious sense in which it is a bigger issue or a different issue for the different English and language readerships of the press. Looking at the entries listed in the aggregating dictionary, again many of the entries are compound – Poverty:Unemployment:Environment – and quite a few refer to problems of water supply and sanitation.

### Local issues

The responses to this question are shown in Table 7.3.3. At the local level, in Britain the two major national issues simply re-surface – Unemployment, and Law and Order. It is possible that most people project from their immediate local concerns to the national level and that the former table is actually mostly a reflection of this one. In the concluding chapter we will refer to this as the idea of 'correspondence' – local verification. In India, the number of 'Don't knows' has come down again, whereas in Britain the number of 'Don't knows' goes up substantially at this local level. (This obviously begs questions about the extent to which people in Britian any longer recognise a local community or identify with one.) In India the top four issues of the three regions are Drainage, etc. (twice), Education (once), Drinking Water (twice), Transportation (three times), Population (once), Pollution (which invariably means atmospheric pollution) (once), Economic (twice). There are also some high figures for parts of the breakdowns shown in the other columns. In the West around dynamic Pune, Population and Poverty are both important for 'No newspaper' group, and Drinking Water is a greater problem both for them and the vernacular group, but not for the English-speaking group.

'Transportation' enters the list of local problems in just as many ways as it would do in Britain. The greatest problem is that there is not enough of it, the roads are too crowded, there are too few buses, but equally and not contradictorily, that there is too much traffic congestion. Much of the pollution is associated with it, particularly in Pune. The impression is of a demand behind this for more 'progress', linked to better drainage and sanitation, better drinking water availability. All these issues could be branded as environmental, but actually arise much more in terms of the developmental.

In Britain there is one direct question asking 'What is the most important environmental problem facing Britain?' which is not replicated exactly in the Indian survey. The answer is Pollution in general, and more specifically air (vehicle) pollution.

The comparison between the two countries almost seems to suggest that many of the problems that Indians face locally have been solved in developed Britain, with one very obvious outstanding exception, which happens to be a problem fast growing in Indian urban areas too – the problem of traffic pollution.

### 7.4 POLICY

Respondents in Britain (Table 7.4.1) were asked how they would like to see a hypothetical 1 per cent increase in tax spent. They were given a list of suggestions, and clearly plumped for four of these, which were, in order, the Health Service (40 per cent), Reducing Unemployment (22 per cent), Education (14 per cent) and improving Law and Order (11 per cent). Other answers had very small responses, including Protection of the Environment (2 per cent). In India a more restricted list of options included Hospitals, Education, Roads, and Protection of the Environment. Education and Health dominate the responses, and on aggregate Protecting the Environment comes out lowest priority equal with road construction. When these figures are broken down by age, the youngest group puts Protection of the Environment second. But for the groups 'No TV' or 'No paper', Education and Hospitals are clearly ranked first, and Protecting the Environment clearly ranked bottom and was insignificant, even considerably below the requirement for new roads.

In response to questions about India's global policy (Table 7.4.2), the 'Don't knows' dominate, overwhelmingly so in the South. There, only the English paper readership have a response, equally weighted between Forests and International Politics (i.e. joining or debating international conventions). In the West the vernacular readership thinks more in terms of Forests but the English readership more in terms of International Politics. In the North the only issue is Forests. These are unequivocally Indian forests.

The level of 'Don't knows' comes down again at the national scale, but again the policy which is most commonly identified is Forests. In the West, the English readership stressed Pollution Control more. In Britain, a similar question shows more 'Don't knows' than in India, but it is not possible to make too many claims for a direct comparison, since the question limits the period of reflection to the 'last year'. Trees get a most cursory and specific mention (Oxleas Wood), perhaps because there is no great public awareness campaign as there has been in India over planting trees, nor has the absence of trees been blamed for anything. (It is the absence of trees *elsewhere* in the world that is seen as a problem from Britain.) If the government has done anything, it is perhaps to have built roads.

The local policy level responses in India almost mimic, again, the national level ones. Forests is the most important response, though in the West mention is also made, particularly by the English readership, of Pollution Control, and Water Management, both reflecting some suggestion that the two problems of Pollution and Drinking Water supply are seen as being addressed. The irony is that the two groups who complained most about the drinking water problem ('Vernacular paper' and 'No paper') are the two which mention Water Management less or not at all – implying perhaps that the sufferers are less convinced that the issue is being addressed.

## 7.5 PERSONAL ACTION

### Membership of environmental organisations

In Britain the vast majority of people (Table 7.5.1) are not members of any environmental or developmental group. Of the 22 per cent that are, the most likely candidates for membership are two of the oldest conservation groups – the National Trust and the Royal Society for the Protection of Birds. This is one of the few questions for which the disaggregation by class and region shows interesting differences. The membership of the National Trust is highly skewed towards the professional classes, and somewhat skewed to southern Britain. Membership of campaigning groups such as Greenpeace and WWF (World-Wide Fund for Nature) is also skewed by class and region. In India, claimed membership of developmental and environmental groups is only 5 per cent, but it is difficult to identify some of the organisations referred to. The majority are local groups and it is evident that

Table 7.4.1 Desired expenditure of extra tax

BRITAIN

The current standard rate of income tax is 25%. If the government were to increase income tax by 1%, for which one of the following areas would you want the money to be used?

| | Total | Sex | | Age | | | | | Class | | | | Region | | | | |
|---|---|---|---|---|---|---|---|---|---|---|---|---|---|---|---|---|---|
| | | Male | Female | 16-24 | 25-34 | 35-44 | 45-64 | 65+ | AB | C1 | C2 | DE | South | Wales | Midlands | North | Scotland |
| Health Service | 40 | 38 | 42 | 33 | 33 | 43 | 45 | 45 | 32 | 41 | 41 | 42 | 37 | | | 43 | 41 |
| Education | 14 | 13 | 15 | 19 | 20 | 19 | 8 | 6 | 20 | 12 | 15 | 11 | 14 | | | 12 | 16 |
| Construction of roads | 1 | 2 | 1 | | 2 | 0 | 0 | 2 | 1 | 2 | 2 | 1 | 1 | | | 1 | 1 |
| Protection of the environment | 2 | 3 | 2 | 5 | 3 | 2 | 1 | 1 | 2 | 4 | 2 | 2 | 2 | | | 1 | 2 |
| Public sector housing | 3 | 3 | 2 | 0 | 3 | 3 | 3 | 1 | 2 | 2 | 2 | 4 | 4 | | | 2 | 2 |
| Reducing unemployment | 22 | 23 | 22 | 23 | 26 | 21 | 21 | 19 | 29 | 20 | 18 | 23 | 22 | | | 18 | 26 |
| Scientific and/or medical research | 2 | 2 | 2 | 4 | 1 | 1 | 0 | 2 | 0 | 2 | 3 | 2 | 1 | | | 3 | 5 |
| Improving law and order | 11 | 11 | 11 | 8 | 8 | 7 | 12 | 20 | 11 | 13 | 13 | 7 | 13 | | | 12 | 8 |
| Other | 2 | 2 | 2 | 1 | 1 | 1 | 1 | 3 | 2 | 2 | 1 | 2 | 2 | | | 2 | 1 |
| None of these | 1 | 1 | 1 | 0 | 1 | 1 | 0 | 0 | 0 | 0 | 1 | 1 | 0 | | | 2 | 0 |
| Don't know | 2 | 2 | 2 | 3 | 3 | 2 | 2 | 2 | 1 | 1 | 3 | 4 | 2 | | | 2 | 2 |

Table 7.4.1  Continued

INDIA

If the government increase tax by 1%, which one of the following would you want the money to be used on? (Choose one)

| | Total | Sex | | Age | | | | | Income | | | No resp | Radio and TV | | | | Newspaper | | |
|---|---|---|---|---|---|---|---|---|---|---|---|---|---|---|---|---|---|---|---|
| | | Male | Female | 16–24 | 25–34 | 35–44 | 45–64 | 65+ | High | Middle | Low | | Radio | No rad. | TV | No TV | Eng | Vern. | None |
| Base | 1623 | 741 | 605 | 260 | 383 | 398 | 297 | 56 | 219 | 569 | 376 | 459 | 1246 | 377 | 1157 | 466 | 430 | 633 | 622 |
| Don't know, no response | 3 | 2 | 1 | 3 | 2 | 2 | 2 | 2 | 2 | 2 | 2 | 4 | 2 | 5 | 2 | 3 | 1 | 4 | 2 |
| Education | 46 | 43 | 49 | 51 | 49 | 46 | 42 | 29 | 60 | 45 | 43 | 42 | 47 | 43 | 48 | 42 | 53 | 46 | 43 |
| Protection of the environment | 14 | 14 | 14 | 20 | 11 | 16 | 10 | 4 | 14 | 21 | 6 | 14 | 14 | 14 | 18 | 5 | 17 | 24 | 3 |
| Hospitals | 23 | 26 | 20 | 17 | 23 | 21 | 29 | 54 | 11 | 16 | 40 | 25 | 23 | 26 | 18 | 38 | 14 | 13 | 39 |
| Roads | 14 | 14 | 16 | 10 | 16 | 15 | 17 | 13 | 13 | 16 | 9 | 15 | 14 | 12 | 14 | 12 | 15 | 14 | 12 |

Table 7.4.2  Public perception of government environmental policies

BRITAIN

| | Total | Sex | | Age | | | | | Class | | | | Region | | |
|---|---|---|---|---|---|---|---|---|---|---|---|---|---|---|---|
| | | Male | Female | 16–24 | 25–34 | 35–44 | 45–64 | 65+ | AB | C1 | C2 | DE | South | Midlands Wales | North Scotland |
| Base | 940 | 453 | 487 | 159 | 200 | 168 | 246 | 166 | 169 | 224 | 266 | 281 | 375 | 236 | 329 |

Can you give any examples of UK government policy since the beginning of this year, which you think will improve or protect the environment?

| | Total | Male | Female | 16–24 | 25–34 | 35–44 | 45–64 | 65+ | AB | C1 | C2 | DE | South | Midlands Wales | North Scotland |
|---|---|---|---|---|---|---|---|---|---|---|---|---|---|---|---|
| VAT on fuel | 3 | 4 | 2 | 2 | 3 | 3 | 1 | 5 | 5 | 3 | 3 | 2 | 3 | 3 | 3 |
| Partnership for Change meeting of NGOs in Manchester in September 1993 | 0 | 0 | 0 | 0 | 0 | 0 | 0 | 0 | 0 | 0 | 0 | 0 | 0 | 0 | 0 |
| Opposed proposals to reintroduce commercial whaling | 1 | 1 | 1 | 1 | 1 | 2 | 0 | 0 | 2 | 1 | 0 | 0 | 1 | 0 | 0 |
| Pledged funding for the Global Environment Facility (GEF) | 2 | 2 | 1 | 2 | 1 | 4 | 1 | 1 | 3 | 2 | 2 | 0 | 3 | 1 | 0 |
| Signed Montreal Protocol – CFCs | 1 | 1 | 1 | 2 | 1 | 2 | 0 | 0 | 2 | 2 | 0 | 0 | 1 | 0 | 1 |
| Stopped plans to build East London crossing at Oxleas Wood | 0 | 0 | 1 | 0 | 1 | 0 | 0 | 0 | 2 | 0 | 0 | 0 | 1 | 0 | 0 |
| Sizewell B | 1 | 1 | 1 | 1 | 0 | 2 | 2 | 1 | 3 | 1 | 1 | 0 | 1 | 2 | 0 |

Table 7.4.2 Continued

*BRITAIN*

Can you give any examples of UK government policy since the beginning of this year, which you think will improve or protect the environment?

| | Total | Sex | | Age | | | | | Class | | | | Region | | | |
|---|---|---|---|---|---|---|---|---|---|---|---|---|---|---|---|---|
| | | Male | Female | 16-24 | 25-34 | 35-44 | 45-64 | 65+ | AB | C1 | C2 | DE | South | Wales | Midlands North | Scotland |
| Base | 940 | 453 | 487 | 159 | 200 | 168 | 246 | 166 | 169 | 224 | 266 | 281 | 375 | 236 | | 329 |
| Adverts on TV about global warming/ energy use/greenhouse effect | 1 | 1 | 0 | 0 | 1 | 1 | 1 | 0 | 0 | 1 | 1 | 1 | 1 | 0 | 0 | 1 |
| Road building | 5 | 5 | 4 | 6 | 5 | 7 | 3 | 3 | 6 | 4 | 5 | 4 | 8 | 2 | 2 | 2 |
| Rail privatisation | 0 | 0 | 1 | 0 | 0 | 1 | 0 | 0 | 1 | 1 | 0 | 0 | 1 | 0 | 0 | 0 |
| Electricity | 0 | 0 | 0 | 0 | 0 | 1 | 0 | 0 | 0 | 0 | 1 | 0 | 1 | 0 | 0 | 0 |
| Pit closures | 1 | 1 | 0 | 1 | 0 | 0 | 1 | 1 | 0 | 1 | 0 | 1 | 1 | 0 | 1 | 1 |
| More pollution tests | 2 | 2 | 2 | 3 | 1 | 2 | 3 | 3 | 3 | 2 | 3 | 2 | 2 | 3 | 3 | 2 |
| Other | 6 | 7 | 5 | 7 | 4 | 4 | 9 | 6 | 5 | 5 | 8 | 6 | 7 | 8 | 8 | 3 |
| Don't know | 82 | 77 | 85 | 79 | 85 | 81 | 80 | 82 | 80 | 81 | 79 | 86 | 76 | 81 | 81 | 88 |

Table 7.4.2 Continued

INDIA

Give examples of government policy in the last year which will improve the global environment.

NORTH

| | Total | Sex | | Age | | | | | Income | | | | Radio and TV | | | | Newspaper | | |
|---|---|---|---|---|---|---|---|---|---|---|---|---|---|---|---|---|---|---|---|
| | | Male | Female | 16–24 | 25–34 | 35–44 | 45–64 | 65+ | High | Middle | Low | Noresp | Radio | Norad. | TV | No TV | Eng | Vern. | None |
| Base | 362 | 69 | 82 | 46 | 39 | 54 | 14 | 3 | 2 | 128 | 97 | 135 | 320 | 42 | 300 | 62 | 69 | 207 | 94 |
| Don't know/no response | 60 | 51 | 66 | 41 | 62 | 69 | 71 | 67 | 0 | 51 | 76 | 58 | 61 | 52 | 57 | 74 | 52 | 45 | 94 |
| Conservation | 1 | 4 | 0 | 2 | 2 | 2 | 0 | 0 | 0 | 1 | 2 | 1 | 2 | 0 | 2 | 0 | 3 | 3 | 0 |
| Education | 2 | 1 | 3 | 2 | 0 | 4 | 0 | 0 | 0 | 3 | 2 | 1 | 2 | 4 | 2 | 0 | 3 | 2 | 2 |
| Forest | 26 | 31 | 21 | 39 | 29 | 14 | 21 | 33 | 0 | 34 | 14 | 27 | 26 | 26 | 28 | 16 | 26 | 36 | 2 |
| International politics | 3 | 4 | 2 | 0 | 2 | 5 | 0 | 0 | 100 | 6 | 0 | 1 | 3 | 0 | 4 | 0 | 8 | 4 | 0 |
| Pollution control | 3 | 5 | 2 | 8 | 2 | 2 | 0 | 0 | 0 | 1 | 2 | 5 | 2 | 9 | 3 | 3 | 5 | 3 | 0 |
| Population | 2 | 1 | 2 | 6 | 0 | 0 | 0 | 0 | 0 | 0 | 0 | 5 | 2 | 4 | 2 | 3 | 0 | 3 | 0 |
| Poverty | 1 | 1 | 0 | 0 | 0 | 2 | 0 | 0 | 0 | 0 | 2 | 0 | 1 | 0 | 1 | 0 | 0 | 0 | 2 |
| Transportation | 1 | 0 | 1 | 0 | 0 | 2 | 0 | 0 | 0 | 1 | 0 | 0 | 1 | 0 | 1 | 0 | 0 | 1 | 0 |
| Other | 2 | 1 | 2 | 2 | 2 | 2 | 7 | 0 | 0 | 3 | 2 | 1 | 2 | 4 | 2 | 3 | 3 | 3 | 0 |

Table 7.4.2 Continued

INDIA

Give examples of government policy in the last year which will improve the global environment.

SOUTH

| | Total | Sex | | Age | | | | | Income | | | | Radio and TV | | | | Newspaper | | |
| --- | --- | --- | --- | --- | --- | --- | --- | --- | --- | --- | --- | --- | --- | --- | --- | --- | --- | --- | --- |
| | | Male | Female | 16–24 | 25–34 | 35–44 | 45–64 | 65+ | High | Middle | Low | Noresp | Radio | Norad. | TV | No TV | Eng | Vern. | None |
| Base | 716 | 340 | 323 | 130 | 192 | 160 | 194 | 34 | 44 | 136 | 279 | 257 | 516 | 200 | 450 | 266 | 182 | 141 | 396 |
| Don't know/no response | 90 | 90 | 89 | 88 | 92 | 86 | 92 | 100 | 77 | 76 | 97 | 92 | 88 | 96 | 86 | 97 | 69 | 94 | 98 |
| Appropriate technology | 0 | 0 | 0 | 0 | 0 | 0 | 0 | 0 | 0 | 1 | 0 | 0 | 0 | 0 | 0 | 0 | 0 | 1 | 0 |
| Conservation | 0 | 0 | 0 | 0 | 0 | 0 | 0 | 0 | 0 | 0 | 0 | 0 | 0 | 0 | 0 | 0 | 0 | 1 | 0 |
| Environment | 0 | 0 | 0 | 0 | 0 | 0 | 0 | 0 | 0 | 1 | 1 | 0 | 0 | 0 | 0 | 0 | 0 | 0 | 0 |
| Forest | 3 | 3 | 3 | 2 | 2 | 6 | 2 | 0 | 9 | 7 | 1 | 2 | 3 | 2 | 4 | 1 | 9 | 1 | 0 |
| Health | 0 | 0 | 0 | 0 | 0 | 0 | 0 | 0 | 0 | 1 | 0 | 0 | 0 | 0 | 0 | 0 | 0 | 0 | 0 |
| International politics | 3 | 3 | 3 | 6 | 1 | 4 | 2 | 0 | 5 | 5 | 1 | 4 | 3 | 2 | 4 | 1 | 10 | 1 | 1 |
| Pollution control | 1 | 1 | 2 | 1 | 1 | 2 | 2 | 0 | 2 | 4 | 0 | 2 | 2 | 0 | 2 | 0 | 4 | 1 | 0 |
| Politics | 0 | 0 | 0 | 1 | 0 | 0 | 0 | 0 | 0 | 0 | 0 | 0 | 0 | 0 | 0 | 0 | 0 | 0 | 0 |
| Population | 0 | 0 | 0 | 1 | 0 | 0 | 0 | 0 | 0 | 1 | 0 | 0 | 0 | 0 | 0 | 0 | 0 | 0 | 0 |
| Poverty | 0 | 0 | 0 | 0 | 0 | 0 | 0 | 0 | 0 | 0 | 0 | 0 | 0 | 0 | 0 | 0 | 0 | 0 | 0 |
| Transportation | 0 | 0 | 0 | 0 | 0 | 0 | 0 | 0 | 2 | 0 | 0 | 0 | 0 | 0 | 0 | 0 | 0 | 1 | 0 |
| Civil violence | 0 | 0 | 0 | 0 | 0 | 0 | 0 | 0 | 1 | 1 | 0 | 0 | 0 | 0 | 0 | 0 | 1 | 0 | 0 |
| Other | 1 | 1 | 2 | 2 | 2 | 1 | 1 | 0 | 5 | 5 | 0 | 0 | 2 | 0 | 2 | 1 | 4 | 1 | 0 |

Table 7.4.2 Continued

INDIA

Give examples of government policy in the last year which will improve the global environment.

WEST

| | Sex | | | Age | | | | | Income | | | | Radio and TV | | | | Newspaper | | |
|---|---|---|---|---|---|---|---|---|---|---|---|---|---|---|---|---|---|---|---|
| | Total | Male | Female | 16–24 | 25–34 | 35–44 | 45–64 | 65+ | High | Middle | Low | Noresp | Radio | No rad. | TV | No TV | Eng | Vern. | None |
| Base | 545 | 332 | 200 | 84 | 152 | 184 | 89 | 19 | 173 | 305 | 0 | 67 | 410 | 135 | 407 | 138 | 179 | 285 | 132 |
| Don't know/no response | 66 | 65 | 67 | 57 | 72 | 66 | 61 | 68 | 59 | 71 | 0 | 55 | 63 | 73 | 56 | 93 | 53 | 57 | 98 |
| Appropriate technology | 0 | 0 | 0 | 0 | 0 | 1 | 0 | 0 | 0 | 0 | 0 | 0 | 0 | 0 | 0 | 0 | 0 | 0 | 0 |
| Conservation | 2 | 2 | 1 | 4 | 1 | 1 | 1 | 0 | 3 | 1 | 0 | 0 | 1 | 2 | 2 | 0 | 4 | 1 | 0 |
| Education | 0 | 0 | 0 | 0 | 1 | 1 | 0 | 0 | 1 | 0 | 0 | 0 | 0 | 0 | 0 | 0 | 1 | 0 | 0 |
| Forest | 16 | 16 | 16 | 12 | 14 | 21 | 14 | 5 | 8 | 20 | 0 | 20 | 19 | 7 | 20 | 6 | 6 | 28 | 2 |
| International politics | 8 | 10 | 5 | 13 | 5 | 6 | 11 | 21 | 16 | 3 | 0 | 7 | 8 | 8 | 10 | 0 | 20 | 6 | 0 |
| Nothing | 2 | 1 | 4 | 3 | 2 | 1 | 3 | 5 | 3 | 1 | 0 | 5 | 2 | 2 | 2 | 1 | 6 | 0 | 0 |
| Pollution control | 4 | 4 | 5 | 8 | 2 | 4 | 6 | 0 | 6 | 2 | 0 | 9 | 5 | 3 | 6 | 0 | 5 | 6 | 0 |
| Population | 0 | 0 | 0 | 1 | 0 | 0 | 0 | 0 | 0 | 0 | 0 | 1 | 0 | 1 | 0 | 0 | 0 | 0 | 0 |
| Other | 2 | 2 | 1 | 1 | 2 | 1 | 4 | 0 | 4 | 1 | 0 | 3 | 1 | 3 | 2 | 0 | 4 | 2 | 0 |

Table 7.4.2 Continued

INDIA

Give examples of government policy in the last year which will improve the global environment.

NORTH

| | Total | Sex | | Age | | | | | Income | | | Noresp | Radio and TV | | | | Newspaper | | |
|---|---|---|---|---|---|---|---|---|---|---|---|---|---|---|---|---|---|---|---|
| | | Male | Female | 16-24 | 25-34 | 35-44 | 45-64 | 65+ | High | Middle | Low | | Radio | Norad. | TV | No TV | Eng | Vern. | None |
| Base | 362 | 69 | 82 | 46 | 39 | 54 | 14 | 3 | 2 | 128 | 97 | 135 | 320 | 42 | 300 | 62 | 69 | 207 | 94 |
| Don't know/no response | 46 | 46 | 41 | 39 | 38 | 54 | 50 | 67 | 0 | 28 | 66 | 50 | 46 | 48 | 43 | 61 | 39 | 29 | 85 |
| Conservation | 1 | 1 | 1 | 2 | 2 | 0 | 0 | 0 | 0 | 2 | 0 | 1 | 1 | 0 | 1 | 0 | 0 | 2 | 0 |
| Economics | 1 | 1 | 0 | 0 | 2 | 0 | 0 | 0 | 0 | 0 | 2 | 0 | 1 | 0 | 1 | 0 | 0 | 0 | 2 |
| Education | 3 | 1 | 5 | 0 | 2 | 3 | 7 | 0 | 0 | 6 | 0 | 2 | 3 | 0 | 3 | 3 | 1 | 5 | 0 |
| Forest | 34 | 28 | 42 | 39 | 40 | 28 | 29 | 33 | 0 | 48 | 23 | 30 | 33 | 43 | 35 | 29 | 50 | 43 | 6 |
| Health | 1 | 1 | 0 | 0 | 0 | 2 | 0 | 0 | 0 | 0 | 2 | 0 | 1 | 0 | 1 | 0 | 0 | 1 | 0 |
| International politics | 1 | 0 | 0 | 0 | 0 | 0 | 0 | 0 | 0 | 0 | 0 | 1 | 1 | 0 | 1 | 0 | 3 | 0 | 0 |
| Pollution control | 4 | 6 | 3 | 9 | 5 | 3 | 0 | 0 | 0 | 4 | 0 | 7 | 4 | 5 | 4 | 3 | 3 | 6 | 0 |
| Politics | 0 | 0 | 0 | 0 | 0 | 0 | 0 | 0 | 0 | 0 | 0 | 1 | 0 | 0 | 0 | 0 | 1 | 0 | 0 |
| Population | 1 | 3 | 0 | 2 | 0 | 2 | 0 | 0 | 0 | 2 | 0 | 1 | 1 | 0 | 1 | 0 | 0 | 1 | 2 |
| Poverty | 1 | 3 | 0 | 0 | 7 | 3 | 0 | 0 | 0 | 2 | 2 | 0 | 2 | 0 | 1 | 0 | 0 | 0 | 4 |
| Water management | 2 | 4 | 1 | 2 | 0 | 0 | 0 | 0 | 0 | 4 | 2 | 0 | 2 | 0 | 3 | 0 | 0 | 4 | 0 |
| Other | 5 | 6 | 6 | 7 | 2 | 5 | 14 | 0 | 100 | 4 | 4 | 6 | 5 | 5 | 6 | 3 | 3 | 9 | 0 |

Table 7.4.2 Continued

## INDIA

Give examples of government policy in the last year which will improve the global environment.

### SOUTH

| | Total | Sex | | Age | | | | | Income | | | Noresp | Radio and TV | | | | Newspaper | | |
|---|---|---|---|---|---|---|---|---|---|---|---|---|---|---|---|---|---|---|---|
| | | Male | Female | 16-24 | 25-34 | 35-44 | 45-64 | 65+ | High | Middle | Low | | Radio | Norad. | TV | No TV | Eng | Vern. | None |
| Base | 716 | 340 | 323 | 130 | 192 | 160 | 194 | 34 | 44 | 136 | 279 | 257 | 516 | 200 | 450 | 266 | 182 | 141 | 396 |
| Don't know/no response | 83 | 87 | 78 | 80 | 83 | 76 | 88 | 100 | 75 | 66 | 93 | 83 | 80 | 91 | 77 | 94 | 62 | 79 | 94 |
| Appropriate technology | 0 | 0 | 0 | 1 | 0 | 0 | 0 | 0 | 2 | 0 | 0 | 0 | 0 | 0 | 0 | 0 | 0 | 0 | 0 |
| Biodiversity | 0 | 0 | 1 | 0 | 1 | 1 | 0 | 0 | 2 | 0 | 0 | 1 | 0 | 0 | 1 | 0 | 2 | 0 | 0 |
| Conservation | 1 | 1 | 2 | 2 | 1 | 1 | 1 | 0 | 2 | 1 | 0 | 2 | 1 | 1 | 1 | 0 | 4 | 0 | 0 |
| Economic | 2 | 1 | 2 | 4 | 0 | 2 | 0 | 0 | 2 | 3 | 2 | 1 | 2 | 0 | 2 | 0 | 4 | 2 | 0 |
| Education | 0 | 0 | 0 | 0 | 0 | 0 | 0 | 0 | 0 | 0 | 0 | 0 | 0 | 0 | 0 | 0 | 0 | 1 | 0 |
| Environment | 0 | 0 | 1 | 1 | 0 | 0 | 0 | 0 | 2 | 1 | 0 | 1 | 0 | 0 | 1 | 0 | 2 | 1 | 0 |
| Forest | 9 | 6 | 12 | 7 | 10 | 14 | 6 | 0 | 10 | 15 | 5 | 9 | 9 | 7 | 12 | 4 | 15 | 15 | 4 |
| Health | 0 | 0 | 0 | 0 | 0 | 0 | 0 | 0 | 0 | 0 | 0 | 0 | 0 | 0 | 0 | 0 | 0 | 0 | 0 |
| International politics | 0 | 0 | 0 | 1 | 0 | 0 | 0 | 0 | 0 | 0 | 0 | 0 | 0 | 0 | 0 | 0 | 0 | 0 | 0 |
| Nothing | 1 | 1 | 1 | 0 | 0 | 0 | 0 | 0 | 2 | 1 | 0 | 1 | 1 | 0 | 2 | 0 | 3 | 0 | 0 |
| Pollution control | 1 | 1 | 0 | 1 | 1 | 2 | 1 | 0 | 0 | 3 | 0 | 0 | 0 | 1 | 0 | 0 | 0 | 1 | 1 |
| Politics | 1 | 1 | 1 | 1 | 1 | 1 | 0 | 0 | 2 | 3 | 0 | 1 | 1 | 0 | 1 | 0 | 2 | 0 | 0 |
| Population | 1 | 1 | 0 | 0 | 0 | 1 | 1 | 0 | 0 | 0 | 0 | 0 | 0 | 0 | 0 | 0 | 0 | 1 | 1 |
| Poverty | 0 | 1 | 0 | 0 | 0 | 0 | 0 | 0 | 2 | 0 | 0 | 1 | 1 | 0 | 1 | 0 | 0 | 0 | 0 |
| Other | 1 | 1 | 2 | 3 | 1 | 2 | 1 | 0 | 2 | 4 | 0 | 1 | 2 | 0 | 2 | 1 | 5 | 1 | 0 |

Table 7.4.2   Continued

INDIA

Give examples of government policy in the last year which will improve the global environment.

WEST

| | Total | Sex | | Age | | | | | Income | | | Noresp | Radio and TV | | | | Newspaper | | |
|---|---|---|---|---|---|---|---|---|---|---|---|---|---|---|---|---|---|---|---|
| | | Male | Female | 16-24 | 25-34 | 35-44 | 45-64 | 65+ | High | Middle | Low | | Radio | Norad. | TV | No TV | Eng | Vern. | None |
| Base | 545 | 332 | 200 | 84 | 152 | 184 | 89 | 19 | 173 | 305 | 0 | 67 | 410 | 135 | 407 | 138 | 179 | 285 | 132 |
| Don't know/no response | 56 | 54 | 59 | 57 | 63 | 54 | 46 | 58 | 45 | 61 | 0 | 58 | 52 | 67 | 44 | 89 | 46 | 42 | 95 |
| Appropriate technology | 0 | 1 | 0 | 0 | 1 | 0 | 0 | 0 | 1 | 0 | 0 | 0 | 0 | 1 | 1 | 0 | 1 | 1 | 0 |
| Conservation | 2 | 2 | 4 | 5 | 2 | 1 | 3 | 0 | 4 | 1 | 0 | 3 | 3 | 2 | 3 | 0 | 5 | 2 | 0 |
| Economic | 1 | 1 | 0 | 2 | 1 | 1 | 0 | 0 | 1 | 1 | 0 | 0 | 0 | 3 | 1 | 0 | 3 | 0 | 0 |
| Education | 1 | 1 | 0 | 3 | 1 | 0 | 0 | 0 | 0 | 1 | 0 | 3 | 1 | 1 | 1 | 1 | 1 | 1 | 1 |
| Forest | 23 | 24 | 22 | 17 | 20 | 27 | 26 | 14 | 14 | 29 | 0 | 25 | 26 | 15 | 29 | 7 | 9 | 40 | 5 |
| International politics | 0 | 0 | 0 | 1 | 0 | 0 | 0 | 0 | 0 | 0 | 0 | 0 | 0 | 0 | 0 | 0 | 0 | 0 | 0 |
| Nothing | 2 | 0 | 5 | 4 | 2 | 1 | 2 | 0 | 2 | 1 | 0 | 4 | 1 | 3 | 2 | 0 | 5 | 0 | 0 |
| Pollution control | 9 | 10 | 6 | 4 | 6 | 9 | 13 | 28 | 20 | 2 | 0 | 5 | 10 | 5 | 11 | 2 | 20 | 7 | 1 |
| Politics | 0 | 1 | 0 | 0 | 1 | 0 | 1 | 0 | 0 | 0 | 0 | 0 | 0 | 1 | 0 | 1 | 1 | 0 | 0 |
| Population | 0 | 0 | 0 | 1 | 0 | 0 | 0 | 0 | 0 | 0 | 0 | 0 | 0 | 0 | 0 | 0 | 0 | 0 | 0 |
| Water management | 4 | 5 | 3 | 4 | 2 | 4 | 7 | 0 | 8 | 2 | 0 | 3 | 5 | 2 | 5 | 1 | 6 | 5 | 0 |
| Other | 1 | 1 | 1 | 2 | 1 | 0 | 2 | 0 | 2 | 1 | 0 | 0 | 1 | 1 | 1 | 1 | 2 | 0 | 0 |

Table 7.4.2  Continued

## INDIA

Give examples of government policy in the last year which will improve the global environment.

### NORTH

| | Total | Sex | | Age | | | | | Income | | | | Radio and TV | | | | Newspaper | | |
|---|---|---|---|---|---|---|---|---|---|---|---|---|---|---|---|---|---|---|---|
| | | Male | Female | 16–24 | 25–34 | 35–44 | 45–64 | 65+ | High | Middle | Low | Noresp | Radio | Norad. | TV | No TV | Eng | Vern. | None |
| Base | 362 | 69 | 82 | 46 | 39 | 54 | 14 | 3 | 2 | 128 | 97 | 135 | 320 | 42 | 300 | 62 | 69 | 207 | 94 |
| Don't know/no response | 58 | 59 | 56 | 50 | 59 | 63 | 64 | 67 | 100 | 48 | 68 | 61 | 58 | 62 | 57 | 65 | 57 | 43 | 89 |
| Conservation | 1 | 1 | 0 | 0 | 2 | 0 | 0 | 0 | 0 | 1 | 0 | 0 | 1 | 0 | 1 | 0 | 2 | 1 | 0 |
| Drainage, garbage, sanitation | 1 | 0 | 0 | 2 | 0 | 0 | 0 | 0 | 0 | 0 | 0 | 1 | 0 | 4 | 1 | 0 | 0 | 1 | 0 |
| Economic | 2 | 3 | 2 | 4 | 0 | 4 | 0 | 0 | 0 | 1 | 2 | 2 | 2 | 0 | 2 | 3 | 0 | 3 | 2 |
| Education | 1 | 0 | 2 | 2 | 0 | 2 | 0 | 0 | 0 | 1 | 0 | 1 | 1 | 4 | 1 | 0 | 2 | 1 | 0 |
| Forest | 20 | 24 | 18 | 22 | 21 | 19 | 14 | 33 | 0 | 30 | 15 | 15 | 21 | 17 | 20 | 23 | 17 | 30 | 0 |
| Garbage | 1 | 0 | 2 | 2 | 0 | 0 | 0 | 0 | 0 | 0 | 0 | 4 | 1 | 4 | 2 | 0 | 0 | 3 | 0 |
| Landslide prevention | 3 | 0 | 6 | 0 | 9 | 2 | 7 | 0 | 0 | 7 | 0 | 1 | 3 | 0 | 4 | 0 | 5 | 3 | 0 |
| Nothing | 2 | 1 | 5 | 2 | 5 | 2 | 7 | 0 | 0 | 1 | 8 | 0 | 3 | 0 | 1 | 10 | 2 | 2 | 6 |
| Pollution control | 6 | 4 | 7 | 9 | 2 | 5 | 0 | 0 | 0 | 4 | 4 | 10 | 7 | 4 | 8 | 0 | 7 | 9 | 0 |
| Population | 1 | 1 | 0 | 2 | 0 | 0 | 0 | 0 | 0 | 0 | 0 | 1 | 1 | 0 | 1 | 0 | 0 | 1 | 0 |
| Poverty | 1 | 1 | 0 | 0 | 0 | 2 | 0 | 0 | 0 | 1 | 0 | 0 | 1 | 0 | 1 | 0 | 0 | 0 | 2 |
| Transportation | 1 | 1 | 2 | 4 | 0 | 0 | 7 | 0 | 0 | 1 | 0 | 2 | 1 | 4 | 2 | 0 | 2 | 2 | 0 |
| Water management | 2 | 3 | 0 | 2 | 2 | 2 | 0 | 0 | 0 | 1 | 2 | 2 | 2 | 0 | 2 | 0 | 4 | 2 | 0 |

Table 7.4.2  Continued

INDIA

Give examples of government policy in the last year which will improve the global environment.

SOUTH

| | Total | Sex | | Age | | | | | Income | | | | Radio and TV | | | | Newspaper | | |
|---|---|---|---|---|---|---|---|---|---|---|---|---|---|---|---|---|---|---|---|
| | | Male | Female | 16–24 | 25–34 | 35–44 | 45–64 | 65+ | High | Middle | Low | Noresp | Radio | Norad. | TV | No TV | Eng | Vern. | None |
| Base | 716 | 340 | 323 | 130 | 192 | 160 | 194 | 34 | 44 | 136 | 279 | 257 | 516 | 200 | 450 | 266 | 182 | 141 | 396 |
| Don't know/no response | 79 | 80 | 77 | 67 | 86 | 73 | 82 | 97 | 75 | 66 | 92 | 73 | 77 | 84 | 73 | 90 | 63 | 76 | 88 |
| Conservation | 1 | 1 | 0 | 1 | 1 | 0 | 0 | 0 | 0 | 1 | 0 | 1 | 0 | 1 | 1 | 0 | 1 | 0 | 0 |
| Drainage, garbage, sanitation | 1 | 1 | 2 | 2 | 0 | 1 | 1 | 0 | 6 | 1 | 0 | 1 | 1 | 1 | 2 | 0 | 3 | 1 | 0 |
| Economic | 1 | 1 | 2 | 2 | 1 | 2 | 1 | 0 | 2 | 2 | 0 | 2 | 2 | 0 | 2 | 0 | 2 | 2 | 1 |
| Education | 0 | 0 | 0 | 1 | 0 | 0 | 0 | 0 | 0 | 1 | 0 | 0 | 0 | 0 | 0 | 0 | 0 | 1 | 0 |
| Environment | 0 | 0 | 0 | 0 | 0 | 0 | 0 | 0 | 0 | 0 | 0 | 0 | 0 | 0 | 0 | 0 | 1 | 0 | 0 |
| Forest | 5 | 5 | 4 | 4 | 3 | 7 | 6 | 3 | 8 | 9 | 2 | 6 | 6 | 2 | 8 | 0 | 9 | 6 | 2 |
| Garbage | 1 | 0 | 1 | 0 | 1 | 1 | 0 | 0 | 0 | 0 | 0 | 1 | 1 | 0 | 0 | 0 | 1 | 1 | 0 |
| Health | 0 | 0 | 1 | 0 | 1 | 0 | 0 | 0 | 0 | 1 | 0 | 1 | 1 | 0 | 1 | 0 | 1 | 1 | 0 |
| Landslide prevention | 1 | 0 | 1 | 1 | 0 | 1 | 0 | 0 | 0 | 1 | 0 | 1 | 1 | 1 | 1 | 0 | 1 | 1 | 0 |
| Nothing | 6 | 8 | 6 | 20 | 2 | 8 | 1 | 0 | 0 | 10 | 5 | 7 | 6 | 8 | 6 | 7 | 6 | 9 | 5 |
| Pollution control | 1 | 1 | 1 | 1 | 0 | 1 | 1 | 0 | 2 | 2 | 0 | 1 | 1 | 0 | 1 | 0 | 2 | 0 | 0 |
| Poverty | 0 | 1 | 0 | 0 | 0 | 1 | 0 | 0 | 0 | 0 | 0 | 0 | 0 | 0 | 0 | 0 | 0 | 0 | 0 |
| Squatter control | 0 | 0 | 0 | 0 | 0 | 0 | 0 | 0 | 0 | 0 | 0 | 1 | 0 | 0 | 0 | 0 | 1 | 0 | 0 |
| Transportation | 0 | 0 | 1 | 1 | 0 | 1 | 0 | 0 | 2 | 0 | 0 | 0 | 0 | 0 | 1 | 0 | 1 | 1 | 0 |
| Unify people | 0 | 0 | 0 | 0 | 0 | 0 | 0 | 0 | 0 | 1 | 0 | 0 | 1 | 0 | 0 | 0 | 1 | 0 | 0 |
| Water management | 1 | 1 | 1 | 0 | 0 | 1 | 1 | 0 | 2 | 1 | 0 | 1 | 1 | 1 | 1 | 0 | 2 | 0 | 0 |

Table 7.4.2 Continued

## INDIA

*Give examples of government policy in the last year which will improve the global environment.*

WEST

| | Total | Sex | | Age | | | | | Income | | | | Radio and TV | | | | Newspaper | | |
|---|---|---|---|---|---|---|---|---|---|---|---|---|---|---|---|---|---|---|---|
| | | Male | Female | 16-24 | 25-34 | 35-44 | 45-64 | 65+ | High | Middle | Low | Noresp | Radio | Norad. | TV | No TV | Eng | Vern. | None |
| Base | 545 | 332 | 200 | 84 | 152 | 184 | 89 | 19 | 173 | 305 | 0 | 67 | 410 | 135 | 407 | 138 | 179 | 285 | 132 |
| Don't know/no response | 52 | 52 | 54 | 52 | 63 | 48 | 45 | 63 | 40 | 60 | 0 | 49 | 50 | 59 | 41 | 87 | 42 | 38 | 93 |
| Appropriate technology | 1 | 1 | 1 | 1 | 0 | 1 | 2 | 0 | 3 | 0 | 0 | 0 | 1 | 0 | 1 | 0 | 2 | 0 | 0 |
| Conservation | 1 | 0 | 0 | 0 | 0 | 1 | 1 | 0 | 2 | 0 | 0 | 0 | 0 | 1 | 1 | 0 | 2 | 0 | 0 |
| Economic | 1 | 1 | 1 | 0 | 1 | 1 | 0 | 4 | 1 | 1 | 0 | 0 | 1 | 1 | 1 | 0 | 2 | 1 | 0 |
| Education | 1 | 1 | 0 | 2 | 0 | 1 | 2 | 11 | 0 | 1 | 0 | 1 | 1 | 1 | 1 | 0 | 0 | 2 | 1 |
| Forest | 26 | 27 | 24 | 19 | 23 | 33 | 25 | 11 | 19 | 30 | 0 | 28 | 28 | 20 | 33 | 7 | 12 | 43 | 5 |
| Garbage | 1 | 1 | 2 | 2 | 1 | 0 | 2 | 0 | 1 | 1 | 0 | 0 | 1 | 1 | 1 | 1 | 2 | 1 | 0 |
| Landslide prevention | 0 | 0 | 0 | 0 | 0 | 0 | 0 | 0 | 0 | 0 | 0 | 0 | 0 | 0 | 0 | 0 | 1 | 0 | 0 |
| Nothing | 2 | 2 | 3 | 7 | 2 | 2 | 0 | 0 | 2 | 0 | 0 | 7 | 1 | 6 | 3 | 1 | 7 | 0 | 0 |
| Pollution control | 6 | 6 | 5 | 9 | 4 | 6 | 6 | 7 | 13 | 2 | 0 | 6 | 6 | 7 | 8 | 1 | 15 | 5 | 1 |
| Population | 1 | 1 | 0 | 1 | 0 | 1 | 0 | 4 | 1 | 1 | 0 | 0 | 1 | 0 | 0 | 0 | 0 | 1 | 0 |
| Transportation | 1 | 0 | 1 | 2 | 1 | 0 | 0 | 0 | 1 | 0 | 0 | 1 | 0 | 2 | 1 | 0 | 2 | 0 | 0 |
| Water management | 7 | 7 | 6 | 3 | 5 | 8 | 12 | 11 | 17 | 2 | 0 | 1 | 9 | 1 | 9 | 2 | 12 | 9 | 1 |
| Other | 1 | 0 | 1 | 1 | 1 | 0 | 2 | 0 | 1 | 0 | 0 | 6 | 1 | 1 | 1 | 1 | 2 | 1 | 0 |

quite a few of those cited, like the Boy Scouts and the YWCA, while no doubt espousing some relevant messages, are not environmental groups in the same sense. Membership of any group in India is biased heavily towards the TV audience and the newspaper readership. Again we repeat the caveat that these categories are good surrogates for class and income – as in Britain, where membership is skewed by class, so it would appear to be here too.

### Environmentally friendly action

Not many people (16 per cent) in Britain (Table 7.5.2) were prepared to say that they 'did nothing', or they 'didn't know' what they did, to help the environment. The commonest cited actions were to use ozone-friendly sprays – but since there are virtually no sprays left which use CFCs, it is not entirely clear what they were claiming. The most favoured other responses were: using recycled paper, saving electricity, and recycling household rubbish, which means taking waste like glass to local collection spots. There is some variation in recycling according to class. In response to the open-ended question about environmentally friendly actions in India, many fewer claim they actually do anything: of those who do, forestry – in this context meaning planting a tree – is the leading answer. But in terms of household action – and it may not be seen intentionally as an environmentally friendly act – the level of recycling is vastly greater than in Britain. This may surprise people who are aware of the litter in India, but it will not surprise anyone who knows how paper is collected, and paper bags for use in the market are then made by cutting-and-pasting by hand, or who sees the way in which rubber and plastic are recycled – turning old tyres into shoe soles, for example. In this labour-abundant but material-short economy, scrap materials command a price which encourages recycling. The reuse of vegetable matter, paper, glass, plastic and cans is put at between 80 and 90 per cent (of respondents) (Table 7.5.3). Selling the scrap is the commonest action, except, logically, for vegetables, which are composted. This is a vastly superior response which contrasts starkly with the British, and it gives credence to one of the ambivalent Indian views that it is the North that produces rubbish and pollution, despite apparent cleanliness.

In terms of social action, more people in Britain claim to have been active in the last two years than in India – the donation of money and the signing of petitions being the most likely actions. In India there was considerable admiration (in Chapter 6) for the British zeal in campaigning. Many fewer do it in India, but campaigning and the donation of money are the most likely actions – and these are very clearly skewed towards the TV audience, the North and West rather than South.

### 7.6 STORIES RECALLED AND KNOWLEDGE ABOUT THE ENVIRONMENT

In India respondents were asked what medium was the main source of information on the environment. Radio, books and organisations do not feature much in the response (Table 7.6.1). The answers emphasise TV, friends, magazines and newspapers. In both countries respondents were asked to name any environmental story they had been aware of in the media, in the case of Britain both in the last week and the last year, and in India in the last year only. In processing the results for India it became evident that quite a few answered with the name of a regular programme slot – like the little cameos on Earthfile on BBC World Service TV broadcast on the Star satellite. These were grouped with the 'No response' category,

Table 7.5.1 Membership of environmental groups

**BRITAIN**

Which of these development or environmental organisations, either local or national, are you a member of?

| | Total | Sex | | Age | | | | | Class | | | | Region | | |
|---|---|---|---|---|---|---|---|---|---|---|---|---|---|---|---|
| | | Male | Female | 16-24 | 25-34 | 35-44 | 45-64 | 65+ | AB | C1 | C2 | DE | South | Midlands Wales | North Scotland |
| Base | 940 | 453 | 487 | 159 | 200 | 168 | 246 | 166 | 169 | 224 | 266 | 281 | 375 | 236 | 329 |
| ActionAid | 1 | 1 | 1 | 1 | 0 | 2 | 2 | 0 | 2 | 2 | 1 | 0 | 2 | 1 | 1 |
| CAFOD | 2 | 2 | 2 | 1 | 0 | 2 | 2 | 3 | 4 | 2 | 2 | 0 | 2 | 2 | 1 |
| Christian Aid | 3 | 3 | 4 | 2 | 1 | 5 | 3 | 4 | 6 | 3 | 3 | 1 | 5 | 3 | 1 |
| CPRE | 0 | 0 | 0 | 0 | 0 | 1 | 1 | 0 | 1 | 0 | 0 | 0 | 1 | 0 | 0 |
| Friends of the Earth | 4 | 4 | 3 | 5 | 4 | 4 | 3 | 2 | 8 | 3 | 3 | 2 | 6 | 2 | 1 |
| Greenpeace | 3 | 3 | 3 | 4 | 3 | 5 | 3 | 1 | 7 | 3 | 2 | 2 | 6 | 2 | 1 |
| Local environment/ nature conservation group | 2 | 1 | 2 | 1 | 2 | 2 | 2 | 2 | 2 | 2 | 2 | 0 | 2 | 4 | 0 |
| National Trust | 7 | 8 | 7 | 4 | 3 | 11 | 6 | 15 | 19 | 7 | 5 | 4 | 10 | 7 | 5 |
| Oxfam | 3 | 3 | 3 | 3 | 4 | 1 | 2 | 4 | 5 | 5 | 2 | 2 | 4 | 3 | 0 |
| RSPB | 5 | 5 | 4 | 4 | 4 | 7 | 4 | 5 | 9 | 5 | 3 | 4 | 7 | 2 | 2 |
| RSPCA | 3 | 4 | 3 | 3 | 2 | 4 | 4 | 5 | 5 | 3 | 4 | 3 | 5 | 2 | 2 |
| Survival International | 1 | 1 | 0 | 0 | 1 | 1 | 1 | 0 | 2 | 1 | 0 | 0 | 1 | 0 | 0 |
| World-Wide Fund for Nature (WWF) | 4 | 3 | 6 | 4 | 4 | 4 | 6 | 4 | 9 | 3 | 3 | 5 | 7 | 4 | 2 |
| Other | 2 | 2 | 3 | 1 | 1 | 6 | 1 | 4 | 6 | 2 | 2 | 1 | 4 | 2 | 1 |
| Don't know | 3 | 3 | 3 | 2 | 1 | 1 | 6 | 3 | 3 | 1 | 1 | 5 | 2 | 2 | 4 |
| None | 78 | 79 | 77 | 83 | 84 | 73 | 76 | 71 | 58 | 80 | 80 | 85 | 74 | 76 | 83 |

*Table 7.5.1* Continued

*INDIA*

*Do you belong to an NGO or other group working to help the environment?*

| | | Sex | | Age | | | | | | TV | | Newspaper | | |
| | Total | Male | Female | 16–24 | 25–34 | 35–44 | 45–64 | 65+ | TV | No TV | Eng | Vern. | None |
|---|---|---|---|---|---|---|---|---|---|---|---|---|---|
| Base | 1623 | 741 | 605 | 260 | 383 | 398 | 297 | 56 | 1157 | 466 | 430 | 633 | 622 |
| No | 95 | 95 | 97 | 94 | 96 | 97 | 97 | 100 | 93 | 99 | 92 | 92 | 99 |
| Yes | 5 | 5 | 3 | 6 | 4 | 3 | 3 | 0 | 7 | 2 | 8 | 8 | 1 |

Table 7.5.2  Environment friendly actions

BRITAIN

| | Total | Sex | | Age | | | | | Class | | | | Region | | |
|---|---|---|---|---|---|---|---|---|---|---|---|---|---|---|---|
| | | Male | Female | 16–24 | 25–34 | 35–44 | 45–64 | 65+ | AB | C1 | C2 | DE | South | Midlands Wales | North Scotland |
| Base | 940 | 453 | 487 | 159 | 200 | 168 | 246 | 166 | 169 | 224 | 266 | 281 | 375 | 236 | 329 |

*What, if anything, do you do now in your day-to-day life to help the environment which you did not do five years ago?*

| | Total | Male | Female | 16–24 | 25–34 | 35–44 | 45–64 | 65+ | AB | C1 | C2 | DE | South | Midlands Wales | North Scotland |
|---|---|---|---|---|---|---|---|---|---|---|---|---|---|---|---|
| Use ozone-friendly aerosols/sprays | 53 | 47 | 57 | 57 | 61 | 58 | 53 | 32 | 61 | 55 | 53 | 45 | 53 | 52 | 53 |
| Use recycled paper products | 40 | 36 | 44 | 40 | 44 | 45 | 39 | 33 | 55 | 38 | 43 | 30 | 42 | 42 | 38 |
| Use car less/use public transport/bicycle/walk more | 16 | 19 | 13 | 16 | 24 | 14 | 13 | 12 | 19 | 18 | 17 | 12 | 18 | 13 | 16 |
| Use 'green' cleaning products | 19 | 18 | 20 | 22 | 21 | 20 | 21 | 12 | 27 | 22 | 19 | 13 | 20 | 21 | 18 |
| Recycle household rubbish | 32 | 29 | 35 | 27 | 33 | 36 | 29 | 35 | 46 | 34 | 31 | 23 | 41 | 29 | 24 |
| Use energy-efficient light bulbs | 8 | 8 | 8 | 7 | 6 | 10 | 12 | 6 | 16 | 11 | 6 | 4 | 9 | 7 | 8 |
| Set central heating thermostat lower | 18 | 15 | 21 | 9 | 21 | 22 | 24 | 11 | 22 | 17 | 20 | 15 | 19 | 14 | 20 |
| Try to save electricity in the home and/or at work | 38 | 38 | 38 | 33 | 45 | 43 | 38 | 29 | 42 | 39 | 38 | 35 | 39 | 33 | 41 |
| Nothing | 12 | 14 | 10 | 12 | 8 | 10 | 12 | 19 | 8 | 13 | 11 | 15 | 12 | 13 | 12 |
| Other | 4 | 3 | 4 | 5 | 4 | 3 | 5 | 3 | 6 | 3 | 3 | 4 | 5 | 4 | 3 |
| Don't know | 4 | 4 | 4 | 3 | 0 | 4 | 4 | 9 | 2 | 2 | 4 | 6 | 3 | 6 | 4 |

Table 7.5.2  Continued

BRITAIN

*Is any of your household rubbish taken to be recycled or do you reuse it yourself in any way?*

| | Total | Sex | | Age | | | | | Class | | | | Region | | |
| | | Male | Female | 16-24 | 25-34 | 35-44 | 45-64 | 65+ | AB | C1 | C2 | DE | South | Midlands Wales | North Scotland |
| Base | 940 | 453 | 487 | 159 | 200 | 168 | 246 | 166 | 169 | 224 | 266 | 281 | 375 | 236 | 329 |
| No, nothing | 38 | 41 | 35 | 37 | 38 | 36 | 42 | 34 | 21 | 32 | 41 | 50 | 31 | 35 | 48 |
| Glass/bottles | 44 | 41 | 47 | 36 | 46 | 47 | 42 | 49 | 65 | 50 | 42 | 29 | 52 | 45 | 34 |
| Paper | 39 | 35 | 43 | 36 | 33 | 43 | 36 | 50 | 55 | 44 | 39 | 26 | 44 | 40 | 33 |
| Plastic/plastic bags | 13 | 10 | 15 | 16 | 17 | 14 | 11 | 8 | 17 | 15 | 10 | 12 | 17 | 11 | 10 |
| Cans | 18 | 17 | 19 | 16 | 18 | 14 | 18 | 23 | 25 | 18 | 18 | 13 | 23 | 22 | 9 |
| Old clothes | 1 | 2 | 1 | 1 | 2 | 2 | 2 | 1 | 3 | 1 | 2 | 0 | 3 | 1 | 1 |
| Other | 2 | 1 | 2 | 4 | 0 | 1 | 1 | 4 | 3 | 0 | 2 | 2 | 3 | 1 | 2 |
| Don't know | 4 | 5 | 4 | 6 | 5 | 4 | 6 | 0 | 6 | 3 | 3 | 6 | 4 | 5 | 4 |

Table 7.5.2 Continued

INDIA

NORTH

What environmentally friendly actions do you carry out now in your daily life which you did not carry out 2 years ago?

| | | Sex | | Age | | | | | Income | | | | Radio and TV | | | | Newspaper | | |
|---|---|---|---|---|---|---|---|---|---|---|---|---|---|---|---|---|---|---|---|
| | Total | Male | Female | 16-24 | 25-34 | 35-44 | 45-64 | 65+ | High | Middle | Low | No resp | Radio | No rad. | TV | No TV | Eng | Vern. | None |
| Base | 362 | 69 | 82 | 46 | 39 | 54 | 14 | 3 | 2 | 128 | 97 | 135 | 320 | 42 | 300 | 62 | 69 | 207 | 94 |
| Don't know/no response | 48 | 64 | 33 | 46 | 41 | 61 | 50 | 0 | 0 | 52 | 51 | 44 | 49 | 43 | 52 | 32 | 35 | 50 | 57 |
| Cleanliness | 1 | 1 | 1 | 0 | 0 | 0 | 0 | 33 | 0 | 0 | 2 | 1 | 1 | 0 | 1 | 3 | 0 | 1 | 2 |
| Conservation | 1 | 1 | 0 | 2 | 0 | 0 | 0 | 0 | 0 | 0 | 0 | 1 | 1 | 0 | 1 | 0 | 3 | 0 | 0 |
| Education | 1 | 1 | 0 | 0 | 0 | 2 | 0 | 0 | 0 | 0 | 0 | 1 | 1 | 0 | 1 | 0 | 0 | 0 | 2 |
| Forest | 17 | 20 | 18 | 26 | 18 | 16 | 7 | 33 | 0 | 17 | 6 | 26 | 15 | 33 | 20 | 6 | 20 | 20 | 8 |
| Garbage | 1 | 0 | 2 | 0 | 3 | 2 | 0 | 0 | 0 | 2 | 0 | 1 | 1 | 0 | 1 | 0 | 0 | 2 | 0 |
| Gardening | 20 | 4 | 39 | 20 | 31 | 16 | 36 | 0 | 100 | 23 | 31 | 9 | 20 | 24 | 16 | 42 | 23 | 18 | 22 |
| Nothing | 11 | 8 | 6 | 7 | 8 | 4 | 7 | 33 | 0 | 6 | 10 | 15 | 12 | 0 | 10 | 16 | 19 | 9 | 8 |

Table 7.5.2  Continued

INDIA

SOUTH

| | Total | Sex | | Age | | | | | Income | | | | Radio and TV | | | | Newspaper | | |
|---|---|---|---|---|---|---|---|---|---|---|---|---|---|---|---|---|---|---|---|
| | | Male | Female | 16–24 | 25–34 | 35–44 | 45–64 | 65+ | High | Middle | Low | No resp | Radio | No rad. | TV | No TV | Eng | Vern. | None |
| Base | 716 | 340 | 323 | 130 | 192 | 160 | 194 | 34 | 44 | 136 | 279 | 257 | 516 | 200 | 450 | 266 | 182 | 141 | 396 |
| Don't know/no response | 42 | 32 | 53 | 46 | 40 | 44 | 42 | 21 | 30 | 37 | 43 | 46 | 39 | 48 | 40 | 44 | 32 | 34 | 49 |
| Animal care | 0 | 0 | 0 | 0 | 0 | 0 | 1 | 0 | 0 | 0 | 0 | 0 | 0 | 0 | 0 | 0 | 0 | 1 | 0 |
| Cleanliness | 4 | 6 | 2 | 4 | 2 | 2 | 7 | 9 | 24 | 5 | 2 | 2 | 4 | 3 | 5 | 3 | 6 | 3 | 3 |
| Conservation | 0 | 0 | 1 | 2 | 0 | 0 | 1 | 0 | 2 | 1 | 0 | 0 | 1 | 0 | 1 | 0 | 2 | 0 | 0 |
| Economic | 0 | 0 | 0 | 1 | 0 | 0 | 0 | 0 | 0 | 1 | 0 | 0 | 0 | 0 | 0 | 0 | 1 | 0 | 0 |
| Education | 0 | 0 | 0 | 0 | 1 | 1 | 0 | 0 | 0 | 1 | 0 | 0 | 0 | 0 | 0 | 0 | 1 | 0 | 0 |
| Environment | 0 | 0 | 0 | 0 | 0 | 0 | 0 | 0 | 2 | 0 | 0 | 0 | 0 | 0 | 0 | 0 | 0 | 1 | 0 |
| Forest | 7 | 11 | 4 | 6 | 6 | 9 | 7 | 14 | 22 | 9 | 5 | 6 | 8 | 6 | 8 | 6 | 11 | 10 | 5 |
| Garbage | 4 | 3 | 5 | 4 | 5 | 5 | 3 | 0 | 4 | 2 | 3 | 6 | 4 | 4 | 6 | 0 | 6 | 6 | 2 |
| Gardening | 7 | 5 | 10 | 9 | 9 | 7 | 4 | 6 | 2 | 12 | 7 | 6 | 7 | 6 | 8 | 5 | 10 | 11 | 4 |
| Nothing | 32 | 37 | 26 | 25 | 33 | 30 | 35 | 48 | 8 | 26 | 38 | 35 | 33 | 31 | 28 | 40 | 27 | 33 | 35 |
| Pollution control | 1 | 2 | 0 | 2 | 1 | 1 | 1 | 3 | 4 | 2 | 1 | 0 | 1 | 1 | 2 | 0 | 3 | 1 | 1 |
| Water management | 0 | 0 | 0 | 0 | 1 | 0 | 0 | 0 | 0 | 1 | 0 | 0 | 0 | 0 | 0 | 0 | 0 | 0 | 0 |
| Other | 2 | 3 | 0 | 0 | 3 | 2 | 2 | 0 | 2 | 4 | 1 | 0 | 2 | 0 | 2 | 1 | 3 | 1 | 1 |

Table 7.5.2 Continued

INDIA

NORTH

|  | Total | Sex | | Age | | | | | Income | | | | Radio and TV | | | | Newspaper | | |
|---|---|---|---|---|---|---|---|---|---|---|---|---|---|---|---|---|---|---|---|
|  |  | Male | Female | 16-24 | 25-34 | 35-44 | 45-64 | 65+ | High | Middle | Low | No resp | Radio | No rad. | TV | No TV | Eng | Vern. | None |
| Base | 545 | 332 | 200 | 84 | 152 | 184 | 89 | 19 | 173 | 305 | 0 | 67 | 410 | 135 | 407 | 138 | 179 | 285 | 132 |
| Don't know/no response | 43 | 43 | 44 | 25 | 55 | 47 | 36 | 26 | 22 | 57 | 0 | 36 | 40 | 53 | 34 | 70 | 16 | 42 | 76 |
| Appropriate technology | 1 | 0 | 1 | 1 | 1 | 0 | 1 | 0 | 1 | 0 | 0 | 1 | 1 | 0 | 0 | 1 | 0 | 1 | 1 |
| Cleanliness | 1 | 1 | 0 | 3 | 1 | 0 | 1 | 5 | 1 | 0 | 0 | 4 | 1 | 1 | 1 | 0 | 2 | 1 | 0 |
| Conservation | 8 | 6 | 11 | 12 | 8 | 5 | 9 | 9 | 16 | 4 | 0 | 1 | 8 | 6 | 9 | 4 | 16 | 7 | 2 |
| Economic | 0 | 0 | 0 | 0 | 1 | 0 | 0 | 0 | 0 | 0 | 0 | 0 | 0 | 0 | 0 | 0 | 1 | 0 | 0 |
| Education | 0 | 0 | 0 | 0 | 1 | 1 | 0 | 0 | 1 | 0 | 0 | 0 | 0 | 0 | 0 | 0 | 0 | 1 | 0 |
| Environment | 0 | 0 | 0 | 0 | 0 | 0 | 1 | 0 | 0 | 0 | 0 | 0 | 0 | 0 | 0 | 0 | 1 | 0 | 0 |
| Forest | 16 | 15 | 17 | 24 | 9 | 19 | 14 | 5 | 11 | 16 | 0 | 34 | 18 | 12 | 21 | 3 | 8 | 27 | 1 |
| Garbage | 9 | 8 | 12 | 12 | 8 | 9 | 9 | 14 | 19 | 4 | 0 | 9 | 9 | 9 | 11 | 4 | 23 | 7 | 2 |
| Gardening | 1 | 1 | 0 | 1 | 0 | 0 | 3 | 0 | 0 | 0 | 0 | 1 | 1 | 1 | 1 | 0 | 2 | 1 | 0 |
| Nothing | 10 | 11 | 9 | 9 | 11 | 9 | 9 | 14 | 8 | 12 | 0 | 4 | 10 | 7 | 7 | 17 | 10 | 5 | 17 |
| Organic | 1 | 2 | 0 | 0 | 1 | 1 | 3 | 9 | 2 | 1 | 0 | 1 | 1 | 1 | 1 | 1 | 1 | 1 | 2 |
| Pollution control | 6 | 9 | 2 | 9 | 1 | 9 | 6 | 18 | 13 | 3 | 0 | 4 | 6 | 6 | 8 | 1 | 14 | 5 | 1 |
| Water management | 2 | 3 | 1 | 1 | 3 | 1 | 8 | 0 | 5 | 2 | 0 | 0 | 3 | 2 | 3 | 0 | 5 | 2 | 0 |
| Other | 1 | 1 | 1 | 2 | 2 | 0 | 0 | 0 | 1 | 0 | 0 | 3 | 1 | 1 | 1 | 0 | 2 | 0 | 0 |

*Table 7.5.3*  Recycling of household waste, India: percentage of total

|                   | Vegetables | Paper | Glass | Plastic | Cans |
|-------------------|-----------|-------|-------|---------|------|
| Burn it           | 15        | 11    | 1     | 5       | 2    |
| Sell it           | 2         | 72    | 64    | 70      | 68   |
| Domestic reuse    | 2         | 4     | 13    | 14      | 20   |
| Compost it        | 70        | 9     | 12    | 5       | 4    |
| Total recycled    | 89        | 96    | 90    | 94      | 93   |
| Total not recycled| 11        | 4     | 10    | 6       | 7    |

as the topic of the story was not known: this exaggerates the 'No response' category, but not by a lot. As ought to happen, the 'No response' is almost 100 per cent for the categories 'No TV' and 'No newspaper'. The equivalent does not occur in the UK. In India the stories that are recalled are overwhelmingly about India: Global Warming and Ozone Depletion appear in the list, but only minimally. There are a few responses about wildlife in Africa – for Indians, too, Africa is more a place of wildlife than economics and politics. But the most commonly remembered stories are about forests/trees, followed by pollution. The Narmada Dam project features, but more for the West, as one might expect, than for other regions, and even there it is recalled less than pollution.

In Britain there are more people who have identified a story – for the 'last week' the two commonest both involve radiation – the dumping of radioactive waste at sea and the possible link between a nuclear power and reprocessing plant at Sellafield in the UK and child leukaemia. This is a story which has featured in the British media for several years, and it features highly in the answers to the next question, asking respondents to mention stories 'within the last year'. A minority of people, 41 per cent, responded with 'Can't remember' or 'Don't know'. The three lead categories include two global (seen from Britain's perspective) stories – destruction of the rain-forests and global warming – but also the *Braer* oil-tanker disaster in the Shetland Islands of Scotland. Leukaemia and Sellafield, the ozone hole and pollution follow.

**Knowledge about specific global environmental problems**

A series of questions were asked in India (Table 7.6.2), but not in Britain, about the ozone hole(s), global warming and acid rain. There is no way we can say what level of knowledge we would expect – such specific questions asked in Britain might well reveal some muddled knowledge for a minority of the general public. As with most issues, the level of 'Don't knows' for any of the questions nears 100 per cent for those who are neither newspaper readers nor with access to TV. With regard to most responses, the discrimination between those who read newspapers and those who do not is a better predictor of likely knowledge than TV viewership or not. In all three regions respondents were more likely to think that ozone holes were a problem of the Northern hemisphere rather than both hemispheres – and in all regions the respondents who read the English press were more likely to respond correctly (in one region by what is an insignificant margin). The very limited knowledge about its precise technical causes – CFCs – is clearly biased towards the English readership in all three regions. Awareness of global warming in two of the three regions is biased towards the English readership, but not in the West, where it seems more equal. In the South greenhouse gases are the lead cause

Table 7.6.1 Environmental news stories recalled

| BRITAIN | Total | Sex | | Age | | | | | Class | | | | Region | | |
|---|---|---|---|---|---|---|---|---|---|---|---|---|---|---|---|
| | | Male | Female | 16–24 | 25–34 | 35–44 | 45–64 | 65+ | AB | C1 | C2 | DE | South | Midlands Wales | North Scotland |
| Base | 940 | 453 | 487 | 159 | 200 | 168 | 246 | 166 | 169 | 224 | 266 | 281 | 375 | 236 | 329 |
| *Within the last week what environmental issues on TV, radio or in the newspapers have you been aware of?* | | | | | | | | | | | | | | | |
| Radioactive waste dumping at sea | 3 | 3 | 4 | 2 | 4 | 4 | 4 | 3 | 3 | 3 | 3 | 3 | 4 | 5 | 2 |
| Water meters leading to health problems | 2 | 2 | 2 | 2 | 2 | 2 | 4 | 3 | 3 | 3 | 2 | 1 | 3 | 2 | 2 |
| National Trust and hunting ban | 2 | 3 | 1 | 0 | 3 | 3 | 1 | 2 | 4 | 2 | 1 | 1 | 4 | 1 | 1 |
| M11 link protesters living in tree | 1 | 2 | 1 | 0 | 3 | 3 | 1 | 0 | 3 | 2 | 2 | 1 | 3 | 0 | 0 |
| Channel tunnel link protests | 1 | 2 | 0 | 2 | 1 | 1 | 0 | 1 | 1 | 0 | 0 | 1 | 2 | 0 | 0 |
| Animal slaughter in Spanish abattoir | 2 | 2 | 2 | 0 | 2 | 2 | 2 | 2 | 2 | 2 | 2 | 1 | 3 | 2 | 0 |
| Birds census | 1 | 0 | 1 | 0 | 0 | 0 | 1 | 0 | 1 | 1 | 1 | 0 | 1 | 1 | 0 |
| National Power to close coal-fired power stations | 2 | 3 | 0 | 0 | 4 | 2 | 2 | 1 | 5 | 1 | 3 | 0 | 4 | 2 | 0 |
| New electric car | 1 | 0 | 1 | 0 | 1 | 0 | 0 | 0 | 1 | 1 | 1 | 0 | 1 | 0 | 0 |
| Balmoral radon/ radiation/gas threat | 0 | 1 | 0 | 0 | 0 | 1 | 0 | 1 | 0 | 0 | 0 | 0 | 1 | 1 | 1 |
| Brazil gold mining | 0 | 0 | 0 | 0 | 0 | 0 | 0 | 0 | 0 | 0 | 0 | 0 | 0 | 0 | 0 |

Table 7.6.1 Continued

BRITAIN

| | Total 940 | Sex | | Age | | | | | Class | | | | Region | | |
|---|---|---|---|---|---|---|---|---|---|---|---|---|---|---|---|
| | | Male 453 | Female 487 | 16–24 159 | 25–34 200 | 35–44 168 | 45–64 246 | 65+ 166 | AB 169 | C1 224 | C2 266 | DE 281 | South 375 | Midlands Wales 236 | North Scotland 329 |
| Base | | | | | | | | | | | | | | | |

*Within the last week what environmental issues on TV, radio or in the newspapers have you been aware of?*

| | Total | Male | Female | 16–24 | 25–34 | 35–44 | 45–64 | 65+ | AB | C1 | C2 | DE | South | Wales | Scotland |
|---|---|---|---|---|---|---|---|---|---|---|---|---|---|---|---|
| Sizewell B computer problems | 1 | 1 | 0 | 1 | 1 | 1 | 1 | 1 | 1 | 1 | 1 | 0 | 1 | 1 | 0 |
| Pakistan nuclear programme halted | 0 | 0 | 0 | 0 | 0 | 0 | 0 | 1 | 0 | 0 | 0 | 0 | 0 | 0 | 0 |
| Seal organs trade | 0 | 0 | 0 | 0 | 0 | 1 | 0 | 0 | 0 | 0 | 0 | 0 | 0 | 0 | 0 |
| Elephants in Zimbabwe | 1 | 1 | 1 | 0 | 0 | 3 | 0 | 2 | 1 | 1 | 1 | 1 | 1 | 2 | 0 |
| Sellafield and leukaemia | 5 | 4 | 6 | 3 | 4 | 8 | 6 | 4 | 3 | 4 | 6 | 6 | 3 | 4 | 8 |
| Indian earthquake | 0 | 1 | 0 | 1 | 1 | 1 | 0 | 0 | 0 | 0 | 0 | 1 | 0 | 1 | 0 |
| Glasgow council house energy usage | 0 | 0 | 0 | 0 | 0 | 0 | 0 | 1 | 0 | 1 | 0 | 0 | 0 | 0 | 0 |
| Mining in Papua New Guinea | 0 | 0 | 0 | 0 | 0 | 0 | 0 | 1 | 1 | 0 | 0 | 0 | 0 | 0 | 0 |
| Any wildlife programme on TV | 3 | 2 | 3 | 1 | 3 | 5 | 2 | 5 | 4 | 3 | 3 | 2 | 3 | 4 | 2 |
| Other | 13 | 17 | 9 | 10 | 11 | 11 | 17 | 12 | 20 | 12 | 12 | 10 | 14 | 14 | 11 |
| None | 62 | 59 | 64 | 75 | 65 | 58 | 57 | 56 | 49 | 63 | 64 | 65 | 58 | 60 | 66 |
| Don't know | 11 | 10 | 12 | 6 | 10 | 10 | 11 | 17 | 14 | 11 | 7 | 12 | 11 | 11 | 11 |

*Table 7.6.1*   Continued

## SOURCES IN INDIA

| Base | Total 1623 | Sex | | Age | | | | | Income | | | |
|---|---|---|---|---|---|---|---|---|---|---|---|---|
| | | Male 741 | Female 605 | 16–24 260 | 25–34 383 | 35–44 398 | 45–64 297 | 65+ 56 | High 219 | Middle 569 | Low 376 | No resp 459 |
| None/no answer | 13 | 14 | 12 | 13 | 16 | 13 | 11 | 14 | 3 | 23 | 6 | 9 |
| TV | 27 | 27 | 26 | 27 | 23 | 27 | 28 | 25 | 31 | 26 | 17 | 32 |
| Magazine | 19 | 16 | 19 | 17 | 18 | 18 | 20 | 20 | 11 | 12 | 39 | 17 |
| Newspaper | 12 | 11 | 14 | 12 | 13 | 12 | 12 | 6 | 13 | 13 | 10 | 13 |
| Friends, etc. | 19 | 20 | 18 | 19 | 19 | 20 | 17 | 13 | 34 | 18 | 11 | 17 |
| Radio | 5 | 4 | 5 | 7 | 4 | 5 | 4 | 5 | 3 | 4 | 7 | 6 |
| Books | 5 | 5 | 5 | 5 | 5 | 4 | 7 | 16 | 2 | 3 | 10 | 5 |
| Organisation | 1 | 2 | 1 | 1 | 1 | 1 | 2 | 1 | 2 | 1 | 0 | 1 |

## STORIES IN INDIA

| Base | Total 1623 | Sex | | Age | | | | | Income | | | |
|---|---|---|---|---|---|---|---|---|---|---|---|---|
| | | Male 741 | Female 605 | 16–24 260 | 25–34 383 | 35–44 398 | 45–64 297 | 65+ 56 | High 219 | Middle 569 | Low 376 | No resp 459 |
| None/no response | 80 | 79 | 82 | 75 | 84 | 79 | 87 | 82 | 57 | 81 | 95 | 78 |
| Biodiversity | 0 | 0 | 0 | 0 | 0 | 0 | 0 | 0 | 0 | 0 | 0 | 0 |
| Bio-gas | 0 | 0 | 0 | 0 | 0 | 0 | 0 | 0 | 0 | 0 | 0 | 0 |
| Development | 2 | 2 | 1 | 3 | 2 | 2 | 0 | 0 | 3 | 2 | 0 | 2 |
| Global warming | 0 | 0 | 0 | 1 | 1 | 0 | 0 | 0 | 1 | 0 | 0 | 0 |
| Int Politics (Rio etc.) | 1 | 1 | 1 | 1 | 1 | 1 | 1 | 0 | 1 | 1 | 0 | 2 |
| Narmada Dam project | 1 | 1 | 2 | 2 | 1 | 2 | 0 | 0 | 2 | 1 | 0 | 2 |
| Ozone | 1 | 1 | 0 | 1 | 1 | 0 | 0 | 0 | 0 | 1 | 0 | 1 |
| Pollution | 3 | 3 | 3 | 2 | 2 | 3 | 2 | 5 | 7 | 3 | 0 | 2 |
| Air pollution | 1 | 1 | 1 | 1 | 0 | 2 | 1 | 2 | 3 | 0 | 1 | 1 |
| Poverty and environment | 0 | 0 | 0 | 0 | 0 | 0 | 0 | 0 | 0 | 0 | 0 | 0 |
| Roads | 0 | 0 | 0 | 1 | 0 | 0 | 0 | 0 | 0 | 0 | 0 | 1 |
| Trees | 6 | 5 | 4 | 6 | 4 | 6 | 3 | 2 | 7 | 5 | 3 | 7 |
| Vermiculture | 0 | 1 | 0 | 0 | 0 | 1 | 1 | 0 | 3 | 0 | 0 | 0 |
| Water management | 2 | 2 | 2 | 1 | 1 | 2 | 3 | 5 | 8 | 1 | 0 | 1 |
| Wildlife | 1 | 1 | 1 | 2 | 1 | 0 | 0 | 2 | 1 | 1 | 0 | 1 |
| Other | 3 | 3 | 2 | 4 | 2 | 2 | 3 | 3 | 5 | 2 | 1 | 3 |

*Table 7.6.1* Continued

## SOURCES IN INDIA

| | Radio and TV | | | | Newspapers | | | Region | | |
|---|---|---|---|---|---|---|---|---|---|---|
| Base | Radio 1246 | No rad. 377 | TV 1157 | No TV 466 | Eng 430 | Vern. 633 | None 622 | North 362 | South 716 | West 545 |
| None/no answer | 9 | 24 | 4 | 34 | 1 | 5 | 26 | 8 | 6 | 24 |
| TV | 27 | 26 | 34 | 3 | 29 | 31 | 20 | 30 | 27 | 24 |
| Magazine | 23 | 3 | 17 | 27 | 15 | 16 | 27 | 23 | 27 | 8 |
| Newspaper | 12 | 12 | 14 | 6 | 16 | 14 | 7 | 12 | 11 | 8 |
| Friends, etc. | 20 | 15 | 22 | 9 | 29 | 24 | 5 | 16 | 14 | 25 |
| Radio | 5 | 6 | 5 | 5 | 6 | 6 | 4 | 7 | 6 | 3 |
| Books | 3 | 13 | 2 | 15 | 3 | 2 | 10 | 2 | 9 | 2 |
| Organisation | 1 | 1 | 1 | 1 | 1 | 1 | 0 | 1 | 0 | 2 |

## STORIES IN INDIA

| | Radio and TV | | | | Newspapers | | | Region | | |
|---|---|---|---|---|---|---|---|---|---|---|
| Base | Radio 1246 | No rad. 377 | TV 1157 | No TV 466 | Eng 430 | Vern. 633 | None 622 | North 362 | South 716 | West 545 |
| None/no response | 78 | 87 | 73 | 97 | 66 | 69 | 99 | 72 | 92 | 69 |
| Biodiversity | 0 | 0 | 0 | 0 | 0 | 0 | 0 | 0 | 0 | 0 |
| Bio-gas | 0 | 0 | 0 | 0 | 0 | 0 | 0 | 0 | 0 | 0 |
| Development | 2 | 1 | 2 | 0 | 2 | 4 | 0 | 2 | 0 | 4 |
| Global warming | 0 | 1 | 0 | 0 | 1 | 0 | 0 | 1 | 0 | 1 |
| Int. politics (Rio, etc.) | 1 | 1 | 1 | 1 | 2 | 2 | 0 | 2 | 1 | 1 |
| Narmada Dam project | 1 | 2 | 1 | 0 | 2 | 1 | 0 | 0 | 1 | 2 |
| Ozone | 1 | 0 | 1 | 0 | 1 | 2 | 0 | 2 | 0 | 1 |
| Pollution | 3 | 1 | 3 | 0 | 4 | 4 | 0 | 3 | 1 | 4 |
| Air pollution | 1 | 1 | 1 | 0 | 1 | 2 | 0 | 1 | 0 | 2 |
| Poverty and environment | 0 | 0 | 0 | 0 | 0 | 0 | 0 | 1 | 0 | 0 |
| Roads | 0 | 0 | 0 | 0 | 1 | 0 | 0 | 1 | 0 | 0 |
| Trees | 6 | 4 | 8 | 1 | 7 | 10 | 0 | 11 | 2 | 6 |
| Vermiculture | 0 | 0 | 0 | 0 | 1 | 1 | 0 | 0 | 0 | 1 |
| Water management | 2 | 0 | 2 | 0 | 4 | 3 | 0 | 1 | 0 | 4 |
| Wildlife | 1 | 1 | 1 | 0 | 3 | 0 | 0 | 0 | 1 | 1 |
| Other | 3 | 2 | 3 | 1 | 6 | 3 | 0 | 5 | 2 | 3 |

*Table 7.6.2*   Knowledge of selected environmental issues (India)

| | Total | Sex | | Age | | | | | Income | | | |
|---|---|---|---|---|---|---|---|---|---|---|---|---|
| | | Male | Female | 16–24 | 25–34 | 35–44 | 45–64 | 65+ | High | Middle | Low | No resp |
| Base | 1623 | 741 | 605 | 260 | 383 | 398 | 297 | 56 | 219 | 569 | 376 | 459 |
| **Have you heard of the ozone hole?** | | | | | | | | | | | | |
| No response | 5 | 4 | 3 | 7 | 2 | 3 | 3 | 4 | 4 | 4 | 2 | 8 |
| No | 58 | 59 | 63 | 50 | 63 | 59 | 69 | 77 | 15 | 54 | 87 | 60 |
| Yes | 37 | 38 | 34 | 43 | 35 | 38 | 29 | 20 | 81 | 42 | 11 | 32 |
| **Where is the ozone hole? *(prompted)*** | | | | | | | | | | | | |
| Don't know/no response | 67 | 68 | 69 | 62 | 70 | 67 | 75 | 84 | 27 | 64 | 91 | 71 |
| Both | 12 | 12 | 11 | 17 | 10 | 12 | 7 | 5 | 24 | 11 | 3 | 14 |
| North | 16 | 16 | 17 | 13 | 15 | 19 | 15 | 11 | 46 | 19 | 5 | 8 |
| South | 5 | 5 | 3 | 8 | 5 | 3 | 2 | 0 | 3 | 6 | 2 | 7 |
| **Why is it talked about?** | | | | | | | | | | | | |
| No response | 55 | 53 | 60 | 58 | 59 | 61 | 50 | 41 | 16 | 59 | 74 | 54 |
| Don't know | 14 | 16 | 12 | 7 | 12 | 9 | 26 | 43 | 12 | 7 | 20 | 19 |
| Climate change | 4 | 5 | 3 | 3 | 4 | 4 | 3 | 4 | 5 | 6 | 2 | 2 |
| Env. harm | 7 | 6 | 8 | 7 | 7 | 8 | 5 | 4 | 11 | 11 | 2 | 6 |
| Health harm | 15 | 16 | 13 | 19 | 12 | 15 | 12 | 7 | 44 | 14 | 2 | 13 |
| Other | 4 | 4 | 5 | 6 | 5 | 3 | 4 | 2 | 11 | 3 | 0 | 6 |
| **What causes it? *(open responses)*** | | | | | | | | | | | | |
| No response | 96 | 96 | 95 | 95 | 95 | 97 | 95 | 96 | 96 | 96 | 98 | 93 |
| Don't know | 2 | 1 | 3 | 2 | 2 | 1 | 3 | 4 | 1 | 1 | 1 | 4 |
| Climate change | 0 | 0 | 0 | 0 | 0 | 0 | 0 | 0 | 1 | 0 | 0 | 0 |
| Pollution | 1 | 1 | 0 | 0 | 1 | 1 | 1 | 0 | 0 | 1 | 1 | 1 |
| Other | 1 | 1 | 2 | 2 | 1 | 1 | 1 | 0 | 2 | 1 | 0 | 2 |
| **What causes it? *(prompted responses)*** | | | | | | | | | | | | |
| Don't know | 67 | 68 | 69 | 60 | 71 | 68 | 76 | 84 | 25 | 63 | 95 | 70 |
| Aerosols | 11 | 7 | 12 | 12 | 11 | 9 | 5 | 2 | 11 | 16 | 2 | 12 |
| CFCs | 21 | 23 | 19 | 27 | 18 | 23 | 18 | 14 | 62 | 21 | 3 | 17 |
| Other | 0 | 1 | 0 | 1 | 0 | 1 | 1 | 0 | 1 | 0 | 0 | 1 |
| **Have you heard of global warming?** | | | | | | | | | | | | |
| No response | 9 | 8 | 6 | 7 | 6 | 8 | 9 | 9 | 6 | 8 | 10 | 12 |
| No | 59 | 61 | 64 | 56 | 67 | 60 | 66 | 73 | 19 | 57 | 82 | 63 |
| Yes | 32 | 31 | 30 | 37 | 28 | 32 | 25 | 18 | 75 | 35 | 8 | 26 |

*Table 7.6.2* Continued

| | Radio and TV | | | | Newspapers | | | Region | | |
|---|---|---|---|---|---|---|---|---|---|---|
| | Radio | No rad. | TV | No TV | Eng | Vern. | None | North | South | West |
| Base | 1246 | 377 | 1157 | 466 | 430 | 633 | 622 | 362 | 716 | 545 |

*Have you heard of the ozone hole?*

| | | | | | | | | | | |
|---|---|---|---|---|---|---|---|---|---|---|
| No response | 4 | 6 | 5 | 4 | 3 | 7 | 3 | 11 | 2 | 3 |
| No | 57 | 64 | 47 | 87 | 25 | 43 | 93 | 42 | 78 | 43 |
| Yes | 39 | 30 | 48 | 9 | 72 | 50 | 5 | 47 | 20 | 53 |

*Where is the ozone hole? (prompted)*

| | | | | | | | | | | |
|---|---|---|---|---|---|---|---|---|---|---|
| Don't know/no response | 64 | 77 | 57 | 93 | 38 | 53 | 97 | 55 | 86 | 51 |
| Both | 13 | 8 | 15 | 4 | 25 | 16 | 1 | 16 | 6 | 16 |
| North | 18 | 10 | 22 | 2 | 29 | 23 | 2 | 19 | 7 | 27 |
| South | 5 | 5 | 6 | 1 | 8 | 8 | 0 | 9 | 1 | 7 |

*Why is it talked about?*

| | | | | | | | | | | |
|---|---|---|---|---|---|---|---|---|---|---|
| No response | 53 | 64 | 47 | 76 | 30 | 49 | 76 | 50 | 61 | 52 |
| Don't know | 14 | 14 | 12 | 19 | 9 | 9 | 22 | 10 | 26 | 1 |
| Climate change | 4 | 2 | 5 | 1 | 5 | 7 | 0 | 4 | 1 | 7 |
| Env. harm | 8 | 6 | 10 | 1 | 14 | 10 | 0 | 12 | 5 | 8 |
| Health harm | 17 | 8 | 20 | 2 | 32 | 20 | 1 | 19 | 4 | 26 |
| Other | 4 | 4 | 5 | 1 | 10 | 5 | 0 | 5 | 2 | 6 |

*What causes it? (open responses)*

| | | | | | | | | | | |
|---|---|---|---|---|---|---|---|---|---|---|
| No response | 96 | 96 | 96 | 97 | 93 | 97 | 97 | 97 | 93 | 99 |
| Don't know | 2 | 2 | 2 | 2 | 2 | 1 | 3 | 2 | 3 | 0 |
| Climate change | 0 | 0 | 0 | 0 | 0 | 0 | 0 | 0 | 0 | 0 |
| Pollution | 1 | 1 | 1 | 1 | 2 | 0 | 0 | 0 | 2 | 0 |
| Other | 1 | 2 | 2 | 1 | 3 | 1 | 0 | 1 | 2 | 0 |

*What causes it? (prompted responses)*

| | | | | | | | | | | |
|---|---|---|---|---|---|---|---|---|---|---|
| Don't know | 65 | 75 | 56 | 94 | 37 | 53 | 98 | 53 | 87 | 51 |
| Aerosols | 11 | 10 | 15 | 1 | 15 | 19 | 0 | 24 | 5 | 10 |
| CFCs | 23 | 15 | 28 | 4 | 48 | 26 | 2 | 21 | 8 | 39 |
| Other | 1 | 0 | 1 | 0 | 0 | 1 | 0 | 1 | 0 | 0 |

*Have you heard of global warming?*

| | | | | | | | | | | |
|---|---|---|---|---|---|---|---|---|---|---|
| No response | 8 | 12 | 9 | 9 | 8 | 12 | 6 | 16 | 8 | 6 |
| No | 59 | 61 | 50 | 82 | 30 | 43 | 91 | 40 | 79 | 46 |
| Yes | 33 | 27 | 41 | 8 | 62 | 45 | 3 | 44 | 13 | 48 |

*Table 7.6.2* Continued

| Base | Total | Sex | | Age | | | | | Income | | | |
|---|---|---|---|---|---|---|---|---|---|---|---|---|
| | | Male | Female | 16–24 | 25–34 | 35–44 | 45–64 | 65+ | High | Middle | Low | No resp |
| | 1623 | 741 | 605 | 260 | 383 | 398 | 297 | 56 | 219 | 569 | 376 | 459 |
| **What will it do? (open responses)** | | | | | | | | | | | | |
| No response | 59 | 64 | 59 | 58 | 65 | 61 | 62 | 75 | 20 | 61 | 75 | 61 |
| Don't know | 9 | 6 | 11 | 4 | 8 | 8 | 13 | 6 | 5 | 6 | 16 | 11 |
| Climate change | 19 | 18 | 19 | 24 | 18 | 18 | 13 | 13 | 48 | 20 | 2 | 17 |
| Disaster | 0 | 0 | 0 | 0 | 0 | 0 | 0 | 0 | 0 | 0 | 0 | 0 |
| Economic harm | 0 | 0 | 0 | 0 | 0 | 0 | 0 | 0 | 0 | 0 | 0 | 0 |
| Env. harm | 3 | 2 | 3 | 3 | 2 | 3 | 2 | 1 | 5 | 4 | 1 | 2 |
| Floods | 6 | 6 | 6 | 7 | 5 | 6 | 7 | 4 | 18 | 5 | 2 | 5 |
| Health harm | 3 | 3 | 1 | 3 | 2 | 3 | 2 | 0 | 3 | 4 | 1 | 4 |
| Other | 1 | 1 | 1 | 1 | 0 | 1 | 1 | 0 | 1 | 0 | 2 | 0 |
| **When will it happen?** | | | | | | | | | | | | |
| No response/don't know | 84 | 81 | 83 | 82 | 85 | 80 | 85 | 84 | 35 | 85 | 98 | 93 |
| Happening already | 5 | 7 | 4 | 6 | 4 | 6 | 6 | 9 | 24 | 3 | 1 | 1 |
| 10 years from now | 6 | 6 | 7 | 6 | 5 | 10 | 6 | 4 | 26 | 5 | 0 | 2 |
| 20 years from now | 1 | 1 | 1 | 2 | 1 | 2 | 0 | 2 | 4 | 2 | 1 | 0 |
| 30 years from now | 1 | 1 | 1 | 1 | 1 | 1 | 1 | 0 | 3 | 0 | 0 | 1 |
| 40 years from now | 0 | 0 | 1 | 1 | 0 | 1 | 0 | 0 | 1 | 0 | 0 | 0 |
| 50 years from now | 1 | 1 | 1 | 2 | 1 | 1 | 0 | 2 | 3 | 1 | 0 | 0 |
| 60 years from now | 0 | 0 | 0 | 0 | 1 | 0 | 0 | 0 | 1 | 1 | 0 | 0 |
| 100 years from now | 1 | 1 | 0 | 1 | 1 | 0 | 0 | 0 | 1 | 0 | 1 | 1 |
| 200 years from now | 0 | 0 | 0 | 0 | 1 | 0 | 0 | 0 | 0 | 1 | 0 | 1 |
| 300 years from now | 0 | 0 | 1 | 0 | 1 | 0 | 0 | 0 | 0 | 1 | 0 | 0 |
| 500 years from now | 0 | 0 | 0 | 0 | 0 | 0 | 0 | 0 | 0 | 0 | 0 | 0 |
| **What causes it?** | | | | | | | | | | | | |
| No response | 71 | 72 | 74 | 66 | 77 | 72 | 78 | 82 | 30 | 70 | 93 | 73 |
| Don't know | 2 | 1 | 1 | 2 | 1 | 1 | 1 | 1 | 3 | 2 | 1 | 3 |
| Climate change | 1 | 1 | 1 | 2 | 0 | 1 | 0 | 0 | 1 | 2 | 0 | 2 |
| Deforestation | 6 | 7 | 5 | 4 | 4 | 7 | 6 | 7 | 21 | 5 | 1 | 2 |
| Disaster | 0 | 0 | 0 | 0 | 0 | 0 | 0 | 0 | 0 | 0 | 0 | 0 |
| Energy use | 1 | 1 | 1 | 1 | 0 | 0 | 1 | 0 | 1 | 1 | 0 | 1 |
| Env. harm | 2 | 1 | 1 | 2 | 2 | 1 | 1 | 1 | 3 | 2 | 1 | 2 |
| Greenhouse gases | 4 | 2 | 4 | 6 | 3 | 3 | 2 | 0 | 4 | 5 | 1 | 5 |
| Ozone | 3 | 4 | 3 | 4 | 5 | 3 | 2 | 1 | 9 | 3 | 0 | 3 |
| Pollution | 9 | 9 | 8 | 11 | 5 | 10 | 8 | 7 | 26 | 9 | 2 | 7 |
| Population | 0 | 0 | 0 | 0 | 0 | 0 | 0 | 0 | 0 | 0 | 1 | 0 |
| Other | 1 | 1 | 1 | 2 | 2 | 1 | 0 | 0 | 2 | 1 | 0 | 2 |

*Table 7.6.2* Continued

| Base | Radio and TV | | | | Newspapers | | | Region | | |
|---|---|---|---|---|---|---|---|---|---|---|
| | Radio | No rad. | TV | No TV | Eng | Vern. | None | North | South | West |
| | 1246 | 377 | 1157 | 466 | 430 | 633 | 622 | 362 | 716 | 545 |

**What will it do?** *(open responses)*

| | | | | | | | | | | |
|---|---|---|---|---|---|---|---|---|---|---|
| No response | 57 | 66 | 51 | 79 | 28 | 45 | 88 | 38 | 74 | 53 |
| Don't know | 10 | 7 | 8 | 13 | 8 | 9 | 9 | 14 | 14 | 0 |
| Climate change | 20 | 16 | 25 | 4 | 36 | 29 | 1 | 27 | 7 | 30 |
| Disaster | 0 | 0 | 0 | 0 | 0 | 0 | 0 | 0 | 0 | 0 |
| Economic harm | 0 | 0 | 0 | 0 | 0 | 0 | 0 | 0 | 0 | 0 |
| Env. harm | 3 | 3 | 4 | 2 | 5 | 5 | 0 | 6 | 1 | 4 |
| Floods | 6 | 6 | 8 | 2 | 15 | 6 | 1 | 4 | 3 | 10 |
| Health harm | 4 | 1 | 4 | 0 | 5 | 4 | 0 | 10 | 1 | 2 |
| Other | 1 | 1 | 1 | 0 | 1 | 1 | 0 | 2 | 1 | 0 |

**When will it happen?** *(open responses)*

| | | | | | | | | | | |
|---|---|---|---|---|---|---|---|---|---|---|
| No response/don't know | 83 | 86 | 78 | 99 | 62 | 76 | 100 | 91 | 98 | 61 |
| Happening already | 5 | 4 | 7 | 0 | 16 | 6 | 0 | 1 | 1 | 12 |
| 10 years from now | 7 | 4 | 8 | 0 | 11 | 10 | 0 | 1 | 0 | 17 |
| 20 years from now | 1 | 2 | 2 | 0 | 2 | 3 | 0 | 1 | 0 | 3 |
| 30 years from now | 1 | 0 | 1 | 0 | 1 | 1 | 0 | 0 | 0 | 2 |
| 40 years from now | 0 | 1 | 1 | 0 | 1 | 0 | 0 | 0 | 0 | 1 |
| 50 years from now | 1 | 1 | 1 | 0 | 2 | 1 | 0 | 1 | 0 | 2 |
| 60 years from now | 0 | 0 | 0 | 0 | 1 | 0 | 0 | 1 | 0 | 1 |
| 100 years from now | 1 | 1 | 1 | 0 | 1 | 1 | 0 | 1 | 0 | 1 |
| 200 years from now | 0 | 1 | 1 | 0 | 1 | 0 | 0 | 2 | 0 | 0 |
| 300 years from now | 1 | 0 | 1 | 0 | 0 | 1 | 0 | 2 | 0 | 0 |
| 500 years from now | 0 | 0 | 0 | 1 | 0 | 0 | 1 | 0 | 0 | 0 |

**What causes it?** *(open responses)*

| | | | | | | | | | | |
|---|---|---|---|---|---|---|---|---|---|---|
| No response | 69 | 77 | 62 | 93 | 44 | 58 | 97 | 53 | 92 | 55 |
| Don't know | 2 | 1 | 2 | 2 | 3 | 2 | 1 | 5 | 1 | 2 |
| Climate change | 1 | 1 | 2 | 0 | 1 | 2 | 0 | 5 | 0 | 1 |
| Deforestation | 6 | 3 | 8 | 1 | 12 | 8 | 0 | 6 | 0 | 12 |
| Disaster | 0 | 0 | 0 | 0 | 0 | 0 | 0 | 0 | 0 | 0 |
| Energy use | 1 | 1 | 1 | 0 | 1 | 1 | 0 | 1 | 1 | 1 |
| Env. harm | 2 | 1 | 2 | 1 | 2 | 4 | 0 | 5 | 0 | 2 |
| Greenhouse gases | 4 | 5 | 5 | 1 | 7 | 4 | 1 | 8 | 4 | 2 |
| Ozone | 4 | 2 | 4 | 1 | 7 | 5 | 0 | 2 | 1 | 7 |
| Pollution | 9 | 9 | 12 | 1 | 19 | 13 | 0 | 12 | 1 | 17 |
| Population | 0 | 0 | 0 | 0 | 0 | 0 | 0 | 1 | 0 | 0 |
| Other | 1 | 1 | 2 | 0 | 3 | 1 | 0 | 3 | 1 | 1 |

*Table 7.6.2*   Continued

| Base | Total 1623 | Sex Male 741 | Female 605 | Age 16–24 260 | 25–34 383 | 35–44 398 | 45–64 297 | 65+ 56 | Income High 219 | Middle 569 | Low 376 | No resp 459 |
|---|---|---|---|---|---|---|---|---|---|---|---|---|
| *What is acid rain? (open responses)* | | | | | | | | | | | | |
| No response | 29 | 35 | 26 | 29 | 33 | 37 | 20 | 21 | 34 | 49 | 5 | 21 |
| Don't know | 54 | 50 | 56 | 43 | 53 | 48 | 70 | 75 | 43 | 29 | 91 | 58 |
| Chemical rains | 12 | 10 | 10 | 20 | 9 | 10 | 4 | 4 | 15 | 14 | 2 | 14 |
| SoxNox rain | 4 | 3 | 5 | 8 | 4 | 2 | 2 | 0 | 5 | 5 | 2 | 5 |
| Other | 2 | 2 | 2 | 1 | 1 | 3 | 3 | 0 | 3 | 4 | 1 | 1 |
| *What will it do? (open responses)* | | | | | | | | | | | | |
| No response | 75 | 75 | 79 | 70 | 81 | 75 | 82 | 82 | 43 | 75 | 95 | 75 |
| Don't know | 4 | 5 | 2 | 3 | 2 | 5 | 3 | 5 | 16 | 1 | 0 | 6 |
| Changes climate | 0 | 0 | 0 | 0 | 0 | 0 | 0 | 0 | 0 | 0 | 0 | 0 |
| Corrodes | 2 | 0 | 2 | 3 | 1 | 1 | 1 | 0 | 0 | 2 | 0 | 3 |
| Deforestation | 0 | 0 | 0 | 0 | 0 | 0 | 0 | 0 | 0 | 0 | 0 | 0 |
| Env. harm | 12 | 12 | 10 | 14 | 10 | 12 | 9 | 11 | 31 | 13 | 3 | 9 |
| Health harm | 6 | 6 | 6 | 9 | 5 | 6 | 4 | 2 | 8 | 9 | 1 | 6 |
| Other | 1 | 1 | 1 | 1 | 1 | 1 | 1 | 0 | 1 | 1 | 0 | 1 |
| *What causes it? (open responses)* | | | | | | | | | | | | |
| No response | 75 | 76 | 78 | 71 | 81 | 75 | 82 | 84 | 41 | 76 | 96 | 75 |
| Don't know | 3 | 2 | 1 | 2 | 1 | 2 | 1 | 0 | 5 | 1 | 0 | 6 |
| Climate change | 1 | 1 | 1 | 0 | 1 | 2 | 1 | 1 | 6 | 0 | 0 | 1 |
| Deforestation | 3 | 4 | 2 | 0 | 3 | 4 | 4 | 3 | 15 | 1 | 0 | 0 |
| Ozone | 1 | 1 | 0 | 0 | 0 | 1 | 0 | 1 | 2 | 0 | 0 | 0 |
| Pollution | 14 | 14 | 12 | 21 | 10 | 13 | 10 | 10 | 29 | 17 | 2 | 14 |
| SoxNox | 3 | 2 | 4 | 4 | 3 | 3 | 2 | 0 | 2 | 5 | 2 | 3 |
| Other | 1 | 0 | 1 | 1 | 0 | 1 | 0 | 0 | 0 | 1 | 0 | 1 |

*Table 7.6.2*   Continued

| Base | Radio and TV | | | | Newspapers | | | Region | | |
|---|---|---|---|---|---|---|---|---|---|---|
| | Radio 1246 | No rad. 377 | TV 1157 | No TV 466 | Eng 430 | Vern. 633 | None 622 | North 362 | South 716 | West 545 |
| *What is acid rain?* | | | | | | | | | | |
| No response | 27 | 33 | 27 | 33 | 20 | 37 | 26 | 15 | 12 | 60 |
| Don't know | 55 | 50 | 50 | 63 | 46 | 39 | 73 | 53 | 81 | 18 |
| Chemical rains | 12 | 12 | 15 | 2 | 22 | 16 | 0 | 23 | 2 | 16 |
| SoxNox rain | 4 | 4 | 5 | 2 | 10 | 4 | 0 | 7 | 4 | 2 |
| Other | 2 | 1 | 3 | 0 | 2 | 4 | 0 | 1 | 1 | 4 |
| *What will it do?* | | | | | | | | | | |
| No response | 73 | 82 | 67 | 95 | 51 | 67 | 97 | 61 | 91 | 64 |
| Don't know | 5 | 2 | 6 | 0 | 7 | 6 | 1 | 7 | 2 | 5 |
| Changes climate | 0 | 0 | 0 | 0 | 0 | 0 | 0 | 0 | 0 | 0 |
| Corrodes | 2 | 2 | 2 | 0 | 3 | 2 | 0 | 4 | 1 | 0 |
| Deforestation | 0 | 0 | 0 | 0 | 0 | 0 | 0 | 0 | 0 | 0 |
| Env. harm | 13 | 9 | 16 | 2 | 26 | 16 | 0 | 17 | 3 | 20 |
| Health harm | 7 | 6 | 8 | 2 | 11 | 9 | 1 | 10 | 3 | 9 |
| Other | 1 | 0 | 1 | 0 | 2 | 1 | 0 | 0 | 1 | 0 |
| *What causes it?* | | | | | | | | | | |
| No response | 73 | 82 | 67 | 96 | 49 | 67 | 98 | 63 | 91 | 63 |
| Don't know | 3 | 2 | 4 | 0 | 4 | 3 | 1 | 7 | 2 | 1 |
| Climate change | 1 | 0 | 1 | 0 | 3 | 2 | 0 | 1 | 0 | 3 |
| Deforestation | 3 | 1 | 4 | 0 | 7 | 4 | 0 | 0 | 0 | 7 |
| Ozone | 1 | 0 | 1 | 0 | 1 | 1 | 0 | 1 | 0 | 1 |
| Pollution | 15 | 11 | 18 | 3 | 27 | 21 | 0 | 22 | 3 | 23 |
| SoxNox | 3 | 2 | 4 | 1 | 7 | 2 | 1 | 4 | 4 | 1 |
| Other | 1 | 0 | 1 | 0 | 1 | 1 | 0 | 2 | 1 | 0 |

*Table 7.6.3* British response to newspaper headlines

I am going to read out a number of headlines which might appear in a newspaper. Assuming that you saw the headline, please would you tell me how likely you would be to read the whole story.

*Example of original scores:*
*Ninth victim of Chinese serial killer found in downtown Shanghai*

| | | Sex | | Age | | | | | Class | | | | Region | | |
|---|---|---|---|---|---|---|---|---|---|---|---|---|---|---|---|
| | Total | Male | Female | 16–24 | 25–34 | 35–44 | 45–64 | 65+ | AB | C1 | C2 | DE | South | Midlands Wales | North Scotland |
| Base | 940 | 453 | 487 | 159 | 200 | 168 | 246 | 166 | 169 | 224 | 266 | 281 | 375 | 236 | 329 |

| | Wts | Percentages | | | | | | | | | | | | | | |
|---|---|---|---|---|---|---|---|---|---|---|---|---|---|---|---|---|
| Very likely | 4 | 7 | 9 | 6 | 16 | 12 | 4 | 5 | 2 | 7 | 7 | 10 | 6 | 8 | 9 | 6 |
| Fairly likely | 3 | 15 | 16 | 14 | 17 | 18 | 19 | 11 | 12 | 16 | 17 | 14 | 14 | 18 | 16 | 12 |
| Unlikely | 2 | 27 | 28 | 27 | 29 | 24 | 33 | 27 | 24 | 27 | 25 | 32 | 24 | 30 | 23 | 27 |
| Very unlikely | 1 | 47 | 44 | 51 | 38 | 44 | 40 | 54 | 59 | 47 | 49 | 41 | 53 | 42 | 49 | 53 |
| Don't know | 0 | 3 | 3 | 2 | 0 | 3 | 4 | 3 | 3 | 3 | 1 | 3 | 3 | 3 | 2 | 2 |

*Stories ranked by weighted score*     Weighted scores

| | Total | Male | Female | 16–24 | 25–34 | 35–44 | 45–64 | 65+ | AB | C1 | C2 | DE | South | Midlands Wales | North Scotland |
|---|---|---|---|---|---|---|---|---|---|---|---|---|---|---|---|
| Two killed in pile-up at local accident black spot | 3.2 | 3.2 | 3.3 | 2.9 | 3.2 | 3.3 | 3.2 | 3.4 | 3.3 | 3.2 | 3.3 | 3.1 | 3.2 | 3.3 | 3.1 |
| Hostages injured in local bank raid | 3.2 | 3.2 | 3.1 | 3.2 | 3.3 | 3.2 | 3.3 | 3.3 | 3.2 | 3.2 | 3.1 | 3.2 | 3.2 | 3.1 | 0 |
| Local company to close with loss of 300 jobs | 3.1 | 3.2 | 3.1 | 2.7 | 3.1 | 3.3 | 3.4 | 3.3 | 3.2 | 3.2 | 3.2 | 3.1 | 3.1 | 3.3 | 3.2 |
| New motorway link will mean destruction of local woodland | 3.0 | 3.0 | 2.8 | 2.9 | 3.1 | 3.1 | 3.1 | 3.2 | 3.0 | 3.1 | 2.8 | 3.0 | 3.1 | 2.9 | 2.9 |

Table 7.6.3 Continued

| | | Sex | | Age | | | | | Class | | | | Region | | |
|---|---|---|---|---|---|---|---|---|---|---|---|---|---|---|---|
| Base | Total 940 | Male 453 | Female 487 | 16–24 159 | 25–34 200 | 35–44 168 | 45–64 246 | 65+ 166 | AB 169 | C1 224 | C2 266 | DE 281 | South 375 | Midlands Wales 236 | North Scotland 329 |
| Weighted scores | | | | | | | | | | | | | | | |
| General election in UK planned for spring | 2.9 | 3.1 | 2.8 | 2.7 | 2.7 | 3.0 | 3.1 | 3.0 | 3.2 | 3.0 | 2.8 | 2.8 | 3.0 | 3.1 | 2.7 |
| Hundreds of acres of rainforest destroyed as Brazil builds highway | 2.8 | 2.7 | 2.7 | 2.8 | 2.9 | 2.7 | 2.8 | 3.0 | 2.7 | 2.8 | 2.6 | 2.8 | 2.9 | 2.5 | 2.5 |
| Local council election results | 2.7 | 2.6 | 2.5 | 2.4 | 2.7 | 2.7 | 3.0 | 2.8 | 2.8 | 2.6 | 2.5 | 2.7 | 2.8 | 2.5 | 2.5 |
| Over 600 feared drowned as Ugandan ferry capsizes | 2.5 | 2.5 | 2.4 | 2.5 | 2.6 | 2.6 | 2.4 | 2.4 | 2.7 | 2.5 | 2.5 | 2.2 | 2.5 | 2.7 | 2.3 |
| Trade embargo brings greater hardship for India's poor | 2.1 | 2.1 | 1.9 | 2.0 | 2.2 | 2.1 | 2.2 | 2.3 | 2.0 | 2.1 | 2.0 | 2.1 | 2.2 | 2.0 | 2.0 |
| French climbdown brings hope to GATT talks | 2.0 | 2.2 | 1.9 | 1.6 | 1.9 | 2.1 | 2.2 | 2.1 | 2.4 | 2.0 | 2.0 | 1.7 | 2.2 | 2.0 | 1.7 |
| Ninth victim of Chinese serial killer found in downtown Shanghai | 1.7 | 1.8 | 1.7 | 2.1 | 1.9 | 1.8 | 1.6 | 1.5 | 1.8 | 1.8 | 1.9 | 1.7 | 1.9 | 1.8 | 1.7 |

Table 7.7.1 Expectation of the future world

## BRITAIN

As far as the environment is concerned, do you think that for the next generation this country will be a better or worse place to live in?

| | Total | Sex | | Age | | | | | Class | | | | Region | | |
|---|---|---|---|---|---|---|---|---|---|---|---|---|---|---|---|
| | | Male | Female | 16-24 | 25-34 | 35-44 | 45-64 | 65+ | AB | C1 | C2 | DE | South | Midlands Wales | North Scotland |
| Base | 940 | 453 | 487 | 159 | 200 | 168 | 246 | 166 | 169 | 224 | 266 | 281 | 375 | 236 | 329 |
| Better | 17 | 19 | 14 | 15 | 18 | 19 | 18 | 13 | 17 | 20 | 14 | 16 | 17 | 17 | 16 |
| Worse | 69 | 67 | 71 | 71 | 75 | 63 | 68 | 68 | 64 | 65 | 73 | 71 | 68 | 67 | 72 |
| Don't know | 14 | 14 | 15 | 14 | 7 | 18 | 14 | 19 | 18 | 14 | 13 | 13 | 15 | 16 | 12 |

## INDIA

Do you think that for the next generation this country will be a better, worse, or same place to live in?

| | Total | Sex | | Age | | | | | Income | | | | Radio and TV | | | | Newspaper | | | Region | | |
|---|---|---|---|---|---|---|---|---|---|---|---|---|---|---|---|---|---|---|---|---|---|---|
| | | Male | Female | 16-24 | 25-34 | 35-44 | 45-64 | 65+ | High | Middle | Low | No resp | No rad. | Radio | TV | No TV | Eng | Vern. | None | North | South | West |
| Base | 1623 | 741 | 605 | 260 | 383 | 398 | 297 | 56 | 219 | 569 | 376 | 459 | 377 | 1246 | 1157 | 466 | 430 | 633 | 622 | 362 | 716 | 545 |
| No answer | 6 | 6 | 6 | 5 | 5 | 6 | 6 | 7 | 1 | 6 | 10 | 6 | 7 | 6 | 5 | 9 | 5 | 9 | 9 | 9 | 7 | 3 |
| Better | 34 | 27 | 39 | 41 | 28 | 32 | 33 | 36 | 19 | 25 | 44 | 44 | 36 | 34 | 32 | 38 | 30 | 35 | 35 | 49 | 45 | 10 |
| Same | 8 | 6 | 11 | 10 | 9 | 8 | 4 | 7 | 6 | 10 | 4 | 9 | 8 | 8 | 8 | 6 | 10 | 7 | 7 | 7 | 7 | 10 |
| Worse | 52 | 61 | 44 | 43 | 57 | 55 | 57 | 50 | 74 | 59 | 42 | 41 | 49 | 53 | 54 | 46 | 55 | 49 | 49 | 36 | 41 | 77 |

identified (by the English readership), whereas in the other regions the causes are more likely to be 'deforestation' and 'pollution'. It will cause 'climate change' and 'floods' (including coastal inundation). In the South and North nobody seems to know when global warming will happen – but on the West around Pune quite clearly it is either already happening (see the comment by Darryl D'Monte in Chapter 4) or it will happen within ten years.

Knowledge about acid rain is limited to the North and West, and is generally thought to be caused by 'pollution' and to do 'environmental harm'. (Possible damage to what may be the world's most famous building, and one of India's major tourist assets – the Taj Mahal – does feature in the media from time to time – and is referred to in the focus group discussion, Chapter 6.) Specific knowledge linking acid rain to sulphur and nitrogen oxides is limited to the English readership.

**Readers' response to headlines in Britain**

There was not an equivalent of this question in India. In Britain respondents were asked to assess their likely interest in 11 different headlines. Four of the 11 stories (Table 7.6.3) concerned developing countries, two of them human-interest stories, one on poverty in India and international trade, and one on rainforests in Brazil. The other seven stories included one which was European/international and six about Britain, one of them on a local motorway-versus-trees issue. The ranking of the weighted overall interest levels is quite clear. The top stories are the local stories, and the bottom stories are the overseas ones, except that the loss of rain-forest in Brazil slips above the local election results in the UK (where local politics is fairly emasculated).

## 7.7 BASIC OPTIMISM AND PESSIMISM ABOUT THE FUTURE NATIONAL ENVIRONMENTS

This can be fairly simply dealt with. In Britain (Table 7.7.1) the question was specifically with regard to the environment, asking if the UK would be a better or worse place for the next generation to live in. The majority think the UK will be a worse place, and this hardly varies by age or sex or region. In India the question was put more simply, just asking whether the country would be a better or worse place to live in, without stressing the environment. The values change considerably by income and by region. The developed and fast developing West (around Pune) think that things will get worse – i.e. those that have 'more development' are more pessimistic. The survey does not say what they are pessimistic about – but that has come out in the other questions asking about the problems facing India and their locality. These involved inadequate drainage and sanitation, and the implication in this is that things will get worse if these issues are not addressed – and to address them means precisely more development in 'civic' areas of the type that the developed countries have already had. The highest scores for things getting better (and there are many more optimists than in Britain) are among the low-income groups and in the North and South – i.e. those least likely so far to 'have development'. The hopes of the poor are therefore more consistent with aspiration, which presumably again is for more development. The fact that the wealthier are more pessimistic is no 'bad thing'. If wealthy nations are better placed to spend money on cleaning up, then the wealthy in India are a useful counterpoise too. But overall, this does reinforce the message that poverty, seen by some as a cause of environmental degradation, is the prime candidate for elimination.

## 7.8 CONCLUSIONS

In India the concerns that most people have are about communalism, the economy, education, population and health – and of course about water and drainage. But these do not fit so easily into a definable category Environment, although the idea of improved sanitation and rubbish disposal, i.e. the cleaning up of urban squalor, is now popularly given a new label, the 'brown agenda', to distinguish it from the 'green agenda'. When pushed specifically for knowledge on 'global environmental issues', the Indian public show little understanding, and such knowledge as there is is more influenced by access to the press than to TV. Precise technical knowledge seems to be most influenced by access to the English press. The biggest discriminators in India are clearly poverty and wealth, and this also mimics the distinction between the illiterate and literate, the less educated and the more educated, those who have access to media sources and those who don't, and perhaps the distinction between the many local publics and the post-nationalist audience which the editors identified and which Gandhi and Nehru strove so hard to implant into India, and which now defeated Congress will no doubt turn to, to sustain India's secularity.

In more homogeneous and more developed Britain, the prime interest of people is also still with their local economy, and particularly with employment. It is not that people are not concerned with environmental issues at all; it is just that they do not have a high priority unless they have a detectable local impact. It is true that people are more aware of problems abroad, but it is unclear whether this relates to their own spheres of possible action.

# 8 Conclusions
## Which worlds are talking?

This book is about the world around us, knowledge about the world around us and the communication of that knowledge, where 'knowledge' can be interpreted in the widest sense to include opinions and attitudes too. Much of it has been empirical, finding out as far as we have been able what the various actors within the overall system of communication think they are saying about the world, how it is received and how this relates to attitudes and behaviour. A little of this chapter will reflect on that empirical content – but we start from a much more abstract point of view, by thinking a little more about the idea of knowledge in a world of communicators. The inspiration for this abstract starting point derives from a paper given by Professor Pranap Sen of the University of Calcutta at the Indian Institute of Advanced Study in May 1995, entitled 'Knowledge, truth and scepticism'.

Professor Sen states that, in trying to find a correct definition of knowledge, epistemologists have standardly laid down three conditions:

> $S$ knows that $p$ if and only if
> (i) $S$ believes that $p$
> (ii) $S$ is justified in believing that $p$; and
> (iii) it is true that $p$.

To take a particular example, Galileo knows that the earth is round if and only if

> (i) Galileo believes that the earth is round;
> (ii) Galileo is justified in believing that the earth is round; and
> (iii) it is true that the earth is round.

The first two are conditions of the subject, but the last is a condition of the object. But,

> To take our example, does it make any sense at all to say that, to know the earth is round, Galileo has to satisfy the condition that it is true that the earth is round? How can that be a condition for Galileo to satisfy?

Sen's conclusion, argued with force and profound consequences, is that the 'truth' condition (iii) cannot be claimed as an equivalent necessary condition of knowledge like the two belief conditions (i and ii).

We know from the history of science that what is established as true by observation in one generation may be replaced by new observations and theories in the next. But to the extent that there is an attempt to marry theory with observation, we can speak of a 'correspondence' between our ideas and an observable world around us.

Professor Sen argues his position from the point of a single observer. But suppose that Galileo does know something. Does this mean that it 'is known'? Intuitively we say no, that for something to 'be known' it has to be known among a community of knowers of some sort. So we extend the model to such a community. Now, in addition to the idea of 'correspondence', that a notion corresponds to observable circumstance and is related to the idea of truth, we have in addition the concept of 'coherence', that the ideas of two or more observers confirm each other. Coherence provides justification for belief. Since there is more than one observer, they have to be able to communicate, and 'coherence' is thus intimately concerned with communication.

Here we explore these issues by means of a computer simulation. What we have set out to do is to create a community of communicating observers, each with some understanding (a 'world-view') of what the world is like. These observers initially know a lot about their own local circumstance but only guess what distant places are like. But we then let them talk to each other by telephone, so that their world-views can be modified by what they hear about distant places. By varying different parameters relating to local correspondence (local observable 'truths') and the way other observers' views are incorporated (the sharing of beliefs) and the rates of communication, we can generate better understanding and further questions about the relationship between communication and knowledge.

What we have postulated is a very simple world, consisting of a $5 \times 5$ board, rather like a chess board. On this board a True World is created by randomly painting each square either black or white. There are of course $2^{25}$ = approx. 34 million possible True Worlds, one of which is randomly selected in any simulation run as the map of the True World. We would never defend the hypothesis that the actual world is finite and knowable in this manner – but for the purposes of this exercise we are happy to use a finite model world to begin with.

Next we hypothesise a sentient person (Senper) inhabiting each of the squares of the True World; each one has its own image of what the True World is like. This image is made up of two parts:

1   *truths* which it can verify about its own square, and the truths it can verify about those neighbours with which it shares a common boundary,[1]
2   *beliefs* about the state of the rest of the world. At the beginning of a simulation for all the 25 boards (windows) of the 25 Senpers, all the non-neighbouring squares are randomly painted either black or white.

The next step is to see if these Senpers can learn the state of the True World by talking to each other. There are two kinds of rules, each with different possible variations. There are Communication Rules which specify whether a Senper can contact neighbours only or distant squares, etc., and Transition Rules which say how Senpers respond to information received.

The simulation begins by using Communication Rules which randomly allow any one Senper, say S located on Cell 3,4, to communicate with any other, for example Q located on Cell 2,5. S compares its map of the world with Q's, and then changes its own map according to the following set of Transition Rules (many variations are clearly possible for both sets of rules). These are called the Sceptic's Rules, and for these the following table shows how S changes its state according to the values of S and Q. Q does not change its state.

| Q<br>S | Black | White | Uncertain |
|---|---|---|---|
| Black | Black | Uncertain | Black |
| White | Uncertain | White | White |
| Uncertain | Uncertain | Uncertain | Uncertain |

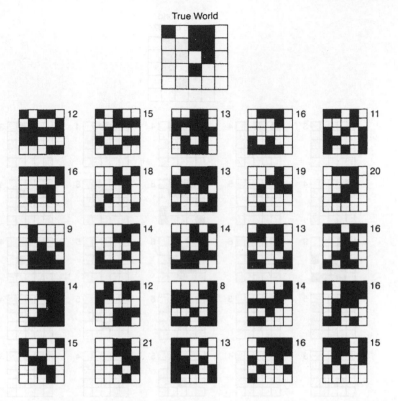

*Figure 8.1a* World views: sceptic's case

*Note:* The top board is the random allocation of black and white which constitutes the True World. The other 25 boards represent the starting world-view of each Senper living on each of the squares of the world. The top left board represents the knowledge and world-view of the Senper living on the top left square of the True World, column 1 – row 1, etc. The Senper knows the state of its own square and the state of the two adjacent contiguous squares (Rook's case). Initially all other sqaures are filled at random – the fantasy of the Senper's highly active imagination. The numbers represent the accuracy of the world-view. Senper 1,1 scores 12 – less than half the squares are imagined correctly – but this is a score the Senper will never 'know' by direct comparison with the True World. On average, the starting scores should be a little more than 13 (i.e. more than 50 per cent of 25), because there is a 50:50 chance of imagining distant squares correctly, plus the island of knowledge about neighbours and self.

There is also one important extra rule (used also in the next set of rules), that at the end of a phone call S will always choose local truth over Q's reports: that is, if Q says one of S's neighbours is White when it is Black, S will still maintain that it is Black. The Sceptic's Rules say that if S thinks that a distant square is Black, but another respondent (who may be anywhere – not necessarily near that square) says that it is White, S becomes Uncertain. If it is Uncertain, and Q tells it something is either Black or White, S is still not sure.

Running this simulation (Fig. 8.1a, b and c)[2] rapidly results in every Senper knowing only its own and its neighbour's states, and everything else is a world of uncertainty depicted in grey. In other words, there are 25 dissimilar windows. The reason for this is obviously in the last line of the transition table – which is clearly

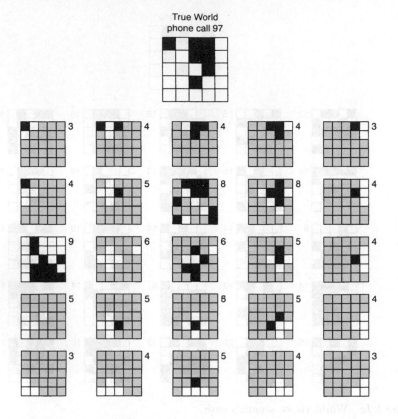

*Figure 8.1b*   World views: sceptic's case

*Note:* Following the Sceptic's Rules, after 97 random phone calls between pairs of Senpers, uncertainty and doubt rapidly invade all minds as conflicting opinions are found, and in this case conflicting opinion ends in doubt.

a sink. Once Uncertainty is attained, Uncertainty is the only possible next result. It derives from the fact that S, who is uncertain about a particular square, will not believe what Q tells him – and indeed there is only a slightly better than average chance that Q actually knows the truth. Scepticism reigns.

Now, in assessing the performance of these Senpers it is possible to calculate at any point how closely their world-view approximates the True World simply by adding up the number of squares for which there is a proper correspondence between the True World board and the Senper's board. These numbers are printed at the side of each Senper board. At the beginning of the first simulation run they all record initial scores of more than 50 per cent, because they know exactly their own and their neighbours' states, and for the other squares they have a 50/50 chance that their random beliefs are correct. But note, most importantly, that none of the Senpers can ever know its own score, because none of them knows what the True World looks like. In this 'Sceptic' simulation run, their knowledge scores rapidly collapse to the minimum possible.

For most of the remaining runs, a new set of Transition Rules is used. In this the sink of Uncertainty is no longer terminal but transient. If S is uncertain about

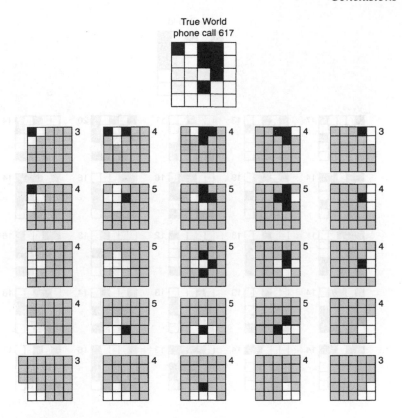

*Figure 8.1c* World views: sceptic's case

*Note:* Continuing with the Sceptic's Rules, after 617 random phone calls there are 25 seperate visions of the world – most of them are clouded by uncertainty and knowledge is minimal.

a square, then he will accept Q's views if Q has a definite belief. This is the case for a Community of Reasonable Believers.

| S | Q | Black | White | Uncertain |
|---|---|---|---|---|
| Black | | Black | Uncertain | Black |
| White | | Uncertain | White | White |
| Uncertain | | Black | White | Uncertain |

We now show two cases of Communication Rules for this table of Transition Rules. First, the world can communicate equally with itself, but the Senpers randomly talk only to their neighbours. This means that long-distance transmission is indirect – but in the end any Senper can 'hear' by implication what any other has said. The Senpers gradually converge on a common understanding of the True World, and ultimately all achieve this understanding. But it is not a smooth process – their knowledge scores go down as well as up along the way, when they pick up a more recent but less correct world-view from a more recent phone call. Convergence on a good degree of truth is rapid (Fig. 8.2a, b, c) – but convergence on complete Truth takes a very long time indeed (more than 40,000 calls on

True World

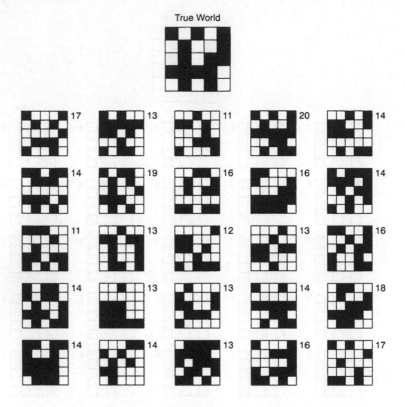

*Figure 8.2a*   World views: reasonable believer's case

*Note:* In this case the Transition Rules are the Reasonable Believer's Rules, and the Communication Rules are that neighbours randomly phone each other. This diagram shows the starting position of knowledge and fantasy for this run.

average). But even then, the point remains that *they never know that they know the True World*, so what criteria could they use for assuming that they do? Our response to this is that it occurs when any Senper finds it is not necessary to change its view of the world, or that the rate of change of its views is approaching zero, then it assumes it is: (a) near the truth, and (b) that its views are held in common with those of most other Senpers.

The next simulation changes the Communication Rules, to conform more to what we think is the actual pattern. The North is represented by the top line of five squares and the South by the rest (Fig. 8.3). This approximates to a 20/80 split in population between developed and developing nations. Any North Senper phones another North Senper 50 times more often than it phones a South Senper. In the South, the Senpers make on average only one call for each 50 made in the North, and phone neighbours only, including North neighbours (Morocco phoning Spain).

This North–South unequal access model produces what for us is initially a counter-intuitive result. Here it useful to distinguish between Belief and Truth. Belief is scored by counting how many squares on any Senpers map are not uncertain (grey). The North very quickly has few squares of uncertainty, because the Northern Senpers

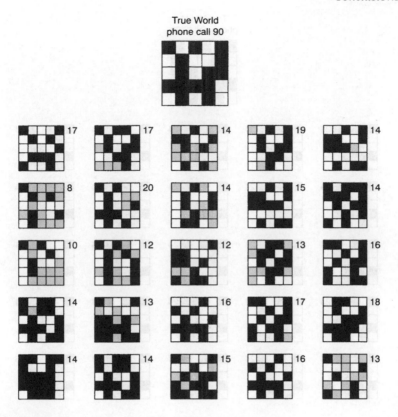

*Figure 8.2b*  World views: reasonable believer's case

*Note:* After 90 random phone calls to neighbours many scores have gone down as Uncertainty sets in over conflicting world-views.

keep adjusting their views by referring to each other – and they actually achieve a True picture of the North (the top line) very fast. In Fig. 8.3b they all agree except for the easternmost Senper. These moments of doubt are so fleeting that it is quite good to have captured one here. Visual inspection will confirm that since most phone calls by this Senper will be to the North, it will soon realign itself with the dominant view. But their True knowledge of the whole world, though they do not know it, is actually considerably less, because they agree with each other about what the South (the other four lines) looks like, but they are often quite wrong (Fig. 8.3b). The problem is that the South is not well connected, and its views do not get represented often enough. The South has both poorer Truth, and a variety of belief (coherence). Some South Senpers have as much certainty as the North, even though it is not True, because they have not had enough communication to cause doubt or modification of their views. The conclusion is that for the World to Know Itself, for Belief globally to converge with global Truth, then all parties must be equally able to communicate. In other words, in some respects unequal access to communication can disadvantage the well connected as much as the less connected. If the world is changing, or if there are linguistic problems blocking communication (neither are simulated here), then presumably convergence is even more difficult.

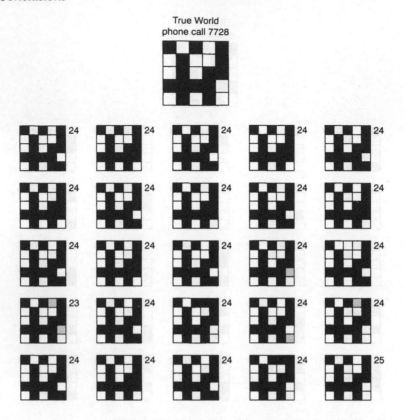

*Figure 8.2c* World views: reasonable believer's case

*Note:* After 7,728 random neighbourly calls there is a high degree of convergence on a common view of the world. The rate of convergence slows rapidly the closer the system gets to a common accurate view, because a mistaken view about one cell might pervade most of the system (good coherence) until it conflicts with a phone call to that cell or its neighbours (local correspondence fails) – and then that has to filter through the rest of the system. This does not happen irreversibly – since the Transition will first cause others to doubt their view of the cell, but they might then find their uncertainty again replaced by the mistaken view of another neighbour. This simulation took 42,556 phone calls before all Senpers agreed on one world vision.

We can go on varying the operating rules in endless ways. Here we will content ourselves with describing two further models. Both of them relax the idea of correspondence – that is to say there is no True World and the Senpers start with nothing but an unconstrained, random world-view, one part of which refers to themselves. In the first 'Pig Head' model, they never doubt their own state; they accept and continue to believe in the state they were 'born' with. The Transition Rules otherwise remain the same. It takes only a little thought to realise that the world-views of all of them will ultimately converge on one universal truth, which is that all accept each other's 'self-proclaimed' state. The second variant is a little more subtle. Again, there is no True World, and all begin with a random world-view. But this time the Transition Rules apply unconstrained: a Senper can be caused to doubt its own state, and can then be persuaded by others to change its own self-image.

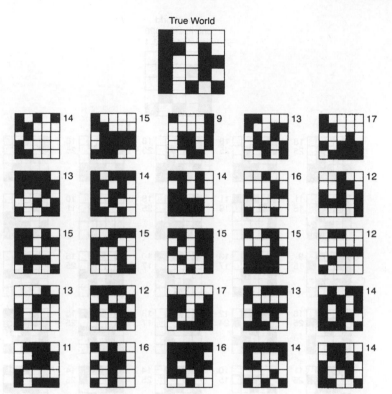

*Figure 8.3a* World views: reasonable believer's case and unequal communication

*Note:* The starting states for a simulation involving Reasonable Believer's Transition Rules, but unequal North-South Communication Rules. The top line of 5 Senpers represents the North. They make 50 phone calls for every one phone call made by any other Senpers in the rest of the world, the South, and they can phone anywhere, but 95 per cent of phone calls are made to other North Senpers. In the South, with the 50 times lower rate, phone calls are to neighbours only (including to North neighbours for the second row).

Given unequal access between North and South as discussed above, there is a very rapid convergence on a World View by the North, which in no sense looks like an average of anything and is impossible to predict from the original North world-views. But this view is not necessarily stable: it may persist for long periods and then go through revolutionary change depending on the sequence of South inputs that may occur. But whatever the dominant North paradigm is at any one time, that is usually what begins to filter into the South, and ultimately one agreed world-view does emerge, dominated by the latest North paradigm.

What can we conclude? In the case of the first models which include corre-spondence:

1   An understanding of the 'true world' is only achieved through accepting the belief of others without knowing whether or not those beliefs are true, and by continually modifying one's own beliefs.

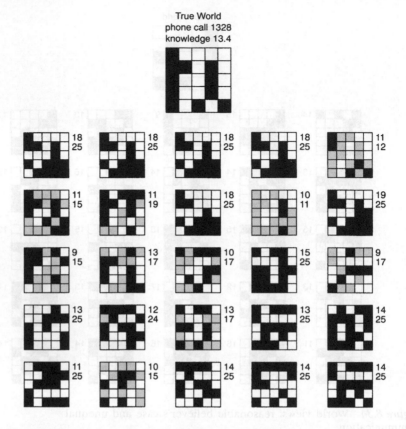

*Figure 8.3b* World views: reasonable believer's case and unequal communication

*Note:* The top of each pair of numbers represents the accuracy of each Senper's world-view. The bottom number represents the Senper's degree of certainty – or lack of doubt. After 1,000 calls the North all share the same world-view and are mostly certain of it, but they are wrong in significant respects. In the South there are islands of certainty – those who have not yet had any communication with anyone else. For others uncertainty is increasing.

2  Because knowledge is achieved through communication in community, the acquisition of knowledge is a historical process in time.

3  It is necessary to have a community of communicators if each can verify as true only a small part of the world.

4  If different communities and cultures, or different disciplines, have different styles of debate and presentation – different forms of communication which may be opaque to varying degrees to other communities – then this may help the internal dialogue in the community in promoting belief, but often inhibits verification against the views of other communities.

For the last model without correspondence (putting aside the 'Pig Head' version):

1  An understanding of the 'true world' is only achieved through accepting the beliefs of others and by continually modifying one's own beliefs.

2  Because knowledge is achieved through communication in community, the acquisition of knowledge is a historical process in time, where the eventual

outcome is the result of the accidental sequence of events in that history – there is no *a priori* known state on which the solution will converge.

3   If different communities have different rates of communication, those with the higher rate are more influential in defining the ultimate agreed 'true world'.

We do not propose that these models fit all that is contained within these pages. We realise that this book is about mass media and the word 'broadcasting' is used quite often. Despite the fact that our models are based on point-to-point communication, that is probably not of great significance – the high rate of communication in the North effectively ends up with rapid broadcasting. But the models do give pause for thought and do suggest some summary images.

First, in terms of asymmetries of communication, the distinction between North and South applies twice in this book, once between the UK and India, and the second time replicated within India itself. This suggests that some sort of convergence between the UK and the 'North-in-India' (post-nationalist) is possible, both of them inadequately aware of the remaining and largest population of the Rest-of-India and its world-views.

Second, there is a shadow hanging over the whole of the book, which is that we have had unrepresentative and limited access to this Rest-of-India, both for physical access reasons (the logistics of mounting forays into the *mofussil* – the interior) and for linguistic reasons. And our book will be published in the UK and India, but in English only, and will therefore itself not make much contact with these other publics either.

Third, although individuals obviously each have a unique world-view, the idea of society is predicated on at least certain common aspects of a world-view, even if they are limited to a respect for some common code concerning the limits of tolerable behaviour or some common goals. These shared aspects of a world-view come through communication. Obviously in the real world communication is not discrete; it occurs through many very different networks, which are infinitely complex and rely on every kind of medium from the nuance of body language, the tone of the spoken word and the style of written text to the formats and conventions of the images and sounds that are carried on radio-signals and cables. At every step there are 'linguistic problems', whether this is the understanding of the symbolism of one form of dance rather than another or whether it is the literal problem of translating abstract concepts from one language to another. In a society with low levels of self-communication, the aspects of the common world-view which define the society may be both simple and 'powerful' to become widely held. There is no doubt that for most of the last 50 years to be Indian has meant to accept the aspiration of 'development' as a dominant part of the world-view. The complex and fractious nature of Indian politics may appear to have a life of its own, but that has mostly revolved around the struggles between communalism (of language, religion and caste) on the one hand and on the other, but tied into them, the methods and rewards of development. The mass media have been used and have in part been content to be used in spreading India's 'world-view' of development within India – although as a monolithic ideal it is challenged by some sectors of society. Of these, the post-nationalist sector is peculiarly well placed in terms of language, wealth and communications technology to imbibe some aspects of other world-views, including some from the North, and to pursue some aspects of anti-development, including aspects of 'environmentalism'.

In Britain there have been two dominant themes in the last 50 years. The first, that all major political parties have espoused in their own way and still do, is the never-ending quest for material prosperity, an unexamined statement that being 'developed' is not a final state but a continuing process. Even if there are degrees of suspicion now that this is a treadmill, because of anxiety about unemployment in a no-growth economy, no political party has yet worked out how to get off it. The second theme has been to redefine its post-colonial international role, which has meant extremely complex and contradictory flirtations with different ideas of what the terms 'European' and 'minor world power' mean.

In the Foreword to this book we said that our project involved three intersecting spheres – of the mass media, global inequalities in development, and the environment. It seems to us clear at the end of it all, that for most of what we have written we could have replaced the third sphere with quite a few other subjects of attention – the economy, for example – and come to broadly similar conclusions. If for a moment we think of the three as three points, then they define a triangle of interest. But the triangle has three pairwise edges. One of these is the relationship between the Environment and North–South inequality and that edge could be the subject of intensive study. The same can be said of the Environment/Media edge. But in the context of these three together, it is the North–South Inequality/Media edge which is the most important for study, because that is the main context of the evolution of world-views within which the 'environment' is embedded. Let us remember that the 'environment' has no independent existence outside human consciousness, that it is a word, no more nor less than that, which is used by some as a category for labelling things, by others as a format for writing about things, by others as the campaigning goal of their lives. Some contemporary commentators are keen to give it concrete form: they stress that the economy is a subset of the 'environment'. Perhaps it is right to redress an imbalance but it does not negate the fact that, seen from the other end, the 'environment' is also a subset of culture. A car is made of refashioned elements of the environment, but only refashioned because of the conscious values and designs of the culture, and it is culture which appraises its handiwork as 'progress' or 'environmental terrorist'.

The pervasive rise of the word 'environment' in the last few decades has been because it is a comprehensive word, a 'whole-thing' word, indicating some holistic understanding of the integration of the *Gestalt*, the antithesis of the atomistic reductionism of science which is perceived to have had dangerous impacts on the world around us. But because 'the environment' is holistic it presents immediate interpretive problems. Within our local lived-world, where we can check our understanding of the environment through correspondence, how do we check our experience against 'the whole thing'? Obviously we cannot, either in knowing all there is to know immediately around us, or, and least of all, in being able to experience the planetary 'whole thing'. Psychologists have an expression for an immature relationship, as when a baby thinks its mother is a breast, or when an adolescent male desires a girlfriend's body before he knows her personality – it is the part–object relation. The whole remains unknown, and a part suffices. Such a part–object relation is only immature if the relationship with the whole could in theory develop. What we have seen in this study is that it is impossible for the relationship with the whole to develop, even if it is given a word such as 'Gaia' or 'The Environment'. The public in Britain and India both express any practical environmental concern in a part–object relation – be it in recycling, in protesting against the siting of new supermarkets, or in worrying about air pollution and asthma – because the whole is too nebulous. They may join an organisation like the National Trust, concerned with the manifest local visual environment and the cultural value of historical artefacts more than anything else, or the Royal Society for the

Protection of Birds. Much of this activity need have no relationship to a grand 'environmentalism' *per se* at all. For some people the quality of life will improve with having better drains and drinking-water, or with having/not having an urban bypass. At this local scale, 'the environment' is a category which observers such as ourselves, as much as the subjects, are inclined to apply to one aspect of the local part of the world-view of individuals. But other categories can equally apply, be it development or conservation, degradation or rehabilitation. Because of wider media exposure and a different educational and cultural history, the British clearly have a stronger acceptance of the idea of a planetary environment 'out there' – beyond the checks of local correspondence. But how they can relate to it is still part–object – mostly through expressed concern for the rainforests, etc.

In the Foreword we proposed a very simplified model of the circulation of ideas through the media gatekeepers, then through the media, to be interpreted by the audience, to be absorbed into society, which is then interrogated by and proposes messages to the media gatekeepers. That model is not good enough to fit with what we have covered in this book, because there are many different circuits within any society, and complex and sometimes tenuous relationships between societies. We also asked what it is that keeps the circle in motion. At the national scales we have worked at, it seems to us that it is the grand projects which societies are pursuing, and the debates around those grand projects. No doubt the media help in propagating and distilling a commoner world-view of these projects, or in giving limited opportunities for propagating the new ideas which may challenge them. (We say limited because, in our simulations discussed above, radical world-views can be established, but they are mostly rapidly extinguished.) So, although we may talk about news and news values and the roles of primary definers, actually none of them are as important as the projects of society. If we are asked what the grounds are for this assertion, we ask the reader to reflect on what (s)he has read in these chapters: there are dominant themes in both societies which resurface in new forms at every stage round the circle. There is no obvious starting point in the cycle of communication, nor any obvious end – whether it is Northern hegemony or rainforests. We subscribe to Shanahan's (1993) citation of the Dominant Social Paradigm (Pirages and Erlich, 1974; Dunlap and Van Liere, 1984), which moulds many of the fundamental assumptions of the media's production.

Yet there is no single global circle of communication. Partly this is because the conventions of message-making in one circuit cannot be readily used in another, partly it is for economic reasons, partly for reasons of political control, etc. But if ever there were such a circle, if ever a global society were established, and if ever 'the environment' or 'sustainable development' became the primary project that society was engaged in, reactively or proactively, then perhaps ...

But for the moment:

### *Eco-politics: a case for equity*[3]
by R.N. Prasher, a senior IAS[4] officer in Haryana

**The Tribune**
Chandigarh, 25 August 1995, page 8

... Meanwhile the rich countries have developed other ways of exploiting them [the developing countries]. The new plan works through environmental phrases, and global warming is the latest bait. ... One does not have to reel out figures to show that the developed countries have been the biggest energy-guzzlers and pollutant-emitters. ... It is equally obvious that during the past decade or so significant success had been achieved by the developed world in reducing the pollution levels in their countries. . . . The most obvious explanation, of course, lies in new technologies which help in curbing pollution.

Industrialised countries cannot deny they have been dumping proportionately outrageous levels of pollutants in air, water, and even space. Now it has been accepted that these were and continue to be the common heritage of all humanity, that poor countries must retard their development to save it. Therefore the debt which the industrialised countries have incurred by utilising the oceans and atmosphere as dumping grounds for their high levels of waste must be recognised. While no industrialised country will be in a position to repay the debt it has incurred to the rest of humanity by abusing the common heritage, it should not prevent us from demanding what it is possible to repay.

As stated earlier, the use of energy in the early phases of development is unclean. If the per capita output of pollutants in India or China reaches the levels that pertained in the USA in the sixties, it will be disastrous for the health of their citizens. However, it will be equally disastrous for the rest of the world, the rich countries included.

In their own interests the rich countries should transfer to the poor nations the technologies required for a cleaner utilisation of fossil fuels. . . . The message is simple. The developing countries shall undergo the process of growth irrespective of the discomfort it causes to them or anyone else. It is up to the rich countries to decide whether they have something at stake in this transition.

Given the inequalities in communication, this is not a message which gets propagated much in the North. And unfortunately in the end it matters less what you know than who else believes you.

## NOTES

1   In spatial statistics and analysis, contact patterns are often referred to by reference to chess moves. Here this is a Rook's case rather than a Queen's case (which would also involve diagonal neighbours touching at the corners only), but the distinction is not of importance to the results.
2   The printed diagrams here do not convey any of the excitement in watching the flickering patterns evolve during an actual run.
3   This is an abridged version.
4   The Indian Administrative Service, successor to the Indian Civil Service, which recruits high fliers into the top ranks of the government bureaucracy.

## REFERENCES

Dunlap, R.E. and Van Liere, K. (1984) 'Commitment to the dominant social paradigm and concern for environmental quality', *Social Science Quarterly* 65: 1013–28.
Pirages, D.C. and Erlich, P.R. (1974) *Ark II: Social Response to Environmental Imperatives*, San Francisco: Freeman.
Shanahan, J. (1993) 'Television and the cultivation of concern', in Hansen, Anders (ed.) *The Mass Media and Environmental Issues*, Leicester: Leicester University Press, Ch. 9, pp. 181–97.

# Appendix I: Coder's handbook[1]

## The mass media and global environmental learning

*The South–North Centre for Environmental Policy*
*SOAS*
*Thornhaugh St, Russell Square, London WC1H 0XG*

## CONTENTS

1 Introduction to the project
2 The data
3 The distinction between backcloth and traffic
4 Data collection
5 Using the coding sheets
6 Addendum 1: Traffic words

## 1 INTRODUCTION TO THE PROJECT

Throughout the world, although to varying extents, mass media play a significant role in creating different 'publics' and in then influencing the opinions of those publics. The mass media are therefore very important in disseminating and moulding attitudes and ideas towards environmental issues, some of which are truly global in nature. In North countries the mass media reach most people, and have a dominant role in informal public education. There has already been some research in North countries on the extent to which environmental issues are covered by the media, but little is known about how or why environmental news or other forms of environmental discourses are produced and disseminated in South countries. In South countries, a majority of the population might not be influenced by mass media in the same way, though the urban and élite groups usually are. It is there-fore important to find out on a comparative basis what members of the public know about environmental issues, and how they come to have that particular knowl-edge. This is a central aim of this project, basing its comparative study on India and the UK.

In spring 1991, the Economic and Social Research Council launched its Global Environmental Change Programme, the largest ever UK social science research programme. Phase II of the programme funds research under four main themes, one of which is International Opinion Formation and Environmental Education. This project has a 2-year grant from the ESRC under this head.

## 2 THE DATA

The part of the project on which you are working involves the collection and analysis of the content of a sample of TV and radio channels and selected newspapers in the UK and India. We are surveying and sampling six weeks (once every two months) of two selected newspapers in both countries, and for the same six weeks the programming on TV and radio.

We are particularly interested in programmes and newspapers stories *with espoused environmental content*. By this, we mean those which include **an explanation of the natural world itself** or which refer to **man's interaction with the natural world**. These definitions should be your guide for deciding which programmes and newspaper stories have espoused environmental content and those which do not.

The principles are the same for radio and the press. For TV the kind of data we are interested in are expressed by such questions as:

What sort of programme is it?
What is the programme about?
How long is the programme?
When was the programme broadcast?
Who broadcast the programme?
Which language was used in the programme?
Who made the programme?
From which country did the programme come?

Some of these questions can be answered easily, and answered consistently – e.g. the date and time of transmission of a programme. Other questions appear much more difficult to answer consistently – the obvious example is 'What sort of programme is it?' However, we want to try and achieve the same degree of consistency in coding programme contents as in coding dates and times.

Most of us think that we can discriminate between different types of television programmes. The labels 'drama' or 'documentary' seem adequate to describe programmes at first sight. Experience shows, however, that this is not always the case. To many people knowledge that something is a 'sports programme' is not very useful. Is it about 'open air downhill skiing' or 'sumo wrestling'? Indeed it is about nineteenth century ladies' tennis, in which case is it about sport or about women's rights? And indeed, which programme is *not* about the environment, in that each one of them is situated in time and at a place and has implicit messages about how society is based on prodigal or conservative use of resources?

The salient features of the scheme that we have adopted are as follows: we ask you to distinguish between the traffic (form of treatment) and the backcloth (content or subject matter) in any programme.

## 3 THE DISTINCTION BETWEEN BACKCLOTH AND TRAFFIC

In our classification scheme, backcloth is subject matter, and traffic is treatment. A programme can use the subject matter of cats with a serious and informative treatment, or a comedy treatment; it may use real cats, or it may use cartoon illustrations. A love story can be comic or it may be the subject of a serious drama.

You will notice that any programme could be about anything, and therefore the space on our data sheets for backcloth is left wide open. But you will also notice that there are two special boxes which have to be filled in for any programme: these are on time and place. The two form a special kind of backcloth – the setting for the rest of the action.

You will see how the same ideas also fit radio and the press when you come to the examples below.

## 4 DATA COLLECTION

### Programme in *outline* and programmes in *detail*

For every TV and radio programme (apart from the Main Evening News) an Outline page must be completed. In addition to this, there are four other options:

(i) Most **programmes have no espoused environmental content**. In this case, no other pages need to be completed.

Although the majority of programmes have no espoused environmental content, they all implicitly relate to a context in time and place. These contexts can be related to such ideas as 'consumerist industrial society' or 'medieval agrarian society'. It is important that we know this, since we hypothesise that although there may be, say, 8 per cent of environmental programming on TV, the other 92 per cent is probably propagating messages predominantly about modern consumer society – e.g. Los Angeles or Miami in the 1990s. For these programmes we need to know the fixed data about duration, origin, language, etc., traffic style (e.g. fictional serial, or documentary, etc.) and the time and place. We do not need detail about the backcloth beyond that. All of the information we require can be recorded using the Outline page only.

(ii) If **some, but not all, of the programme has environmental content**, then a Detail page should be completed for every relevant part.

(iii) If the **programme consists of several different parts, all of which have environmental content** then a Detail section should be completed for every part.

(iv) If the **whole programme is of environmental concern**, then a Detail page should be completed for the whole programme.

For those programmes with espoused environmental content, we want to know not only all the information normally recorded using the Outline page but also the details of the backcloth. The Detail page is intended for just that purpose.

### Newspapers

The principles are the same as for the other data collected. However, the unit of measurement is obviously not duration, but column centimetres. The traffic words are also slightly different – since now we have words such as 'obituary' and 'editorial' as well as 'report' and 'in-depth feature'.

## 5 USING THE CODING SHEETS

### Outline page

For radio and television programmes – apart from the Main Evening News – you should first fill out the small box in the top left-hand corner with either an **R** for radio or a **T** for television.

You should then continue by filling in the boxes which record the **Date**, your **Name** (an abbreviation or your first name will do) and the **Channel** or **Station** which you are monitoring. All data recorded will be typed onto computer, so please write clearly, using BLOCK CAPITALS for any words which might be unusual or unfamiliar.

When you are recording the **Date**, please write it in the order: year, month, day, e.g. 921114 is 14 November 1992. Note that you *do not* put in full stops (92.11.14 is wrong) but you *do* write zeros – e.g. 930305 is 5 March 1993.

The **Start Time** should be recorded in the order hours, minutes, seconds, e.g. 225432 is 32 seconds past 6 minutes to 11 p.m. Again use zeros. 6 a.m. is 06. Often you may be able to provide the hours and minutes only, in which case you should fill out the *hour* and *min* boxes only, leaving the *sec* box blank. For the Main Evening News we do require data to the second.

**New/Repeat**: A programme is *New* if it has not been transmitted on this channel before. It is a *Repeat* if it has already been transmitted on this channel.

The **End Time** should be recorded in the same way as the **Start Time**.

The box for **Adverts** is for you to record the total time given to adverts during the entirety of the programme, including those which immediately follow the end of the programme. Where you do not actually watch the programme, you will have to estimate the amount of time given to adverts.

To understand how to code the sections **Title** and **Subtitle**, it is useful to know the difference between what we call series, serials, unique and linked programmes. A **unique** programme is a simple one-off that does not relate to any of the following categories. A **linked** programme is one where the actual programme itself has no continuity (for example of either the story or the characters or the setting) with any others, but the time slot identifies a continuity of the programme scheduling, or the broadcasting company indicates the link: e.g. Film on Four or Panorama, Play for Today. **Series** are programmes where there is continuity of characters but each episode is self-contained and the sequence of showing the episodes could be rearranged without damage – e.g. The Trainer, M*A*S*H, The Magic Roundabout. **Serials** have continuity of character and one long story split into different episodes, e.g. Brookside, Eastenders, Howard's Way.

For unique programmes, the **Title** is the **Title**, and there is no **Subtitle**. For linked programmes the **Title** is the **Title** of the film or play, etc. and the **Subtitle** is Film on Four, Play for Today, etc.

For series and serials, the **Title** is the **Title** of the series e.g. The Trainer, Brookside and the **Subtitle** is the episode title.

Country/ies of Origin refers either to the country in which the programme was made – in the case of TV and radio programmes – or the country from which the story is being filed – in the case of newspaper stories. See Addendum I in the full version of the Handbook for a list of abbreviations for countries.

The **Company/ies of Origin** is the name of the production company which made the programme. This information will not always be available, but note it down if possible.

All programmes have an **Original Language**. Many will be broadcast in this. **Dubbed into** should be used in cases where the original soundtrack has been replaced by a soundtrack in another language. Indicate this other language. **Subtitled into** should be used when the original soundtrack can still be heard, but a written translation in this new language is provided on the screen to coincide with the original text.

A **Voice-over** is the main soundtrack which can be heard over any other speech. Sometimes, a voice-over will act as a direct translation into the new language of what is being said in the original spoken language; at other times the voice-over will be providing a summary or commentary on what is being said.

The box for **Traffic Words** is for recording the treatment of the subject matter. For a list of the words to choose from please see Addendum 2 in the full version of the Handbook. All **Traffic** words are abbreviated to three letters and separated from each other by colons.

**Place/Time** Although for many programmes the place/time location of the programme is of little interest, for others it is very important. Place should be given as specifically as possible, e.g. Nairobi, Madrid, but sometimes it is not possible to be more precise than, say, South America. Sometimes the time setting of a programme is very precise, e.g. Bastille Day 1989, while at other times it is not, e.g. Moghul Period. When the programme is set in the present, we simply use 'now'. C17, C18 represent 17th century, 18th century, etc.

## Detail page

The first box to be completed refers to whether or not the information you are recording relates to a whole programme or **Part** of a programme. If you are recording information on a programme which is entirely devoted to the environment, then you will not need to tick this box. However, if – as is often the case – you are filling out a Detail page for some of a programme, only part of which has environmental content, then you should tick here.

The **Start Time** and **End Time** should be recorded in exactly the same way as for the Outline page, in the order hours, minutes, seconds.

**New/Repeat** for a Detail page may be slightly different from that for an Outline page in that the Part will occasionally be a Repeat, e.g. a film clip when the rest of the Programme is New – and vice versa. In order to ascertain this information, you will usually need to be told it by the broadcaster.

The **Subtitle** is the title of this Part of the programme.

The instructions for completing sections **Country/ies, Original Language, Dubbed into, Subtitled into, Voice-over in, Traffic, Place** and **Time** remain the same for the Detail page as for the Outline page. Where any of these eight sections is the same for the Part as for the whole programme, you are not required to fill them out again. Instead, simply put two sets of inverted commas – "" – or write *ditto* to indicate that they are unchanged.

In the **Backcloth** box you record what the programme is about. You do this by writing in as many elements of the programme as you think necessary. An element can be a word or a combination of words, but all elements are separated by a colon (:). In general it is better to use single word elements or elements with a few words rather than a lot – to help us in our data processing. Here are two examples:

> Education: English Curriculum: Imaginative Approach: 1988 Education
> Reform Act: Dominant White Culture [Taken from a News Story]
> Seal Culling: Environmental Activists: Fish stocks: Fishermen: Politics
> [Taken from a TV programme]
> **Main Evening News** only

You should start by filling in the boxes which record whether you are listening to **Radio** or watching **TV**, the **Date**, your **Name** (an abbreviation or your first name will do) and the **Channel** or **Station** which you are listening to or watching. All data recorded will be typed onto computer, so please write clearly, using BLOCK CAPITALS for any words which might be unusual or unfamiliar.

The **Time** the programme starts should then be recorded in the order hours, minutes, seconds, e.g. 225432 is 32 seconds past 6 minutes to 11 p.m. Again, following this is a box for you to record the total number of minutes and seconds devoted to **Adverts** during and immediately following the programme. Please make every attempt to record all these data to the nearest 5 or 10 seconds.

The **Time headlines start** should be recorded in the same manner.

In the section for **Headlines** for the news programmes, one section of the box has been allowed for each headline. If you find that the headline you are recording is more than one line long, then you can squeeze more lines of information into each section. However, please only use one section per headline – and make sure it all still remains legible!

For the **Main Evening News**, at the end of each story you have the option **Time Edition Ends**. This is for you to record the time at which the whole news programme comes to an end. In the case of channels and stations where there is commercial advertising, you will need to write in the time the last item ends and then move to the next story, record ADV in the traffic, forget the backcloth, and then record time item starts and time edition ends.

### Newspapers only

In the case of **Newspapers**, you are asked to record slightly different information from that recorded for radio and TV.

After recording the **Date** of the edition and your **Name**, there is a box marked **Page Number**. Here you should first record the number of the page on which the item begins, followed by a colon and then the total number of pages in that edition of the newspaper or – in the case of newspapers which have several sections to them, such as Sunday editions – the total number of pages in that particular section, using section numbers or letters where given, e.g. A34, H21. For magazine colour supplements use the letter M, e.g. M3:M25 means an article which starts on the third page of a 25-page magazine. H20:H25 means page H20 out of 25 in section H, 17:32 means page 17 of 32, and so on ...

For each item, you are then asked to complete a box for **Title**.

The **Country** box refers to the country (or countries) from which the story has been filed and not necessarily, but quite possibly, the country or countries about which the story is written. Information on the latter will be recorded under **Backcloth**, e.g. a report about British Gas exploration in the Amazonian forests of Brazil, written by a reporter based in Washington would have USA in the **Country** box, 'Brazilian Amazonia' appearing in the Place box (or both Brazil and UK in the Place box separated by a colon, if some of the report is also about headquarters in the UK), and 'Gas Exploration: Rainforest' somewhere under **Backcloth**.

The **Agency** box is for recording the name of the provider of the story. Sometimes, you will be able to name the **Agency** (e.g. Reuters); sometimes you will have the name of the *reporter*, and whether or not he or she is *staff*. This information should be recorded where possible. In some cases, **Agency** information will be available where **Country** information is not. For instance, a story from Reuters about the Sardar Sarovar Dam project in India would have India in the **Backcloth** box, but not in the **Country** box. Reuters would appear under **Agency**.

**Column cm** is for you to record the amount of space given by a newspaper to a particular story. You should work out the **total** number of column cm by measuring the length of all the columns devoted to the items and adding them together. Then, within this space measure how many of these cm – if any – are taken up by **photographs** and/or other **graphics** and/or **tables**.

## 6 ADDENDUM 1

**Traffic words**

*Note:* Not all words will apply to TV, radio and newspapers. Which is normally applied in which context is signalled by T, R and N. You might find contexts in which you use a word in an appropriate but uncommon way, a T word in a newspaper context, for example.

Words have been grouped together – simply to help you identify when to use them. Each group is headed by a *key word*, given in brackets.

USE THE 3-LETTER ABBREVIATION FOR EACH WORD ON THE CODING SHEETS. E.G. **ADV** or **CIN**

YOU ARE AT LIBERTY TO INVENT OTHER WORDS, BUT PLEASE DO SO ONLY IF YOU FIND IT **ABSOLUTELY NECESSARY**, AND THEN LET US KNOW WHAT YOU HAVE DONE WHEN YOU HAND YOUR DATA SHEETS IN.

### Group A (Information)

| | | |
|---|---|---|
| Demonstration DEM | TRN | Illustrates a physical skill or process with the intention that the viewer could gain the skill by imitating. |
| Explanation EXP | TRN | Imparts in the recipient a capacity to understand something. The attempt to illuminate or convince or give understanding is by reasoned presentation, leading the viewer along. |
| Instruction INS | TRN | Imparts in the recipient a capacity to undertake a given process. The presentation is formal or procedural, the broadcaster expects to convince or give understanding by authority. |

### Group B (Talk)

| | | |
|---|---|---|
| Biography BIO | TRN | Biographical treatment of a person which attempts to give a reasonably comprehensive view of the person. |
| Commentary COT | TR | Simultaneous verbal comment and additional information. The talk is unscripted, although the commentator may refer to previously scripted material if it becomes necessary during the commentary. Examples include sports commentaries, arts commentaries and public events commentaries. |
| Discussion DIS | TR | Two or more people of equivalent status talking about the subject matter. The theme or subject may be guided or bounded by a chairman. Ideas are batted backwards and forwards, the talk is not |

scripted or known precisely in advance and may take the form of a chat. Combinations with Ethical Concern, Light, and Serious will be important.

| | | |
|---|---|---|
| Interview INV | TRN | An interviewer asks questions of an interviewee who answers them. Also includes cases where only the responses to the interviewer's questions are given. |
| Narration NAR | TRN | Telling the story, not necessarily synchronised voice and vision. Essentially one leading voice throughout with prepared script. Usually associated with non-fiction. (Excludes theatrical narration, i.e. reading or recitation.) In Newspapers may be equivalent to Extract. |
| Report REP | TRN | Statement of what is reported to be fact or assertion of unbiased appraisal by the broadcaster of some-thing, implicitly of interest to or concern to the public. (*Note:* This is always applied to news programmes.) |
| Results RES | TRN | Report when applied to the resolution of a defined competition, e.g. a sport report or election result. |
| Solo dissertation SOL | TR | As Discussion except a single person weighs the various points of view, in scripted or unscripted discussion/elaboration/public lecture/chat. Combinations with light, serious, ethical concern, etc. will be important. |

### Group C (Performance)

| | | |
|---|---|---|
| Competition COP | TRN | Carries the idea of winning or losing on the subject matter, with an ordered set of results, or at least one designated winner. |
| Exhibition EXH | T | Of an assembly of visual artistic artefacts. This EXH is a visual analogue of performance. It applies to static displays. |
| Performance PER | TR | Of an artistic production, e.g. the performance of a piece of music in a concert, the performance of a ballet, etc. |
| Quiz QUI | TR | Involves someone asking prepared questions and someone answering them to define a level of attainment; a performance of the quiz game subject matter. |
| Reading REA | TR | Refers to the theatrical reading of poetry or prose. |

Recitation    T    A static performance by a single person of a
REC          learnt piece of prose or poetry.

### Group D (Drama)

| Animation | TN | Picture, Puppet, or Object animation. Includes cartoons and human puppets, e.g. Playdays. |
| ANI | | |

| Cinema Movie | T | Drama made independently of broadcasters for general release and non-theatrical (=club type) release. (It subsequently comes to TV where broadcasters usually purchase a number of showings.) |
| CIN | | |

| Drama | TR | Commissioned and/or made by broadcaster for themselves and for future sale. May be Series/Serial/Unique e.g. The Bill, Eastenders, The Archers. |
| DRA | | |

| Fantasy and Surreal | TRN | Creation of phenomena outside normal experience, e.g. the stretching of the bodies of cartoon characters, Popeye's fantastic strength, Dumbo's ability to fly, Pinocchio's growing nose. ... Many science fiction programmes will carry this traffic. |
| FAN | | |

| TV Movie | T | Movie programmes made for television outside of a broadcasting organisation. |
| TVM | | |

### Group E (Tone)

| Comedy | TR | Attempt to be amusing and/or make the audience laugh. |
| COM | | |

| Light | TRN | Frivolous presentation with the intention of entertaining. |
| LIG | | |

| Serious | TRN | Non-trivial presentation, thought-provoking and requiring some concentration. |
| SEI | | |

### Group F (Enrichment)

| Aesthetic Enrichment | TRN | Helps the audience or reader to understand and appreciate the aesthetics of x. This enriches the future enjoyment of x. Often this will occur with Instruction. (In general a performance does not achieve this, even though it may be aesthetically enjoyable.) |
| AES | | |

| Scientific Enrichment | TRN | To help the audience understand and appreciate Science. |
| SCI | | |

| Social Scientific<br>Enrichment<br>SOC | TRN | To help the audience understand and appreciate human behavioural science. |

### Group G (Ethics)

| Ethical Concern<br>ETH | TRN | With the explicit intent of stimulating an awareness of social and moral problems. |
| Spiritual<br>Guidance and<br>Worship<br>SPI | TRN | A theological/religious presentation to guide the viewer along a prescribed theological path, or supposing precepts by which people can guide their lives. Includes church services and formal worship. |

### Group H (Fact/Fiction)

| Non-fiction<br>NON | TRN | The broadcaster puts the programme or item out as being non-fiction, i.e. purports to be about actual phenomena past or present. |
| Fiction<br>FIC | TRN | The broadcaster does not put the programme or item out as factual. (Note:   Some Drama series may be both Non-Fiction and Fiction.) |

### Group I (Programme Relationships)

| Linked<br>LIN | TRN | Many programmes or newspaper features are linked together, often by a set time slot, e.g. Kilroy is broadcast every weekday morning at the same time. The Environment Guardian appears every Friday. |
| Series<br>SER | TRN | The programme or item is one of a set sharing the same characters and/or subject matter. The episodes can be permuted, i.e. it does not matter in which order they are viewed since each is self-contained. |
| Serial<br>SEL | TRN | The programme or item is one of a set sharing the same characters and subject matter and/or setting with a common theme or story. The episodes cannot be permuted since they follow a definite showing order which must be followed if the individual episodes are to make sense. |
| Unique<br>UNI | TR | A unique programme is one made by itself for itself which is not part of a series or serial. |

### Group J (Participation)

| Charity Appeal<br>CHA | TRN | |

| Follow-up<br>FOL | TRN | The programme or article includes or is immediately preceded or followed by a device which enables the viewer, listener or reader to gain or share additional information about the subject(s) covered. This may for example take the form of a phone-in, help-line, write-in or make reference to CEEFAX or ORACLE. |
|---|---|---|
| Home<br>Participation<br>HOM | TR | Interactive audience participation at home, which might include such elements as expressing opinion, asking/answering questions/pledging money by telephone/fax. |
| Letters<br>LET | TRN | Part of newspaper or programme in which readers' or audience letters are published/read and answered. |
| Party Political<br>Broadcast<br>PAR | TRN | Regulated broadcast on behalf of a political party. In newspapers this would be coupled with Advert. |
| Studio<br>Participation<br>STU | TR | Audience participation in the studio, e.g. Kilroy, Question Time. |

### Group K (Overall Type)

| Advert<br>ADV | N | Newspapers only |
|---|---|---|
| Documentary<br>DOC | TR | |
| Editorial<br>EDI | TRN | |
| Feature Article<br>FEA | N | |
| Game Show<br>GAM | TR | |
| Magazine<br>MAG | TR | In magazine programmes the parts of the programme are linked by a presenter or presenters, e.g. Good Morning ... with Anne and Nick on BBC1. |
| News<br>NEW | TR | |
| Obituary<br>OBY | TRN | |

| Weather<br>WEA | TRN | |
|---|---|---|

### Group L (Target Audience)

| Children<br>CHI | TRN | Specifically targeted at children. |
|---|---|---|
| Disability<br>DIB | TR | Specifically targeted at people with a disability, including the visually or hearing impaired. |
| Elderly<br>ELD | TR | Specifically targeted at the elderly. |
| Schools<br>SCH | TR | Specifically produced for use as a teaching aid in schools. |
| Teenage<br>TEE | TRN | Specifically targeted at teenagers. |

### Group M (Parts)

| Anchor<br>ANC | TR | Portion of – or entire news programme in which the newscaster appears on the screen or on mike. Often this will occur in conjunction with Graphics, Report and/or Roving Reporter. |
|---|---|---|
| Extract<br>EXT | N | Part of a book or other work published elsewhere, reprinted in the paper. |
| Graphics<br>GRA | TN | Any visual aid such as a graph, still photograph, moving or still diagram. |
| Music<br>MUS | TR | Recorded music. |
| Roving Reporter<br>ROR | R | Portion of news programme in which a reporter – other than the Anchor – files a Report, carries out an Interview or provides Commentary from a location outside the studio or from within the studio itself. |
| Vox Pops<br>VOX | TR | Continual string of short comments or answers to a specific question, in which the individual responses of the interviewees are not interspersed with the voice of the interviewer. |
| Review<br>REV | TRN | Review of books, cinema, theatre, videos, TV programmes . . . |

### NOTE

1   The Handbook has been abridged for this appendix.

# Appendix II: The dictionary to aggregate from B¹ to B²

*Note:* Most of the Place and Time words are in another dictionary: some coders have used some words in the general backcloth here.

495," ,dictionary entries after new words, ",0

| | |
|---|---|
| accident | mal |
| acid rain | plu:mal:air:lan |
| advertising | com:val |
| afforestation | mit:lan:dev |
| Africa | afr |
| agriculture | ecn:lan:rur |
| air pollution | air:plu |
| Albania | cis |
| alternative energy | tec |
| Amazon | sca |
| Amsterdam | oec |
| Andhra pradesh | ind |
| animal | ecl |
| animals | ecl |
| animals domestic | ecl:ecn |
| animal rights | ecl:val |
| animal rights | ecl:val |
| Antarctic | poa |
| Antarctica | poa |
| Antartica | poa |
| Appalachia | nam |
| aquatic life | wat:ecl |
| arctic | poa |
| armed revolt | pol:mal |
| Artist | val:hum |
| Asia | oas |
| Australia | oas |
| Autstralia | oas |
| Avalanche | dis:lan |
| Bangladesh | oas |
| Bhopal | ind:urb |
| Biodiversity | ecl |
| biosphere | lan:wat:ecl |
| birds | ecl:air |

| | |
|---|---|
| blood sports | ecl:val |
| BNFL | unk:ecn:res |
| Bombay | ind:urb |
| BP | unk:res:ecn |
| Brazil | sca |
| Brunei | oas |
| Budapest | cis |
| buddhism | val |
| building | lan:ecn |
| buildings | lan:ecn |
| Canada | nam |
| cancer | hea |
| CDN | nam |
| Central America | sca |
| Centra America | sca |
| CFC | plu:air |
| Channel Tunnel | com:ecn:oec:unk |
| Chemical-industry | ecn |
| chemical industry | ecn |
| chemical pollution | plu |
| chemical poluution | plu |
| chemical weapons | mal:plu:pol |
| Chernobyl | mal:plu:hea |
| children | hum |
| China | oas |
| CIS | cis |
| CISEE | cis |
| CITES | ecl:int:mit |
| City | urb |
| cleaning | mit |
| CO | plu |
| $CO_2$ | plu |
| Coal power | res:ecn |
| coastal | lan:wat |
| coastal development | lan:wat:ecn |
| coastal pollution | lan:wat:plu |
| coastal salinity | lan:wat |
| Coasts | lan:wat |
| Cochin | ind |
| Colombia | sca |
| company investment | ecn |
| composting | con |
| Congo | afr |
| Conservation | con |
| consumerism | val:ecn |
| corrosion | mal |
| Costa rica | sca |
| costs | ecn |
| counter urbanisation | urb:val |
| culinary | cul |
| culinary habits | cul |

| | |
|---|---|
| cycling | com |
| cylcling | com |
| Dam | wat:lan:ecn |
| damage | mal |
| danube | wat:oec |
| deforestation | lan:mal |
| deforetation | lan:mal |
| Dehra dun | ind |
| desertification | ecl:mal |
| disaster | dis |
| disease | hea |
| displaced persons | hum:pol:mal |
| DK | oec |
| dog pollution | plu:ecl:urb |
| domestic animals | ecl:ecn |
| drought | lan:wat:dis |
| early warning system | pol:tec |
| East Europe | cis |
| East Germany | cis:oec |
| EC | oec |
| ecology | ecl |
| economics | ecn |
| education | val |
| Egypt | afr |
| El Salvador | sca |
| endangered-species | mal |
| endangered species | mal |
| energy | res |
| energy conservation | res:mit |
| energy conservation | res:mit |
| Energy consevration | res:mit |
| energy exploration | res:ecn |
| energy tax | res:pol:ecn |
| England | unk |
| English Heritage | ins:unk:val |
| env awareness | val |
| env-harmony | val |
| env-management | con |
| env awareness | val |
| env conformity | val |
| env education | val |
| env legislation | val:pol |
| env mamangement | con |
| env management | con |
| env schem | mit:con:dev |
| env scheme | mit:con:dev |
| env schemes | mit:con:dev |
| env sheme | mit:con:dev |
| erosion | mal |
| Ethiopia | afr |
| Europe | oec |

| | |
|---|---|
| evolution | ecl |
| famine | mal:hea |
| farming | ecn:lan:rur |
| fauna | ecl |
| finance | ecn |
| Finland | oec |
| fishing | ecn:wat |
| fishing industry | wat:ecn |
| flooding | dis:wat |
| floods | dis:wat |
| flora | ecl |
| food | hea:hum |
| food pollution | plu:hea |
| forest fire | mal:lan |
| forest fires | mal:lan |
| forestry | ecl:lan |
| forests | ecl:lan |
| France | oec |
| frankfurt | oec:urb |
| fresh water habitat | ecl:wat:lan |
| fundraising | ecn:val |
| G7 | oec |
| Gales | dis |
| Ganges | wat:ind |
| garden | ecn:lan:con |
| gardening | ecn:lan:con |
| gardens | ecn:lan:con |
| gas | res |
| GATT | com:int:pol:ecn |
| GB | unk |
| genetic eng | tec |
| genetic engineering | tec |
| geology | tec |
| Germany | oec |
| global warming | mal:glo:air |
| golf | ecn:lan:val |
| government | pol |
| government enquiry | pol |
| government ineptitude | pol:mal |
| govt | pol |
| govt legislation | pol:val |
| Greenham | unk |
| greenhouse effect | glo:mal:air:int |
| Greenland | oec |
| Green car | tec:mit |
| green consumersim | val |
| ground water | wat:lan |
| Gujarat | ind |
| Gulf | oas |
| habitat | ecl |
| habitat loss | ecl:mal |

| | |
|---|---|
| habitat loss | ecl:mal |
| Hawaii | oec |
| HCFC | plu:mit |
| HCFCs | plu:mit |
| health | hea |
| Herbicide natural | plu:mit |
| Himalayas | ind:oas |
| homeopathy | hea:hum:val |
| Hong Kong | oas |
| human life | hum:val |
| human rights | hum:pol:val |
| Hungary | cis |
| hunting | val:ecl |
| hurricane | dis:air |
| IAEC | pol:tec:int:ins |
| Iceland | oec |
| IMF | ins:int:ecn |
| India | ind |
| indigenes | hum:val |
| indigenous peoples | hum:val |
| industrial accident | mal |
| industrial action | mal:pol |
| industrial compensation | mal:ecn |
| industrial pollution | mal:plu |
| industry | ecn |
| insects | ecl |
| institution | ins |
| institutions | ins |
| intensive farming | lan:ecn:tec |
| inter black market | int:val:ecn |
| intergenerational equity | hum:val |
| international | int |
| international debt | ecn:int |
| International politics | int:pol |
| international pressure | int:pol |
| Iraq | nme |
| Ireland | oec |
| irrigation | lan:wat:ecn |
| Italy | oec |
| Jamaica | sca |
| Japan | oec |
| Kenya | afr |
| Kerala | ind |
| Kuwait | nme |
| lake | lan:wat |
| lakes | lan:wat |
| Lak District | unk |
| land | lan |
| land degradation | lan:mal |
| land ownership | lan:val:ecn |
| land reclamation | lan:mit |

| | |
|---|---|
| land use | lan:ecn |
| land-use | lan:ecn |
| Latvia | cis |
| law breaking | val |
| Leeds | unk |
| legislation | val |
| leisure | val |
| lifestyle | val |
| lifestyles | val |
| Litter | plu |
| local council | pol:ins |
| local government | pol:ins |
| london | unk:urb |
| Madhya Pradesh | ind |
| Maharashtra | ind |
| Malaysia | oas |
| Manchester | unk:urb |
| marine-life | wat:ecl |
| marine life | wat:ecl |
| marine pollution | wat:plu |
| marshland drainage | wat:lan:tec |
| Mauritius | oas |
| media | com |
| medical science | tec:hea |
| medicines | tec:hea |
| Mediterranean | oec |
| Mexico | sca |
| military | pol |
| mineral exploration | res:ecn |
| mining | res:ecn |
| Mississippi | nam:wat |
| monsoons | air:wat |
| morality | val |
| moral values | val |
| mosquito | ecl |
| mosquitos | ecl |
| movies | com |
| MPs | pol |
| Naramada Dam | wat:ind:pol:ecn:ecl:hum |
| Naramada dam project | ind:wat:pol:ecn:ecl:hum |
| Narmada Dam | wat:ind:pol:ecn:ecl:hum |
| NASA | ins:tec:nam |
| National consumer cou | ins:unk |
| National Rivers Aut | unk:ins |
| National Trust | ins:unk:con |
| natural disaster | dis |
| natural disasters | dis |
| natural dister | dis |
| nature | ecl |
| nature trails | lan:ecl:con:val |
| Nepal | oas |

| | |
|---|---|
| Netherlands | oec |
| New Zealand | oas |
| NGO | ins |
| NGOs | ins |
| Niger | wat:afr |
| Nile | wat:afr |
| noise pollution | plu |
| norfolk | unk |
| North sea | oec:wat:int |
| Norway | oec |
| NPT | pol:mit:int |
| nuclaer safety | hea |
| nuclear accident | mal:tec:plu:hea |
| nuclear power | res:ecn |
| nuclear reprocessing | res:ecn:tec |
| nuclear safety | res:mit:tec |
| nuclear waste | plu:tec |
| nuclear weapons | pol:tec |
| OAS | oas |
| OECD | oec |
| oil pollution | plu |
| OPEC | ins:int:res |
| organic | con |
| organic farming | lan:rur:ecn:con |
| organic farming | lan:rur:ecn:con |
| organic gardening | lan:con |
| outer space | spa |
| overfishing | wat:ecl:mal:ecn |
| oxford | unk |
| Oxford University | unk:ins:val |
| Ozone | air |
| ozone depletion | air:mal |
| Ozone hole | air:mal:poa |
| Pakistan | oas |
| Papua New Guinea | oas |
| parks | con |
| Peru | sca |
| pesticide | tec |
| pesticides | tec |
| pesticide pollution | plu |
| Phill | oas |
| physical deformity | hea:mal |
| planning permit | pol:lan |
| plantlife | ecl |
| plant life | ecl:lan |
| politics | pol |
| pollution | plu |
| pollution control | plu:mit |
| Polynesia | oas |
| population | hum |
| population growth | hum |

| | |
|---|---|
| positive pollution | plu:mit |
| poverty | ecn:hum:mal |
| prehistory | cul |
| premature death | hea:mal |
| prem death | hea:mal |
| prem death animal | ecl:mal:hea |
| prem death human | hum:mal:hea |
| prem death insects | ecl:mal |
| privatisation | pol:val:ecn |
| protest | pol |
| provatisation | pol:ecn:val |
| pygmies | hum:afr |
| quarrying | res:ecn:lan |
| quebec | nam |
| radioactive waste | plu |
| radioactivity | plu |
| rainforest | ecl:lan |
| Rain forest | ecl:lan |
| rare birds | ecl |
| rare fauna | ecl |
| rare flora | ecl |
| recycling | res:mit |
| recyling | res:mit |
| regulation | pol:val |
| rehabilitation | hum:val |
| remote sensing | tec |
| reneawable resources | res:con:mit |
| Renewable energy | res:con:mit |
| renewble energy | res:con:mit |
| repair costs | ecn:mit |
| reptiles | ecl |
| research investment | ecn |
| resettlement | hum:mit |
| ressetlement | hum:mit |
| ROK | oas |
| RSA | afr |
| RSPB | ins:unk:con:mit |
| Russia | cis |
| sanitation | hea |
| satellite | tec |
| science | tec |
| science education | tec:val |
| scientic research | tec |
| scientifc research | tec |
| scientific research | tec |
| scietific research | tec |
| Scotland | unk |
| sea | wat |
| sea pollution | wat:plu |
| seismic monitoring | tec |
| self sufficiency | tec:val |

| | |
|---|---|
| sellafield | unk:res:ecn:tec |
| shetland | unk |
| shias | val |
| siltation | mal:lan:wat |
| Slovakia | cis |
| soft loans | ecn:pol:int |
| soil erosion | mal:lan |
| solar system | spa |
| somerset | unk |
| South America | sca |
| South China Sea | oas:wat |
| space | spa |
| space exploration | tec |
| space expolration | tec |
| Spain | oec |
| spiritual values | val |
| sport | val |
| submarine | wat:tec |
| subsidence | lan:mal |
| subsistence farming | val:ecl:ecn |
| Sudan | afr |
| sustainable development | con:ecn |
| sustainable forestry | ecl:lan:con:mit |
| Sweden | oec |
| Sydney | oas |
| symbiosis | ecl |
| Taiw | oas |
| Taiwan | oas |
| technology | tec |
| technology transfer | tec:ecn:pol |
| Thail | oas |
| Tibet | oas |
| tigers | ecl |
| tourism | hum:ecn |
| toursim | hum:ecn |
| town planning | ins:lan:val |
| toxic | waste plu |
| toxic waste | plu |
| traffic accident | mal:com |
| traffic accidents | com:mal |
| transport | com |
| trees | ecl:lan |
| tribal life | cul:hum |
| tribal living | cul:hum |
| Turkey | nme |
| UK | unk |
| UKenv awareness | unk:val |
| UKNGO | unk:ins |
| Ukraine | cis |
| UKscientific research | unk:tec |
| UN | ins:int:pol |

| | |
|---|---|
| UNCED | ins:int:pol:mit |
| unemployment | ecn:mal:hum |
| United Nations | ins:int:pol |
| urbanisation | urb |
| urban migration | rur:urb:hum:ecn |
| US | nam |
| USA | nam |
| values | val |
| vehicles | com |
| vehicle pollution | com:plu |
| vermiculture | lan:ecl |
| Vietnam | oas |
| volcanic eruption | dis |
| volcano | tec |
| volunteer action | hum:val |
| Wales | unk |
| war | pol:mal |
| war threat | pol:int:mal |
| waste | plu:res:mal |
| waste disposal | plu:mit |
| waste dosposal | plu:mit |
| water | wat |
| water cycle | wat:tec |
| water pollution | wat:plu |
| water shortage | wat:mal |
| wealth | ecn:val |
| West-Africa | afr |
| West Europe | oec |
| West Indies | sca |
| WHO | hea:int:ins |
| wildelife | ecl |
| wilderness | val:ecl:rur |
| wildlife | ecl |
| wild life | ecl |
| Wiltshire | unk |
| World bank | ins:int:ecn |
| WWF | ins:int:con |
| Yem | nme |
| YMCA | ins:int:hum:val |
| yorkshire | unk |
| Zambei | wat:afr |
| zoo | con |
| zoos | con |

# Appendix III:
# Monitoring of broadcasts

Outline page filled out for periods marked 'x'.
Detail page filled out for items within time period 'M'.

## A. NON-NEWS BROADCASTS

Channel BBC1
```
              00000000000111111111122222222222
              01234567890123456789012345678
Mon  93Feb    .....x..xxxxxxxxxxxxxxxxx....
Tue  93Feb    .....x..xx..xxMMMxMxxxxxx...
Wed  93Feb    .....x..xx..xxxMxxMxMxMxx....
Thu  93Feb    .....xMxxxxxxMxMMxxMxxxx....
Fri  93Feb    .....x..xx..xMxMMMxMxxxx....
Sat  93Feb    ......xxxxMxxxxxxxxxxxx...
Sun  93Feb    ......MxxxxMxxxxxxxxMx.....
```

Channel BBC1
```
              00000000000111111111122222222222
              01234567890123456789012345678
Mon  93Apr    .....xxxMMMxxxxxMxxxxxx.....
Tue  93Apr    .xx..xxxxxxxxMMMMxxxxx....
Wed  93Apr    .xx..xMMxMMxxxMMMxMxxxxx....
Thu  93Apr    .x...xxxMxxxx.xxMMxxMxxx....
Fri  93Apr    .....xxxxxxxxxMMMxxxxxx.....
Sun  93Apr    ......MMMxMMxxxxxxxxxx......
```

Channel BBC1
```
              00000000000111111111122222222222
              01234567890123456789012345678
Mon  93Jul    xxx..xMMMxMxxxMMMxxxxxx....
Tue  93Jul    .xx..xxxxMxMxxxxxMxxxxx.....
Wed  93Jul    .xx..xMMxxxMxxxMMMMxxxx.....
Thu  93Jul    .....xxxxxxMxxxMMxxxxxx....
Fri  93Jul    x....xxxxxxMxxxxxxxxx....
Sat  93Jul    x.....xxxxxxxxxxxxxxxx....
Sun  93Jul    x.x..xxxxxMxxxxxxMxxxxxx...
```

Channel BBC1
```
              00000000000111111111122222222222
              01234567890123456789012345678
Mon  93Sep    xxx..xMMxxxxxxxMMMMxxxxx...
Tue  93Sep    x.x..xxxMxMMxxxxxMxxxxx.....
Wed  93Sep    .xx..xMxxxMxxxxxMMxxMxxxx....
Thu  93Sep    ..x..xxxxxMxxxxxxMMxMxxx....
Fri  93Sep    xx...xMxxxMxxMxMxMxxxxxxx...
Sat  93Sep    x.....xxxxxxxxxxxxxxx....
Sun  93Sep    ......MMxxxMxxxxxxxxxx....
```

Channel BBC1
```
              00000000000111111111122222222222
              01234567890123456789012345678
Mon  93Nov    ..x..xMxxxxxxxxxxxxMxxx....
Tue  93Nov    xxx..xxxxxxxxxxxxxxxxxxx...
Wed  93Nov    xxxxxxMxxxxxxxxxMxxxxxx....
Thu  93Nov    ..x..xxxxxxxxxxxxxMMxxx....
Fri  93Nov    xx...xxxxxxxxxxxxMxxxxx...
Sat  93Nov    x.....xxxxxxxxxxxxxxxx...
Sun  93Nov    ......xxxxxMMxxxxxxxxxx....
```

Channel BBC1
```
              00000000000111111111122222222222
              01234567890123456789012345678
Mon  94Jan    .xx..xMxxMxxxxMMMMxxxxx....
Tue  94Jan    .xx..xxxMxxMxxMMMMxMxxxx....
Wed  94Jan    .xx..xMxxxxxxMMMxxxMxxx....
Thu  94Jan    ..x..xxxxxxxxMMMMxMxMM.....
Fri  94Jan    .....xMMxMMxxxxMMxMxxxxx....
Sat  94Jan    ......xxMxxxxxxxxxMMxxx....
Sun  94Jan    ......xxMMMMMxxxxxxxxx.....
```

```
Channel  BBC2                                  Channel  BBC2
             00000000000111111111122222222                  00000000000111111111122222222
             01234567890123456789012345678                  01234567890123456789012345678
Mon  93Feb   ......MMxxMMxxxxMMMMxxx.....      Mon  94Jan   x......xxxxMMxxxxxxxxxxx....
Tue  93Feb   .....xMxMxMxxxxxxxxMMxxxM....     Tue  94Jan   x......xxxMMMMxxxxxxxxxM.....
Wed  93Feb   .....xxxxMMxMMxxMMMMxxx.....      Wed  94Jan   x......xMxxxMxxxxxxxxxx....
Thu  93Feb   .....xMMxxMMxxxxxxxxxx.....       Thu  94Jan   x......xMMxMxxxxxxMMMxx....
Fri  93Feb   .....xxxMxxxMMxxxxxMMxxx....      Fri  94Jan   x......xMxMxxxxxxxxMxxxx....
Sat  93Feb   .....xxxMxxxxxxxxxxxxxx...        Sat  94Jan   x.....MMxxxxxxMxxxxxxx....
Sun  93Feb   .....xxxMxxxxxxMxxxxxxx...        Sun  94Jan   ......xxMMMMxxxxxMxxxxxx....

Channel  BBC2                                  Channel  ITV
             00000000000111111111122222222                  00000000000111111111122222222
             01234567890123456789012345678                  01234567890123456789012345678
Mon  93Apr   .....xMMxxxxMMxxxxxMxxxx....      Mon  93Feb   xxxxxxxxxxxxxxxxxxMxxxxx....
Tue  93Apr   .....xMMMxxxMMxxxxxMMxxx....      Tue  93Feb   xxxxxxxxxxxMMMMM.xxxxxx....
Wed  93Apr   .....xxxxxxxMMxxxMMxxxxx....      Wed  93Feb   xxxxxxxxxxxMMxx.xxxxxxx...
Thu  93Apr   .....xMMxxxMMMxxxxxMMxxx....      Thu  93Feb   xxxxxxxxxMxMMMMMxMxxxxxx...
Fri  93Apr   xx...xxxxxxxMMxxxxMMMxxx....      Fri  93Feb   xxxxxxxxxxMMMxxxxxxMMxx....
Sat  93Apr   .....MMMxxxxMMMxxxxxxxxx...       Sat  93Feb   xxxxxxxMMMxxx..xxxxxxxx...
Sun  93Apr   .....xxxxxMxxxxxxMMxxxxx...       Sun  93Feb   xxxxxxxxxxxxxxxxxxxxxxx....

Channel  BBC2                                  Channel  ITV
             00000000000111111111122222222                  00000000000111111111122222222
             01234567890123456789012345678                  01234567890123456789012345678
Mon  93Jul   ......xxxxxxMxxxxxxMMMx.....      Mon  93Apr   xxxxxxxxxMMMMMxMMMxxxxxx....
Tue  93Jul   .....xxxxxxxxxMxxxxMxxx.....      Tue  93Apr   xxxxxxxxxxxxMMxMM.xxxxxx....
Wed  93Jul   x.....xxxxxMMxxxxxMMxxx.....      Wed  93Apr   xxxxxxxxxxxMMMxMMMxxxxxx....
Thu  93Jul   .....xxxxxxxMxxxxxMxMMx.....      Thu  93Apr   xxxxxxxxxxxMMMxxxMMxxxxx....
Fri  93Jul   x....xxxxxxxMMxxxxMMMMxx....      Fri  93Apr   xxxxxxxxxMMMMMxMMMxxxxx.....
Sat  93Jul   xx...xxxMMMMMxxxxxxxxxx....       Sat  93Apr   xxxxxxxxxxxxxxxxxxxxxxx....
Sun  93Jul   x....xxxxxMMxxxxxxMMxxxx....      Sun  93Apr   xxxxMMMMxxxxxM..MMxxxxMM....

Channel  BBC2                                  Channel  ITV
             00000000000111111111122222222                  00000000000111111111122222222
             01234567890123456789012345678                  01234567890123456789012345678
Mon  93Sep   ......xxxxxMxxxxMxxxxxx.....      Mon  93Jul   xxxxxxxxxxxxxxxxxxMxxMxxx....
Tue  93Sep   .....xMMMxxxxxxxxxxMMMxx....      Tue  93Jul   xxxxxxxxxxxxxxxxxxMxxxxxx....
Wed  93Sep   .....xxMMxxMxxxxxMMxxxx....       Wed  93Jul   xxxxxxMMxxxxxxxxxMMxxxxxx...
Thu  93Sep   .....xxxxxxMxxxxxxxxxxxx....      Thu  93Jul   xxxxxxxxxxxxMxxxMMxxxxxx...
Fri  93Sep   .....xxxxxxMMxxxxxxx.Mxxx...      Fri  93Jul   xxxxxxxxxxxxxxxxxxxxxxx....
Sat  93Sep   x....xxxMxxxxxxxxxxxx....         Sat  93Jul   xxxxxxxxxxxxxxxxxxxxxxx....
Sun  93Sep   .....xxxMMMxxxxxxxxMMxxxx....     Sun  93Jul   xxxxxxxxxxxMxxxxxxxxxxxx...

Channel  BBC2                                  Channel  ITV
             00000000000111111111122222222                  00000000000111111111122222222
             01234567890123456789012345678                  01234567890123456789012345678
Mon  93Nov   xxxx...xMMxMxxxMMxMMxxxx....      Mon  93Sep   xxxxxxxxxxMxxxxxxMxxxxxx....
Tue  93Nov   xxxx...xxxxMxMMxxxxxxxxx.....     Tue  93Sep   xxxxxxxxxxxMxxMxMxMxxxxxx....
Wed  93Nov   xxxx...xxMxxMMxxxxxxxxxx....      Wed  93Sep   xxxxxMMMMxxxxxxxxxxxxxxx....
Thu  93Nov   x......xxxxMMMxxxxxxxxxx....      Thu  93Sep   xxxxxxxxxxxxxxxxxMxxxxxx....
Fri  93Nov   ......xxMMxMxxxxxxMxxxx....       Fri  93Sep   xxxxxxxxxxxxxxxxxxxxxxx....
Sat  93Nov   ......MMMxxxxxxMMxxxxxxx...       Sat  93Sep   xxxxxxxxxxxxxxxxxxxxxxx...
Sun  93Nov   ......xxxMMxxxxxxxxxxxxx....      Sun  93Sep   xxxxxxxxxxxxxxxxxxxxxxxx....
```

```
Channel ITV                                    Channel C4
       0000000000111111111122222222            0000000000111111111122222222
       0123456789012345678901234578            0123456789012345678901234578
Mon 93Nov  xxxxxMMxxxxxxMxMMMxxxxxx....   Mon 93Sep  x....xxxxxxxxxMMxxMMMMxx....
Tue 93Nov  xxxxxxxxxxMxxxxxxMxxxxxx....   Tue 93Sep  x....xxxxxxxxxxxxxxxxx....
Wed 93Nov  xxxxxMMMxxxxxxxxxxMxMMxxxx...  Wed 93Sep  xxx..xxxxxxxxxxxxxxxxxx....
Thu 93Nov  xxxxxxxxxxxMMxxxxxMxxxxxx....  Thu 93Sep  .....xxxxxxxxxxxxxxxxxxxx..
Fri 93Nov  xxxxxxxxxxxxxxxMMxxxxxx....    Fri 93Sep  xx...xxxxxxxxxxxxxxxxx...
Sat 93Nov  xxxxxxxxxxxxxxxxxxxxxxxx....   Sat 93Sep  x....xxxxxxxMxxxxxxxxxx....
Sun 93Nov  xxxxxMMxxxxxxxxxxxxxxxxx....   Sun 93Sep  xxx..xxxxMxxxxxxxxMMxxxx....

Channel ITV                                    Channel C4
       0000000000111111111122222222            0000000000111111111122222222
       0123456789012345678901234578            0123456789012345678901234578
Mon 94Jan  xxxxxxxxxxxxxxxxxxMMxxxxx....  Mon 93Nov  x...xxxxxMMxxxMMxxxxxxxx....
Tue 94Jan  xxxxxxxxxxxxxxxxxMxxxxxx....   Tue 93Nov  ....xxxxMMxMxxxxxxxxxxx....
Wed 94Jan  xxxxxxxxxxxxxxxxxxxxxxx....    Wed 93Nov  x...xxxxxxxMxxxxxxxxMxxx....
Thu 94Jan  xxxxxxxxxxxxxxxxxxxxxxxx...    Thu 93Nov  x...xxxxxMxMxxxxxxxxMxxx....
Fri 94Jan  xxxxxxxxxxxxxxxxxxxxxxx....    Fri 93Nov  xx...xxx.xxxMxxxxxxxMxxx....
Sat 94Jan  xxxxxxxxxxxxxxxxxxxxxx....     Sat 93Nov  xx..xxxxxxxxxxxxxxxxxx...
Sun 94Jan  xxxxxxxxxxxxMxxxxxxxxxx....    Sun 93Nov  xx..xxxxxxxxxxxxxxxxxxx....

Channel C4                                     Channel C4
       0000000000111111111122222222            0000000000111111111122222222
       0123456789012345678901234578            0123456789012345678901234578
Mon 93Feb  ....xxMxMxMxxxMMxxxxxxMMx...   Mon 94Jan  xx..xxMxMMxxxxMMxxxxxxxx....
Tue 93Feb  x...xxxxMxMMxxMMxxxMxxxx....   Tue 94Jan  xx..xxxxxMMMxxxxxxMMxxxx....
Wed 93Feb  xxx.xxxxMMMxxxxxxxxxxxxx....   Wed 94Jan  xx..xxMxxxxxxxxxxxMMxxx....
Thu 93Feb  x...xxxxxMMxxxxxxxxxxxx...     Thu 94Jan  xx..xxxxxMxxxxxxxxxxxx....
Fri 93Feb  xx..xxxxMMMxxMMMxxxMxMxxx....  Fri 94Jan  xx..xxMMxxxxxxxxxxMxxx....
Sat 93Feb  xxx..xMMxxxMMxxxxMMMxxxx....   Sat 94Jan  xxx.xxxxxxxMxxxxxMxxxxx...
Sun 93Feb  xx...xMxMxMxMMxxxxxMxxMMM....  Sun 94Jan  xx..xxxxxxxMMxxxxMxxxxx....

Channel C4                                     Channel DODA
       0000000000111111111122222222            0000000000111111111122222222
       0123456789012345678901234578            0123456789012345678901234578
Mon 93Apr  .....xMMMxxMMxMMxxxxxxxx....   Mon 93Feb  .....xxxxxxxxxxxxxxxxx.....
Tue 93Apr  x....xxxMMxMMxMMxxMMMMxx....   Tue 93Feb  .....xxxxxxxxxxxMMxxxx.....
Wed 93Apr  xxx..xxxxxxxxxxxxxxxxxx....    Wed 93Feb  .....xxxxx..xMxxxxxxMxx.....
Thu 93Apr  .....xxxMMxMMxxxxxxxxxx....    Thu 93Feb  .....xxxx...xxx.xxxxxxx.....
Fri 93Apr  x....xMxxxxxxxxxxxxxxxx...     Fri 93Feb  .....xxxxxxxxxxxMxxxxxxx...
Sat 93Apr  x....xxxxxxxxxxxxMMMxxx....    Sat 93Feb  .....xxxxx..xxxMxxxxxxx.....
Sun 93Apr  x....xxxxxxxxxxxxMxxxxx....    Sun 93Feb  .....xxMMxxxxxxxxxxxxxx.....

Channel C4                                     Channel DODA
       0000000000111111111122222222            0000000000111111111122222222
       0123456789012345678901234578            0123456789012345678901234578
Mon 93Jul  x....xxxxMxMxxxxMxxxxxxx....   Mon 93Jul  .....xxxxx..xxxxxxMxxxx.....
Tue 93Jul  xx...xxxMMxxMxxxxxxxxxx....    Tue 93Jul  .....xxx....xxxxxxxxxxx.....
Wed 93Jul  xxx..xMxMMxxxMxxxxMxxxxx....   Wed 93Jul  .....xxxx...xxxxxxxxxxxx...
Thu 93Jul  .....xxxMMxxxMxxxxxxxxx....    Thu 93Jul  .....xMMx..xxxxxxMMxx......
Fri 93Jul  xx...xxxMMxxMMxxxxxMxxx....    Fri 93Jul  .....xxxx...xxx.xxxxxxxxx...
Sat 93Jul  xx...xxxxxxxxxxxxxxxxx....     Sat 93Jul  .....xxxx...xxxMxxxMxxxx.....
Sun 93Jul  .....xxxxxxxxxxxxxxxxxx...     Sun 93Jul  .....xxxxxxxxxxxMxxxxxxx.....
```

```
Channel  DODA                                      Channel  RAD4

         00000000001111111111222222222                     00000000001111111111222222222
         01234567890123456789012345678                     01234567890123456789012345678
Mon  93Sep  .....xxxxx..xxxxxxMxxxx.....       Mon  93Jul  .....xxMxxxxxxxxxMMxxxxxx....
Tue  93Sep  .....xxxxx..xxxxxxxxxxx.....       Tue  93Jul  ....xxxxxxxxxxxxMxxxxxxx....
Wed  93Sep  .....xxxxx..xxxxxxxxxxx.....       Wed  93Jul  .....xMxxMMxxxxxxxMxxxMx....
Thu  93Sep  .....xxxxx..xxxxxxMxxxx.....       Thu  93Jul  .....MMxxxxxxxxxxxxxxxxx....
Fri  93Sep  .....xxxxx..xxxxxMMxxxx.....       Fri  93Jul  .....MxxxMMxxxMxMMMxxxxx....
Sat  93Sep  .....xxxx...xxxxxxxxxxx.....       Sat  93Jul  ....xxxxxxxxxxxMMxxxxxx....
Sun  93Sep  ...........................        Sun  93Jul  .....xMMMxMxxxxxxxxxxxxx....

Channel  DODA                                      Channel  RAD4

         00000000001111111111222222222                     00000000001111111111222222222
         01234567890123456789012345678                     01234567890123456789012345678
Mon  93Nov  .....xxxxxxxxxxxxxMxxxxx....        Mon  93Sep  .....xMxxxxxxxxxxxxxxxxx....
Tue  93Nov  .....xxxxx..xxxxxxxxxxx.....        Tue  93Sep  .....xxMxxxxxxxxxxxxxxxx....
Wed  93Nov  .....xxxxx..xxxxxxxxxxx....         Wed  93Sep  ....xxxxxMxxxxxxxxxxxxx....
Thu  93Nov  .....xxxxx..xxxxxxxxxxx.....        Thu  93Sep  ....xxMxxxxxxxMxxxxxxx....
Fri  93Nov  .....xxx.x..xxxxxMMxxxxx....        Fri  93Sep  ....xxxxxMxxxxxxMxxxxxx....
Sat  93Nov  .....xxxxxxxxxxxxxxxxxx....         Sat  93Sep  ....xxxxxxxxxMMxxxxxxx....
Sun  93Nov  .....xxxxxxxxxxxxxxxxxx.....        Sun  93Sep  ....xxxxMMMxxxxxxxxxxx....

Channel  DODA                                      Channel  RAD4

         00000000001111111111222222222                     00000000001111111111222222222
         01234567890123456789012345678                     01234567890123456789012345678
Mon  94Jan  .....xxxxxxxxxxxxxxxxxx....         Mon  93Nov  .....MMMxxxxxxxxxxxxxxx....
Tue  94Jan  .....xxxxx...xxxxxxxxxx.....        Tue  93Nov  .....MMMxxxxxxxMxxMxxxx....
Wed  94Jan  .....xxxxxx.xxxxxxMxxxxxx...        Wed  93Nov  .....MMMxxxMxxxxxxMxxxx....
Thu  94Jan  .....xxxxxxxxxxxxxxxxxx.....        Thu  93Nov  .....MMMxMMxxxMMMxxMxxxx....
Fri  94Jan  .....xxxxxxxxxxxxMxxxMMxx...        Fri  93Nov  .....MMMxxMxxxMxxMxxxxxx....
Sat  94Jan  .....xxxxxx..xxxxxxxxxxx....         Sat  93Nov  ....xxxxxxxMMxMMxxxxxxx....
Sun  94Jan  .....xxxxxxxxxxxxxxxxxx.....        Sun  93Nov  .....xMxMMxxxMMxxxxxxxxx....

Channel  RAD1                                      Channel  RAD4

         00000000001111111111222222222                     00000000001111111111222222222
         01234567890123456789012345678                     01234567890123456789012345678
Mon  93Apr  xxxxxx..xxxxxxxxxxxxxxxx....        Mon  94Jan  .....MMMxxMMxxxxMxMxxxM.....
Tue  93Apr  xxxxxx..xxxxxxxxxxxxxxxx....        Tue  94Jan  .....xxxxxxMxxxxxxMMxxx.....
Wed  93Apr  xxxxxx..xxxxxxxxxxxxxxxx....        Wed  94Jan  .....xMMxxxMxxxxMxxxxxx.....
Thu  93Apr  xxxxxx..xxxxxxxxxxxxxxxx....        Thu  94Jan  .....xxMxxxMxxxxMxxMxxx.....
Fri  93Apr  .xxxxxxxxxxxxxxxxxxxxxxxx..         Fri  94Jan  .....xMxxMMxxxxxxxxxxxx.....
Sat  93Apr  .xxxxxxxxxxxxxxxxxxxxxxxx..         Sat  94Jan  .....xxxxxMxxxMxxxxxxx.....
Sun  93Apr  xxxxxxxxxxxxxxxxxxxxxxxx...         Sun  94Jan  .....xxxxxxxxxxxMxMxx.....

Channel  RAD4                                      Channel  AIR

         00000000001111111111222222222                     00000000001111111111222222222
         01234567890123456789012345678                     01234567890123456789012345678
Mon  93Apr  .....xMxMMxMxxxxxxMMxxxx....        Mon  93Feb  ....xxMxxxxxxxxxxxxxx.....
Tue  93Apr  .....x...xxxxxxxxxxxxxxx....        Tue  93Feb  ....xxMxxxxxxx.xxxxxxx....
Wed  93Apr  .....xx.xxxxxxxxxxMxxxxx....        Wed  93Feb  ....xxMxxx.xxxx.xMxxxxx....
Thu  93Apr  .....xMMxxxMxxxxxxxxxxxx....        Thu  93Feb  ....xxMxxx.xxxx.xxxxxxx....
Fri  93Apr  .....x..xMMMxxxxxxxxxxxx....        Fri  93Feb  ....xxMxxxx.xxxx.xxxxxxx....
Sat  93Apr  ....xxxxxMxxxxxMxxxxxxx....         Sat  93Feb  ....xxxxxx.xxxxxxxxxxx....
Sun  93Apr  ....xxxxxxxxMxxMxxxxxxx....         Sun  93Feb  ....xxxxxx.xxxx.xxxxxxx.....
```

```
Channel AIR
            00000000001111111111222222222
            01234567890123456789012345678
Mon   93Jul  ....xxMMxx.xxx..xxxxxxxx.....
Tue   93Jul  .....xMxxx.xxx..xxxxxxxx.....
Wed   93Jul  .....xMxxx.xxx..xMxxxxxx.....
Thu   93Jul  ....xMxxx.xxx..xxxxxxxx.....
Fri   93Jul  ....xMxxx.xxx..xxxxxxxx.....
Sat   93Jul  ....xMMxx.xxx..xxxxxxxx.....
Sun   93Jul  .....xMxxx.xxxx.xxxxxxxx.....

Channel AIR
            00000000001111111111222222222
            01234567890123456789012345678
Mon   93Sep  ....xxMxxx.xxxx.xxxxxxxx.....
Tue   93Sep  ...xxMxxx.xxxx.xxxxxxxx.....
Wed   93Sep  ...xxMxxx.xxxx.xxxxxxxx.....
Thu   93Sep  ...xxMxxx.xxxx.xxxxxxxx.....
Fri   93Sep  ...xxMxxx.xxxx.xxxxxxxx.....
Sat   93Sep  ...xxMxxx.xxxx.xxxxxxxx.....
Sun   93Sep  ....xxMxxx.xxxx.xxxxxxxx.....

Channel AIR
            00000000001111111111222222222
            01234567890123456789012345678
Mon   93Nov  ....xxMxxx.xxx..xxxxxxxx.....
Tue   93Nov  ....xxMxxx.xxx..xxxxxxxx.....
Wed   93Nov  ....xxMxxx.xxx..xxxxxxxx.....
Thu   93Nov  ....xxMxxx.xxx..xxxxxxxx.....
Fri   93Nov  ....xxMxxx.xxx..xxxxxxxx.....
Sat   93Nov  ....xxMxxx.xxx..xxxxxxxx.....
Sun   93Nov  ....xxMxxx.xxxx.xxxxxxxx.....

Channel AIR
            00000000001111111111222222222
            01234567890123456789012345678
Mon   94Jan  ....xxMxxx.xxxx.xxxxxxxx.....
Tue   94Jan  ....xxMxxx.xxxx.xxxxxxxx.....
Wed   94Jan  ...xxMxxx.xxxx.xxxxxxxx.....
Thu   94Jan  ....xxMxxx.xxxx.xxxxxxxx.....
Fri   94Jan  ....xxMxxx.xxxx.xxxxxxxx.....
Sat   94Jan  ....xxMxxx.xxxx.xxxxxxxx.....
Sun   94Jan  ....xxMxxx.xxxx.xxxxxxxx.....
```

## B. NEWSBROADCASTS

Complete monitoring of all programmes located by 'x'.

```
Channel BBC1
            00000000001111111111222222222
            01234567890123456789012345678
Mon   93Feb  ...................x.......
Tue   93Feb  ...................x.......
Wed   93Feb  ...................x.......
Thu   93Feb  ...................x.......
Fri   93Feb  ...................x.......
Sat   93Feb  ...................x.......
Sun   93Feb  .....................x.......

Channel BBC1
            00000000001111111111222222222
            01234567890123456789012345678
Mon   93Apr  ..................x..x.......
Tue   93Apr  ...................x.......
Wed   93Apr  ...................x.......
Thu   93Apr  ...................x.......
Fri   93Apr  ...................x.......
Sat   93Apr  ...................xx......
Sun   93Apr  ...................xx......

Channel BBC1
            00000000001111111111222222222
            01234567890123456789012345678
Mon   93Jul  ...................x.......
Tue   93Jul  ...................x.......
Wed   93Jul  ...................x.......
Thu   93Jul  ...................x.......
Fri   93Jul  ...................x.......
Sat   93Jul  ....................x.......
Sun   93Jul  ...................x.......

Channel BBC1
            00000000001111111111222222222
            01234567890123456789012345678
Mon   93Sep  ...................x.......
Tue   93Sep  ...................x.......
Wed   93Sep  ...................x.......
Thu   93Sep  ...................x.......
Fri   93Sep  ...................x.......
Sat   93Sep  ....................xx.....
Sun   93Sep  ...................xx.......

Channel BBC1
            00000000001111111111222222222
            01234567890123456789012345678
Mon   93Nov  ...................x.......
Tue   93Nov  ...................x.......
Wed   93Nov  ...................x.......
Thu   93Nov  ...................x.......
Fri   93Nov  ...................x.......
Sat   93Nov  ....................xx.....
Sun   93Nov  ...................xx.......

Channel BBC1
            00000000001111111111222222222
            01234567890123456789012345678
Mon   94Jan  ...................x.......
Tue   94Jan  ...................x.......
Wed   94Jan  ...................x.......
Thu   94Jan  ...................x.......
Fri   94Jan  ...................x.......
Sat   94Jan  ....................x......
Sun   94Jan  ...................xx.......
```

```
Channel BBC2                                    Channel BBC2
          0000000000111111111122222222            0000000000111111111122222222
          0123456789012345678901234 5678           0123456789012345678901234 5678
Mon  93Feb  .....................xx.....       Mon  94Jan  .....................xx.....
Tue  93Feb  .....................xx.....       Tue  94Jan  .....................xx.....
Wed  93Feb  .....................xx.....       Wed  94Jan  .....................xx.....
Thu  93Feb  .....................xx.....       Thu  94Jan  .....................xx.....
Fri  93Feb  .....................xx.....       Fri  94Jan  .....................xx.....
Sat  93Feb  ...........................       Sat  94Jan  ....................x.......
Sun  93Feb  ...........................       Sun  94Jan  ...........................

Channel BBC2                                    Channel ITV
          0000000000111111111122222222            0000000000111111111122222222
          0123456789012345678901234 5678           0123456789012345678901234 5678
Mon  93Apr  .....................xx.....       Mon  93Feb  ................x...x......
Tue  93Apr  .....................xx.....       Tue  93Feb  ................x...x......
Wed  93Apr  .....................xx.....       Wed  93Feb  ................x...x......
Thu  93Apr  .....................xx.....       Thu  93Feb  ................x...x......
Fri  93Apr  .....................xx.....       Fri  93Feb  ................x..........
Sat  93Apr  ..................xx.........       Sat  93Feb  ..............x............
Sun  93Apr  ...........................       Sun  93Feb  ...................x........

Channel BBC2                                    Channel ITV
          0000000000111111111122222222            0000000000111111111122222222
          0123456789012345678901234 5678           0123456789012345678901234 5678
Mon  93Jul  .....................xx.....       Mon  93Apr  ....................x......
Tue  93Jul  .....................xx.....       Tue  93Apr  ....................x......
Wed  93Jul  .....................xx.....       Wed  93Apr  ....................x......
Thu  93Jul  .....................xx.....       Thu  93Apr  ....................x......
Fri  93Jul  .....................xx.....       Fri  93Apr  ....................x......
Sat  93Jul  ..................x.........       Sat  93Apr  ...................x.......
Sun  93Jul  ...........................       Sun  93Apr  ...........................

Channel BBC2                                    Channel ITV
          0000000000111111111122222222            0000000000111111111122222222
          0123456789012345678901234 5678           0123456789012345678901234 5678
Mon  93Sep  .....................xx.....       Mon  93Jul  ....................x......
Tue  93Sep  .....................xx.....       Tue  93Jul  ....................x......
Wed  93Sep  .....................xx.....       Wed  93Jul  ....................x......
Thu  93Sep  .....................xx.....       Thu  93Jul  ....................x......
Fri  93Sep  .....................xx.....       Fri  93Jul  ...................xx......
Sat  93Sep  .................x..........       Sat  93Jul  ....................x......
Sun  93Sep  ...........................       Sun  93Jul  ....................x......

Channel BBC2                                    Channel ITV
          0000000000111111111122222222            0000000000111111111122222222
          0123456789012345678901234 5678           0123456789012345678901234 5678
Mon  93Nov  .....................xx.....       Mon  93Sep  ....................x......
Tue  93Nov  .....................xx.....       Tue  93Sep  ....................x......
Wed  93Nov  .....................xx.....       Wed  93Sep  ....................x......
Thu  93Nov  .....................xx.....       Thu  93Sep  ....................x......
Fri  93Nov  .....................xx.....       Fri  93Sep  ....................x......
Sat  93Nov  .................x..........       Sat  93Sep  ..................xx.......
Sun  93Nov  ...........................       Sun  93Sep  ....................x......
```

Channel ITV
```
          0000000000111111111222222222
          0123456789012345678901234567 8
Mon   93Nov  .....................x......
Tue   93Nov  .....................x......
Wed   93Nov  .....................x......
Thu   93Nov  .....................x......
Fri   93Nov  .....................x......
Sat   93Nov  ....................x........
Sun   93Nov  .....................x......
```

Channel ITV
```
          0000000000111111111222222222
          0123456789012345678901234567 8
Mon   94Jan  ....................x......
Tue   94Jan  ....................x......
Wed   94Jan  ....................x......
Thu   94Jan  ....................x......
Fri   94Jan  ....................x......
Sat   94Jan  ...................x.......
Sun   94Jan  ....................x.....
```

Channel C4
```
          0000000000111111111222222222
          0123456789012345678901234567 8
Mon   93Feb  .................x.........
Tue   93Feb  .................x.........
Wed   93Feb  .................x.........
Thu   93Feb  .................x.........
Fri   93Feb  .................x.........
Sat   93Feb  ..........................
Sun   93Feb  ..........................
```

Channel C4
```
          0000000000111111111222222222
          0123456789012345678901234567 8
Mon   93Apr  ..................x.........
Tue   93Apr  ..................x.........
Wed   93Apr  ..................x.........
Thu   93Apr  ..................x.........
Fri   93Apr  ..................x.........
Sat   93Apr  ..........................
Sun   93Apr  ..........................
```

Channel C4
```
          0000000000111111111222222222
          0123456789012345678901234567 8
Mon   93Jul  ..................x........
Tue   93Jul  ..................x........
Wed   93Jul  ..................x........
Thu   93Jul  ..................x........
Fri   93Jul  ..................x........
Sat   93Jul  ..........................
Sun   93Jul  ..............x...........
```

Channel C4
```
          0000000000111111111222222222
          0123456789012345678901234567 8
Mon   93Sep  .................x.........
Tue   93Sep  .................x.........
Wed   93Sep  .................x.........
Thu   93Sep  .................x.........
Fri   93Sep  .................xx........
Sat   93Sep  ..........................
Sun   93Sep  ..........................
```

Channel C4
```
          0000000000111111111222222222
          0123456789012345678901234567 8
Mon   93Nov  .................x.........
Tue   93Nov  .................x.........
Wed   93Nov  .................x.........
Thu   93Nov  .................x.........
Fri   93Nov  .................x.........
Sat   93Nov  .................x.........
Sun   93Nov  ...............x...........
```

Channel C4
```
          0000000000111111111222222222
          0123456789012345678901234567 8
Mon   94Jan  .................x.........
Tue   94Jan  .................x.........
Wed   94Jan  .................x.........
Thu   94Jan  .................x.........
Fri   94Jan  .................x.........
Sat   94Jan  .................x.........
Sun   94Jan  ..............x...........
```

Channel DODA
```
          0000000000111111111222222222
          0123456789012345678901234567 8
Mon   93Feb  ................x.xx......
Tue   93Feb  ................x.xx......
Wed   93Feb  ................x.xx......
Thu   93Feb  ................x.xx......
Fri   93Feb  ................x.x.......
Sat   93Feb  ................x.xx......
Sun   93Feb  ................x.xx......
```

Channel DODA
```
          0000000000111111111222222222
          0123456789012345678901234567 8
Mon   93Apr  ...............x.xx......
Tue   93Apr  ...............x.xx......
Wed   93Apr  ...............x.xx......
Thu   93Apr  ...............x.xx......
Fri   93Apr  ...............x.xx......
Sat   93Apr  ...............x.xx......
Sun   93Apr  ...............x.xx......
```

```
Channel DODA
              0000000000111111111122222222 2
              0123456789012345678901234567 8
Mon  93Jul    ...................x.x.......
Tue  93Jul    ...................x.xx......
Wed  93Jul    ...................x.x.......
Thu  93Jul    ...................x.x.......
Fri  93Jul    ...................x.xx......
Sat  93Jul    ...................x.xx......
Sun  93Jul    ...................x.xx......

Channel DODA
              0000000000111111111122222222 2
              0123456789012345678901234567 8
Mon  93Sep    ...................x.xx......
Tue  93Sep    ...................x.xx......
Wed  93Sep    ...................x.xx......
Thu  93Sep    ...................x.xx......
Fri  93Sep    ...................x.xx......
Sat  93Sep    ...................x.xx......
Sun  93Sep    ...................x........

Channel DODA
              0000000000111111111122222222 2
              0123456789012345678901234567 8
Mon  93Nov    ...................x.xx......
Tue  93Nov    ...................x.xx......
Wed  93Nov    ...................x.x.......
Thu  93Nov    ...................x.xx......
Fri  93Nov    ...................x.xx......
Sat  93Nov    ...................x.xx......
Sun  93Nov    ...................x.x.......

Channel DODA
              0000000000111111111122222222 2
              0123456789012345678901234567 8
Mon  94Jan    ...................x.xx......
Tue  94Jan    ...................x.xx......
Wed  94Jan    ...................x.xx......
Thu  94Jan    ...................x.x.......
Fri  94Jan    ...................x.xx......
Sat  94Jan    ...................x.xx......
Sun  94Jan    ...................xx......

Channel RAD1
              0000000000111111111122222222 2
              0123456789012345678901234567 8
Mon  93Apr    ..................x..........
Tue  93Apr    ..................x..........
Wed  93Apr    ..................x..........
Thu  93Apr    ..................x..........
Fri  93Apr    ..................x..........
Sat  93Apr    ............................
Sun  93Apr    ............................
```

```
Channel RAD1
              0000000000111111111122222222 2
              0123456789012345678901234567 8
Mon  93Nov    ..................x..........
Tue  93Nov    ..................x..........
Wed  93Nov    ..................x..........
Thu  93Nov    ..................x..........
Fri  93Nov    ..................x..........
Sat  93Nov    ..................x..........
Sun  93Nov    ............................

Channel RAD4
              0000000000111111111122222222 2
              0123456789012345678901234567 8
Mon  93Apr    ............................
Tue  93Apr    ............................
Wed  93Apr    ............................
Thu  93Apr    ............................
Fri  93Apr    ............................
Sat  93Apr    ...................x........
Sun  93Apr    ...................x........

Channel RAD4
              0000000000111111111122222222 2
              0123456789012345678901234567 8
Mon  93Jul    ..................x.........
Tue  93Jul    ..................x.........
Wed  93Jul    ..................x.........
Thu  93Jul    ..................x.........
Fri  93Jul    ..................x.........
Sat  93Jul    ............................
Sun  93Jul    ............................

Channel RAD4
              0000000000111111111122222222 2
              0123456789012345678901234567 8
Mon  93Sep    ..................x.........
Tue  93Sep    ............................
Wed  93Sep    ..................x.........
Thu  93Sep    ..................x.........
Fri  93Sep    ..................x.........
Sat  93Sep    ..................x.........
Sun  93Sep    ..................x.........

Channel RAD4
              0000000000111111111122222222 2
              0123456789012345678901234567 8
Mon  93Nov    ............................
Tue  93Nov    ............................
Wed  93Nov    ............................
Thu  93Nov    ............................
Fri  93Nov    ............................
Sat  93Nov    ..................x.........
Sun  93Nov    ..................x.........
```

```
Channel RAD4
              00000000001111111111222222222
              012345678901234567890123456678
Mon  94Jan    .................x.........
Tue  94Jan    .................x.........
Wed  94Jan    .................x.........
Thu  94Jan    .................x.........
Fri  94Jan    .................x.........
Sat  94Jan    .................x.........
Sun  94Jan    .................x.........

Channel AIR
              00000000001111111111222222222
              012345678901234567890123456678
Mon  93Feb    .....................x.....
Tue  93Feb    .....................x.....
Wed  93Feb    .....................x.....
Thu  93Feb    .....................x.....
Fri  93Feb    .....................x.....
Sat  93Feb    .....................x.....
Sun  93Feb    .....................x.....

Channel AIR
              00000000001111111111222222222
              012345678901234567890123456678
Mon  93Apr    ....................x......
Tue  93Apr    ....................x......
Wed  93Apr    ....................x......
Thu  93Apr    ....................x......
Fri  93Apr    ....................x......
Sat  93Apr    ....................x......
Sun  93Apr    ....................x......

Channel AIR
              00000000001111111111222222222
              012345678901234567890123456678
Mon  93Jul    ....................x.......
Tue  93Jul    ....................x.......
Wed  93Jul    ....................x.......
Thu  93Jul    ....................x.......
Fri  93Jul    ....................x.......
Sat  93Jul    ....................x.......
Sun  93Jul    ....................x.......
```

```
Channel AIR
              00000000001111111111222222222
              012345678901234567890123456678
Mon  93Sep    ....................x......
Tue  93Sep    ....................x......
Wed  93Sep    ....................x......
Thu  93Sep    ....................x......
Fri  93Sep    ....................x......
Sat  93Sep    ....................x......
Sun  93Sep    ....................x......

Channel AIR
              00000000001111111111222222222
              012345678901234567890123456678
Mon  93Nov    ....................x......
Tue  93Nov    ....................x......
Wed  93Nov    ....................x......
Thu  93Nov    ....................x......
Fri  93Nov    ....................x......
Sat  93Nov    ....................x......
Sun  93Nov    ....................x......

Channel AIR
              00000000001111111111222222222
              012345678901234567890123456678
Mon  94Jan    ....................x......
Tue  94Jan    ....................x......
Wed  94Jan    ....................x......
Thu  94Jan    ....................x......
Fri  94Jan    ....................x......
Sat  94Jan    ....................x......
Sun  94Jan    ....................x......
```

# Appendix IV:
# Public opinion in India
## *QUESTIONNAIRE*

*Note:* Anonymity of the respondent is maintained.

1. What do you think are the most important issues affecting the world today?
2. What do you think are the most important issues affecting India today?
3. What do you think are the most important issues affecting your town or village today?
4. If the government were to increase tax by 1 per cent, for which one of the following areas would you want the money to be used? (Choose only one)

    Hospitals
    Education
    Construction of roads
    Protection of the environment
    (e.g. afforestation, pollution control)

5. Give any examples of Government of India policy in the last year which will improve the global environment.
6. Give any examples of Government of India policy in the last year which will improve the national environment.
7. Give any examples of Government of India policy in the last year which will improve the local environment of your town or village.
8. Do you belong to any NGO or other group working on development or environmental issues (e.g. Friends of Earth, Oxfam) or any such local level group in your town or village?
    If yes, which:
9. What environmentally friendly actions do you carry out now in your daily life which you did not carry out two years ago?
10. What happens to your household waste?

(Put 'X' in the appropriate boxes in the table below):-

|  | Vegetables and fruit peelings | Paper | Glass | Plastic | Cans |
|---|---|---|---|---|---|
| Burn it |  |  |  |  |  |
| Sell it |  |  |  |  |  |
| Reuse it in the house |  |  |  |  |  |
| Compost it |  |  |  |  |  |

11. Have you in the last two years done any of the following?

    * Talked to MP/MLA/District Official/Panchayat Official about environmental issues?
    * Written a letter to a MP/MLA/District Official/Panchayat Official about environmental issues?
    * Taken part in a campaign about environmental issues?
    * Given money to or raised money for environmental NGOs?
    * None of the above.

12. What is your main source of information on the environment?

    1. TV
    2. Magazine
    3. Newspaper
    4. Friends/Family/People I work with
    5. Radio
    6. Books
    7. Being a member of a concerned organisation

13. To which radio stations do you listen most regularly?

    1. AIR
    2. Overseas radio
    3. Vividh Bharati
    4. Foreign radio
    5. Other (specify)
    6. Never listen to radio

14. Which TV channel do you watch most frequently?

    1. DD-1
    2. DD-2
    3. STAR TV
    4. Cable TV
    5. Other (Specify)
    6. Never watch television

15. Which newspaper do you read most frequently?

    1. _____
    2. Never read newspapers.

16. Have you in the last year listened to/read/watched any environmental newspaper stories, television or radio programmes?

    YES    NO
    If yes, which:

17. Do you think that for the next generation this country will be a

    better
    worse or
    same
    place to live in?
    What can our generation do about it?

18. In shopping for consumer items and household items (i.e. soap, soap powder, cosmetics etc.) which *ONE* of the following factors most influences your choice?

    1. Price
    2. Attractiveness of packaging
    3. Reusability of packaging
    4. Durability
    5. The product is not harmful to the environment
    6. The product is not harmful to health.

19. Have you heard of the ozone hole?

    YES    NO

    If yes, is it in the:

    1. Northern Hemisphere
    2. Southern Hemispheres
    3. Both Hemispheres

    Why is it talked about?

    Is it caused by:
    CFCs
    Aerosols/Dangerous Gases
    Other (Specify)
    Don't know

20. Have you heard of global warming?

    YES    NO
    If yes,
    What will it do?
    When will it happen?
    What causes it?

21. What is acid rain?

    What does it do?
    What causes it?

*Information about the respondent:*

Name: _____

Address: _____

Age: 16–24 (   ),    25–34 (   ),    35–44 (   ),    45–64 (   ),

    65 & above (   )

Sex: Male (   ),    Female (   )

Occupation:_____

Monthly income: _____

or tick category

        Below 1000           10,000–12,000
        3000–5000            12,000–14,000
        5000–8000            14,000 & above
        8000–10,000

# Index